Game-theoretic Models of the
Political Influence of Interest Groups

Game-theoretic Models of the Political Influence of Interest Groups

by

RANDOLPH SLOOF
University of Amsterdam

KLUWER ACADEMIC PUBLISHERS
BOSTON / DORDRECHT / LONDON

Distributors for North, Central and South America:
Kluwer Academic Publishers
101 Philip Drive
Assinippi Park
Norwell, Massachusetts 02061 USA

Distributors for all other countries:
Kluwer Academic Publishers
Distribution Centre
Post Office Box 322
3300 AH Dordrecht, THE NETHERLANDS

ISBN 978-1-4419-5050-5

Library of Congress Cataloging-in-Publication Data

Sloof, Randolph.
    Game-theoretic models of the political influence of interest groups / by
Randolph Sloof.
       p. cm.
    Includes bibliographical references and index.

    1. Pressure groups—Mathematical models. 2. Game theory.
I. Title
JF529.S6   1998
324'.4'015193—dc21   98-26317
                                                       CIP

Printed on acid-free paper

Printed in the United States of America

CONTENTS

ACKNOWLEDGEMENTS

This book is based on my doctoral dissertation, which I wrote during my four year stay as a Ph.D. student at CREED, the Center for Research in Experimental Economics and political Decisionmaking of the University of Amsterdam. In completing this research project I have received substantial encouragement and valuable support from a number of people and institutions. The main providers of scientific support were certainly Frans van Winden and Jan Potters. As my supervisor Frans performed an excellent job. He inspired my interest in interest group behavior in the first place, and continued to overwhelm me with new research questions and interesting ideas during the whole project. Both the fourth and fifth chapter originate from our joint work. I enjoyed doing research under Frans' supervision. As a co-author of my first two (joint) articles, on which the Chapters 2 and 4 in this book are largely based, Jan learned me how to write readable papers. Especially when we started writing the survey article (see Section 2.1) he must have asked himself in despair whether anything useful could come out of it. My stubbornness in regard to how the survey should be arranged certainly must have annoyed him from time to time. Besides his practical guidance in writing papers, Jan's thesis appeared to be one of the main sources of inspiration for my own work.

Other colleagues at CREED provided scientific support as well. Isidoro Mazza read some parts of the book and made quite a number of helpful comments. Being my room mate Jörgen Wit made more consequential remarks than he may realize himself. With our desks opposite to each other it was always possible to throw some theoretical or practical (computer) problem to the other side of the room in order to be solved immediately. Joep Sonnemans now and then provided the reassurance I asked for when I was solving a mathematical problem. Endless discussions with Theo Offerman always made me very angry, but may have affected my opinion about research methodology more than I am happy to admit.

Also a number of scholars outside CREED provided useful comments on earlier versions of the book or parts of it. Mentioning them all by name does not fit within the limited scope of these acknowledgements. Therefore, I confine myself to displaying my gratitude towards the members of the dissertation committee – David Austen-Smith, Arnoud Boot, Eric van Damme, Rick van der Ploeg and Claus Weddepohl – for their valuable input.

Besides making an inspiring research environment, my colleagues at CREED appeared to be even more able in keeping my mind off doing research. Joep managed to connect any casual topic that came up for discussion, typically introduced by Theo, to an amazing story about some member of his, apparently immense, family. Hessel Oosterbeek caused a lot of amusement by typically first expressing to Jörgen and me his deeply felt agitation about some evil in- or outside the university. Both Frans van Dijk and Arthur Schram convinced me that being somewhat blunt has a very charm of its own, and is very convenient in putting off (Ph.D.) students that badger the life out of you. Finally, Tanga McDaniel and Mark Olson were so kind to inform me on the American way of life during numerous mensa meals.

My parents, family and friends provided strong moral support. They always showed a sincere interest in my personal well-being, and I have much appreciated their kindness not to inquire too persistently after the progress and the content of this book. Probably, in completing this book they did the most important job by showing me that there is more than just doing research, and by distracting my attention towards these aspects of life. Finally, I received substantial financial support from the Nederlandse Organisatie voor Wetenschappelijk Onderzoek (NWO), Shell and the Tinbergen Institute to visit conferences in often very nice places.

Amsterdam, May 1998.

1 INTRODUCTION

In this chapter the topic of this book is introduced. Section 1.1 provides a brief and rather general motivation for the scientific project undertaken here. Interest groups are a very popular object of scientific inquiry, and they received already considerable research attention from scholars in political science, as well as from researchers in economics. Necessarily, then, this book adds to a literature which is already quite developed. A detailed positioning in this literature of the theoretical material presented in this monograph will be given in Chapter 2. This second chapter will also, by means of a review of the empirical literature, provide a more general overview of the issues deemed to be important when studying the influence of interest groups on public policy. The outline of the entire book is described in greater detail in Section 1.2. As most issues involved are more easily presented in later chapters, this introductory chapter is kept brief.

1.1 MOTIVATION

Substantial political power is often attributed to interest groups. Examples abound in both the economics and political science literature, as well as in journalistic accounts and popular publications. On many occasions the authors express concerns about the negative impact of interest groups on the democratic quality of government. "The interests of a small group are served at the expense of the interests of the general public, the taxpayers!", is an often heard popular complaint. Apart from crippling the democratic process, it is frequently asserted that interest groups are detrimental to society as a whole because their influence on governmental decision-making reduces economic growth, fosters inflation, and increases unemployment (cf. Olson, 1982). Allegedly, influence activities of interest groups are typically directed towards the redistribution of existing wealth ("rent-seeking"), rather than to inducing policies that lead to the creation of new wealth. The redistribution policies that result from the influence of interest groups are distortionary, and welfare (efficiency) losses are the outcome. Moreover, the expenditures on these influence activities are socially wasteful as well. Using a somewhat provocative terminology, these political expenditures are part of the "parasite economy", and do not belong to the realm of the productive economy (cf. Rauch, 1994).

Less frequently do the activities of interest groups and their resulting political influence receive a more positive judgment. A variety of arguments are brought to the fore in this respect. For instance, it is sometimes argued that interest groups make effective participation of ordinary citizens in politics easier. Interest groups play an important role as a way of expressing and representing interests, supplementary to political parties and elections. In this view interest groups are an important means by which the people can let government know what it wants. The representativeness of this people's voice may be questioned, though. Indeed, a number of scholars claim that the early observation of Schattschneider (1960, p. 35) that "The flaw in the pluralist heaven is that the heavenly chorus sings with a strong upper-class accent." is as yet still true.[1] Another now and then mentioned advantage of the influence of interest groups is that it precludes the "tyranny of the majority". The preferences of minorities that are most intensely concerned with specific issues are then still acknowledged by government. Yet another claim is that the very existence of a large number of interest groups leads to political stability, not only because policies are more moderate then, but also because "the isolation and alienation of the individual" can be avoided (cf. Wilson, 1981). Lastly, an argument underlining the positive role of interest groups that is increasingly put forward is that they serve a useful purpose by providing the government with (policy) relevant information.[2] Even though it is very likely that this information is colored by the specific interests of the information supplying group, when the government takes this into account the information may still be valuable to the government.

With these different and opposing views at hand it seems almost impossible to provide a definitive assessment of whether interest groups are, in general, good or bad for society. Such *normative* questions are still important to address, however, especially on a less general level when policy issues are publicly discussed that directly affect the way in which interest groups try to influence governmental decision-making. Examples of the type of issues that have recently received considerable attention in a number of countries include questions like: Should interest group lobbying be restricted?, Should lobbyists publicly register?, Should political parties and/or candidates be required to disclose information about the donations they received?[3] Rather than basing these policies on popular shibboleths concerning the alleged positive or negative impact interest groups have,[4] such policy decisions are preferably based on a well-informed trade-off between the pros and cons of interest group influence. To sharpen our ideas about these potential positive and negative effects interest groups may have, basic knowledge and understanding of "how and when" they influence governmental policies is needed. Only when such *positive* issues are to a reasonable extent resolved, normative aspects can be satisfactorily addressed.

In this book we will mainly focus on positive questions concerning the influence of interest groups. Normative questions are only occasionally touched upon, for instance in Chapter 4 where we briefly discuss the desirability of full disclosure laws for campaign contributions. Even then a comprehensive and in-depth overview of all possible arguments is not given. In the end it is only hoped that, on a more general level, the positive analysis presented in this monograph contributes to a better and more balanced judgment.

Even when one limits the attention to positive questions concerning the "how and when" of interest group influence, a wide variety of issues can be addressed, and advanced by means of different methods of scientific inquiry. For instance, questions that one may want to address are: Are lobbying activities of interest groups influential and if so, why are they influential?, Does interest group influence vary over the type of interest group involved?, Are government agencies captured by the industries they try to regulate?, What kind of interest groups are most likely (not) to be formed? To answer these types of questions one may rely on verbal reasoning, anecdotal evidence, well-elaborated case studies, formal modeling, and empirical (econometric) models. The scope of this monograph necessitates that the analysis is delimited in the number of questions that are addressed. Moreover, we have chosen one specific way to analyze them. The remainder of this section elaborates on these limitations with respect to both the questions posed and the methods employed.

First, the questions addressed in this monograph all focus on the *interaction* between interest groups, governmental agents, and voters. We start from the assumption that interest groups want "something" from the government, favorable regulation for instance, and therefore try to influence the behavior of governmental agents and/or individual voters. The general purpose of this research project is to shed light on how interest groups try to affect the behavior of these agents, and why and when these influence activities are likely to be effective. The analysis is directed towards the strategic issues involved in the behavior of interest groups and the agents they want to influence. To keep the analysis both well-focused and rather general at the same time, we abstract away from exactly what types of interest groups are involved and what exactly they want from the government.[5] Moreover, the existence of interest groups is taken as given, and they are assumed to behave as unitary actors. Consequently, group formation and collective action problems (cf. Olson, 1965) will not directly be addressed.

Second, *formal modeling* is used in an attempt to increase our basic understanding of interest group influence. At the danger of abstracting away too much, formal models allow a rigorous derivation of results from a set of assumptions. By scrutinizing the set of assumptions it can be determined what assumption drives a particular result. For instance, careful selection and inspection of assumptions made in models of interest group behavior increases our insight in the mechanisms by which their influence is effectuated. Another advantage of a formal approach is that theories based on explicit formal models are more likely to yield clear-cut competing hypotheses than theories based on verbal reasoning. Therefore, such theories typically can be more rigorously tested (experimentally, or in the field) than the latter.

As we focus on the strategic interaction between interest groups, on the one hand, and governmental agents and individual voters, on the other hand, game theory seems to be the appropriate analytical tool. Indeed, a number of recent studies use game-theoretic models in order to obtain an explanation of the impact of interest groups on public policy. These game models have increased our understanding of the means by which, and under what circumstances, interest groups may affect government policy. To give just one example, models based on the canonical signaling game show how interest groups may provide policy relevant information to politicians through lobbying. However, these models also indicate that, only when the preferences of the interest groups and the politicians are sufficiently aligned, this information may be trusted. The formal game-theoretic analysis employed in these studies allows to make explicit what "sufficiently" means in this context, and thus makes the theory testable.

In our view, game-theoretic models of interest group behavior have yielded valuable results. This monograph aims to contribute to the research in this area. In particular, a number of game-theoretical models are presented that extend previous models. These extensions are not only meant to make these earlier models more realistic, they are particularly used to examine issues largely ignored in this literature so far. The directions in which the existing models are extended are guided by observations obtained from empirical studies. The main underlying characteristic is that, in contrast to most earlier studies, the choice interest groups have between various means of influence is studied. The specific choices that are analyzed are introduced in the next section when we discuss the outline of the book.

By building on existing models of a more or less well-demarcated subfield the contribution of this monograph is necessarily modest in nature. Our study can be best seen as a follow-up of Potters (1992). It has served its purpose if its contribution appears to be of some significance to a better understanding of the phenomena at hand.

1.2 OUTLINE OF THE BOOK

The next chapter starts with a comprehensive overview of the empirical literature concerned with the influence of interest groups on the formation of public policy. A vast number of studies belongs to this literature, and a wide range of issues are addressed. In fact, most topics alleged to be important when studying the influence of interest groups received at least some attention in this literature. Consequently, the overview of the empirical results obtained up till now not only yields some basic knowledge about the "how and when" of interest group influence, it also provides a general background for the topic of this monograph. It demarcates, albeit somewhat roughly, the relevant field of study. The second part of Chapter 2 provides a very rough and general review of existing theoretical models used to obtain a basic understanding of interest group influence. It is observed that some topics that received considerable attention in empirical work are only marginally addressed by means of theoretical modeling. Specifically, they include the choice an interest group has between various means of influence, differences between groups in their potential for using different methods, and interaction effects between distinct means of influence. The questions addressed by the theoretical models presented in this monograph mainly relate to these topics.

In all game-theoretical models to be presented in Chapters 4, 5 and 6, the distinctive feature of the influence of interest groups is the strategic transmission of information. Chapter 3 therefore first elaborates upon the basic asymmetric information game that underlies all these models. In the context of this basic (signaling) game the mechanisms underlying influence through strategic information transmission are explained. Subsequently, the applications of this game that are used in later chapters are introduced, as well as some extensions that are of particular interest in view of these applications. The extensive discussion of the basic signaling game presented in Chapter 3 enables us to introduce and analyze the extended models of later chapters more swiftly, without having to discuss the basic ingredients in full depth.

Chapter 4 first shows how campaign contributions can be modeled as indirect endorsements revealing relevant information to the electorate when deciding whom to vote for. Subsequently, it studies the question when interest groups will choose to support a candidate through campaign contributions, and when they opt for reaching out to the public directly by means of direct endorsements. Chapter 5 focuses on the query when to use "words" (lobbying) or "actions" (pressure) to influence public policy. Interesting relationships between the "reputation" of the interest group and the use of the two means are disentangled, as well as interaction effects between lobbying and pressure. Chapter 6 addresses the choice of an interest group between different levels of government. At the same time, the model presented there studies the problem of delegation of policy authority from politicians to bureaucrats in the (potential) presence of interest group influence.

Finally, Chapter 7 summarizes and evaluates the main findings reported upon in this monograph. Based on the latter evaluation a number of promising avenues for future research are indicated.

In presenting the theoretical material we have tried to keep the main text as descriptive and interpretative as possible. Although all theoretical results reported follow from a rigorous analysis of formal game-theoretic models, they are generally presented in the body of the text in an informal and, hopefully, intuitive way. Formal derivations of all the theoretical results obtained, as well as discussion of issues which are mainly of technical interest only, are relegated to the appendices at the end of the chapters in question. It is hoped that in this way the results and intuitions obtained are also accessible to readers less familiar with game theory and to readers not interested in technical details.[6]

NOTES

1. See, for instance, Danielian and Page (1994) and Schlozman and Tierney (1986).

2. For instance, in a special contribution on lobbying a Dutch newspaper (de Volkskrant, October 5, 1996, Nobody calls himself a lobbyist) writes that: "Bureaucrats in The Hague have to obtain their information somewhere, and lobbyists are mostly welcome." Similarly, in a weekly newsmagazine (HP/De Tijd, October 11, 1996, Shadows around the Inner Court) it is stated that: "He [a member of parliament] obtains his knowledge from reports and conversations with experts. Lobbyists can be important sources of information. The more confidence they are able to inspire, the larger their influence will be."

3. A strong call for restrictions on lobby-activities in the U.S., especially on the lobby activities of former federal officeholders by order of foreign clients, can for instance be found in Choate (1990) and Phillips (1994). A number of countries have already introduced the registration of lobbyists. Among them are Canada, Germany and the United States (cf. Rush, 1994). Also lobbyists of the European Parliament working in Brussels have to publicly register. By now, over 10.000 lobbyists are registered (de Volkskrant, October 5, 1996, Nobody calls himself a lobbyist). In the spring of 1996 proposals to further regulate the lobbying activities of "Euro-lobbyists" were rejected by the European Parliament. Lastly, rules to disclose information about political donations received were recently discussed in the Netherlands and in Britain (see Chapter 4).

4. Shibboleths are slogans that take the place of hard thinking. In a provocative article in *The Economist* Paul Krugman recently warned against the public discussion of monetary policy being increasingly dominated by shibboleths (Krugman, 1996).

5. A particular interest group in our models could for instance represent a trade association, an environmental group, an individual firm, or a consumer organization. Regarding the subject of influence such a group could for instance vie for protectionist measures, environmental legislation, government subsidies, or product safety requirements.

6. Full appreciation of most of the technical material presented in the appendices only requires knowledge of non-cooperative game theory at the intermediate text book level (e.g., Gibbons, 1992).

REFERENCES

Choate, P., 1990, Agents of influence, (Simon and Schuster, New York).

Danielian, L.H. and B.I. Page, 1994, The heavenly chorus: Interest group voices on TV news, *American Journal of Political Science* 38, 1056-1078.

Gibbons, R., 1992, A primer in game theory (Harvester Wheatsheaf, New York).

Krugman, P., 1996, Stable prices and fast growth: just say no, *The Economist*, August 31.

Olson, M., 1965, The logic of collective action (Harvard University Press, Cambridge).

Olson, M., 1982, The rise and decline of nations (Yale University Press, New Haven).

Phillips, K.P., 1994, Arrogant capital: Washington, Wall Street, and the frustration of American politics (Little Brown, New York).

Potters, J., 1992, Lobbying and pressure. Theory and experiments (Thesis Publishers, Amsterdam).

Rauch, J., 1994, Demosclerosis. The silent killer of American government (Times Books, New York).

Rush, M., 1994, Registering the lobbyists: Lessons from Canada, *Political Studies* 42, 630-645.

Schattschneider, E.E., 1960, The Semisovereign people (Holt, Rinehart and Winston, New York).

Schlozman, K. and J. Tierney, 1986, Organized interests and American democracy (Harper and Row, New York).

Wilson, G.K., 1981, Interest groups in the United States (Clarendon Press, Oxford).

2   REVIEW OF THE LITERATURE

In this chapter the existing literature on the political influence of interest groups will be reviewed and discussed. Section 2.1 provides a comprehensive overview of the empirical literature. This overview yields an elaborate assessment of the actual importance of interest groups for the formation of governmental policies. There appears to be ample empirical evidence that interest groups affect the political decision-making process significantly. Since a vast number of empirical studies is reviewed, covering a wide variety of issues, this overview also indicates the themes in the literature concerned with the positive analysis of interest group influence. In that way also most of the supposedly relevant determinants of the behavior of interest groups and their influence are enumerated.

Our review of the empirical literature points at some specific empirical observations which suggest a number of interesting avenues worthy of further theoretical investigation. In particular, these results indicate that interest groups have several means of influence at their disposal which may differ in their effectiveness, and between which a mutual dependence may exist with respect to this effectiveness. In Section 2.2 it is observed, though, that theoretical models addressing the issue of the choice an interest group has between different means of influence are relatively scarce. These observations motivate the direction of the theoretical work presented in subsequent chapters. Section 2.2 completes the positioning of our research in the existing literature by briefly reviewing the few theoretical studies that do focus on the endogenous choice an interest group has between several means of influence. More generally this section provides the theoretical context in which the formal models presented in the remainder of this monograph have to be placed. Finally, Section 2.3 summarizes.

2.1   THE POLITICAL INFLUENCE OF INTEREST GROUPS: EMPIRICAL EVIDENCE[1]

Allegations of the importance of interest groups for the formation of government policies have long been based on anecdotal evidence and casual empirics. Only in the seventies quantitative historical records in combination with more rigorous econometric analysis started to get more widely used. Since then there has been a vast literature, using empirical models to assess the influence of interest groups on public policymaking. Each of these empirical studies typically gives insights on certain confined aspects of interest group politics. In order to get a comprehensive assessment of the political influence of interest groups, an overview of this empirical literature is needed. This section provides such a survey.

Specifically, in this section an overview is given of the results obtained from studies that use quantitative data and an empirical model to address the 'how and when' of interest group influence on public policy. The goal is to present a stylized picture, rather than being very precise (and painstakingly long), or selecting a carefully defined subfield that leaves out much material that is closely related.[2] By taking such a broad perspective, many of the details and nuances of the individual studies have to be glossed over, on both the conceptual, theoretical and methodological level. For instance, some studies deal with organized interests, others with unorganized groups that are merely characterized by a common interest. Some authors derive hypotheses from underlying theoretical models, others simply refer to common sense arguments. Some use ordinary least squares, others employ more refined estimation techniques. The differences in approach are often subtle and sometimes important, but we refrain from fully spelling these out. As a consequence, we will have to be somewhat vague in the use of notions like interest group, influence, and public policy. The advantage of being less precise in these aspects is that it enables us to discuss, in a concise way, most issues that are said to be important when determining the influence of interest groups on public policy.

Our focus is on studies that explicitly relate interest group variables to public policy variables. Studies that relate interest group variables to other variables which are not (fully) determined by public policymakers, like public opinion, economic growth or inflation, are left aside.[3] To assess the influence of interest groups on political decision-making, generally, an equation is estimated in which the dependent variable represents a policy instrument that the interest group is hypothesized to influence. Roughly, two sets of dependent variables can be distinguished. One set concerns the behavior of individual political decisionmakers. The second set relates to policy outcomes.

The empirical analysis of individual political decisionmakers focuses predominantly on roll call voting by members of the U.S. Congress. Occasionally, voting in committees is considered, and only very rarely other, non-voting activities of individual legislators. Typically, the legislator-directed studies focus on one vote issue or a set of related issues. Popular issues for empirical inquiry are labor related issues and issues concerning industry specific regulation, like price supports, protective regulation, licensing, etc.. Another feature is that, generally, cross-section data are used to compare voting by different legislators. Only rarely time series data are used to analyze changing vote patterns of individual legislators (see, for instance, Bronars and Lott, 1997, and Grenzke, 1989a).

The studies that focus on policy outcomes investigate the extent to which interest groups have influenced the ultimate outcome of the political decision-making process. The kinds of policy variables that are investigated are quite diverse. Examples are state outlays on agricultural research, federal government expenditures, restrictions on trucking weights, and regulated sugar prices. Again, cross-sectional data are more often used than time series. The situation in the U.S., where the individual states all have significant policy freedom, provides an excellent opportunity to use data of inter-state differences of both policy variables and interest group characteristics. Compared to data from samples across countries or across industries, inter-state data have the advantage of a relatively high degree of comparability. Data on inter-state differences in occupational and industrial regulation, as well as social transfers, have been particularly popular for empirical testing. Cross-country and cross-industry studies, on the other hand, mainly deal with protection and tax policies.

Both sets of endogenous variables are important and useful for the empirical assessment of interest group influence. Proponents of strict methodological individualism probably prefer the use of data on individual policymakers. When policy outcome variables are used, the complicated issue of how to derive the aggregate policy outcome from decisions by individual legislators is usually not addressed. Even though the influence of interest groups is mediated by the actions of individual politicians, it is still an important question whether this influence is detectable in the ultimate outcome of the political process. On the other hand, if the influence on individual politicians is hardly detectable in the data, it is still possible that many small influences add up to a significant influence on the aggregate policy outcome. The use of policy variables may also be useful when it is not clear which activity of a legislator is being influenced by the interest group. Apart from roll call voting, activities like lobbying other congressmen, formulating amendments and shaping the details of bills, are important for determining legislative outcomes. Data on these activities are hard to come by, but interest group influence on these activities might be detectable in the aggregate policy outcomes.

Public policy is the outcome of a trade-off between different interests. Due to the voters' electoral control and the reelection motive of politicians, public policy in democratic societies is to a certain, possibly even great, extent guided by the preferences of the electorate. For that reason, it is important to control for the preferences of the (relevant) constituency when estimating the impact of interest groups on either legislators' behavior or public policy outcomes. When, for instance, the interests of the constituency and the interests of a specific interest group are largely aligned, and constituency variables are omitted as explanatory variables in the estimation, the influence of the interest group is most likely to be overstated (cf. Smith, 1995). Almost all studies reviewed here in some way control for constituency interests. The way in which the preferences of the relevant constituency should be measured, however, is subject to debate. Similar remarks apply to the legislators' own goals apart from reelection, especially their intrinsic valuation of policy outcomes (ideology). These personal objectives may affect policy choices, and therefore should be controlled for.[4] Because the interests of constituents and the ideology of legislators are hard to measure, an assessment of an interest group's relative influence becomes very difficult. (cf. Potters and Sloof, 1996).

In the sequel, we discuss various types of interest group related variables that have been tested on their capacity to explain political decision-making and policy outcomes. A distinction is made between variables that directly relate influence to political *activities* of interest groups (Subsection 2.1.1), and those that relate influence to structural *characteristics* of interest groups and their environment (Subsection 2.1.2).[5] Contrary to the observable political activities, like lobbying and contributing to campaigns, the structural and environmental characteristics of interest groups are usually not legislator-specific. Therefore, the distinction between activities and characteristics largely runs parallel to the use of data on individual legislators and the use of data on policy outcomes, respectively. Subsection 2.1.3 summarizes the results.

For convenience a schematic representation of the setup of the overview is presented in Figure 2.1 (see the next page). The terms used in this figure will be made precise when discussing the issue(s) at hand.

2.1.1   Political activities of interest groups

What do interest groups do to influence governmental policy? A comprehensive list would include activities like influencing and mobilizing the electorate, financing electoral campaigns, lobbying politicians and bureaucrats, or going to court. Much has been said and written about the use and relative success of each of these techniques on the basis of more or less detailed case studies. However, relatively few conclusions have been based on systematic and rigorous empirical inquiry. The single most systematically studied activity of interest groups is contributing to election campaigns. This, of course, is not surprising. Campaign contributions are easily quantifiable and data on both source and target are (at least for some countries) readily available. However, as will be seen, the evidence concerning the impact of campaign contributions on politicians' behavior is rather equivocal. The fact that there are different contribution strategies may be part of the explanation of these (mixed) results. These strategies will be discussed in the second part of this subsection. Thereafter, we review the empirical evidence regarding the impact of activities other than contributing to campaigns. Although the evidence is scarcer here, there are some interesting results concerning 'lobbying'.

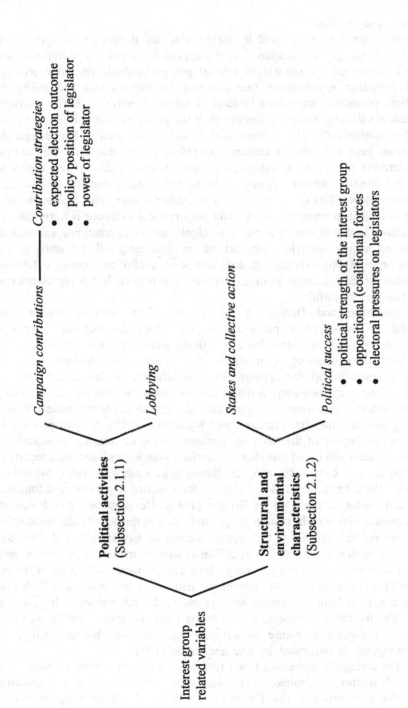

Figure 2.1 Schematic representation of the setup of the overview

Campaign contributions

The empirical research is inspired by early theoretical papers on the investment approach to campaign contributions.[6] In this approach campaign contributions are envisaged as explicit investments by interest groups to obtain favorable policy.[7] Because campaign expenditures (are assumed to) increase the probability of (re)election, politicians are willing to serve an interest group's wishes to a certain extent, and are thereby willing to deviate from the policy position preferred by the voters (or themselves). These theoretical models thus predict that campaign contributions have a discernible influence on public policy. Empirical studies that try to determine the impact of campaign contributions on public policy typically address the question whether money — in the U.S. mainly donated by political action committees (PACs) — can buy a legislator's vote. The answer to this question is subject to extensive debate, and the empirical evidence is fairly mixed. Some authors find that contributions have significant and sometimes substantial influence on voting behavior, whereas others find moneyed influence to be moderate or even non-existing.[8] It does not seem useful to give an extensive presentation of the different studies, methods and results, but a representative sample may be insightful.

Silberman and Durden (1976) is one of the earliest studies that investigates the impact of campaign contributions. They find that total campaign contributions of organized labor have significant influence on a representative's voting behavior concerning a minimum wage issue. Contributions from an opposing business group also appeared to be significant, but less influential. An omission of the study, however, is that it did not control for the legislator's party affiliation, and thus was likely to overestimate the impact of contributions if PACs give along partisan lines (cf. Frendreis and Waterman, 1985). A usual practice is to evaluate the impact of the labor (or business) share of total contributions on voting. Significant effects of this share on various vote issues have been reported, even when there is control for party affiliation (e.g., Coughlin, 1985, McArthur and Marks, 1988, Peltzman, 1984). Evidence for a limited impact of contributions on roll call voting is provided by Wright (1985). He examines five PACs of national associations which, though judged to belong to the set of most influential PACs, were not able to change the voting outcome on any specific bill. Grenzke (1989a) investigates some 120 PACs, affiliated with 10 major organizations, and finds that contributions do not generally affect House members' voting patterns. Langbein (1993) studies a specific ideological issue, i.c. gun control, and finds that contributions from both proponents and opponents did not convert a legislator's opinion. The limited influence of contributions from ideological PACs on votes which are ideological in nature (concerning, e.g., abortion, homosexuality, or nuclear weapons) is confirmed by Kau and Rubin (1993).

The ambiguity of results is not restricted to roll call voting. Though the number of studies is limited here, also committee voting and legislative involvement give mixed results. For instance, Wright (1990) investigates voting on two issues in the Agricultural and the Ways and Means committee, and finds little influence of contributions. Contrarily, Schroedel (1986) finds some evidence for the impact of contributions on committee voting. Hall and Wayman (1990) hypothesize that PAC money is allocated to enhance legislative involvement of a group's supporters and to demobilize the opposition, rather than buying votes. They investigate participation of committee members — e.g., attendance of committee meetings, handing in of amendments, negotiating behind the scenes — on a specific issue for three different committees and find strong support for the mobilization hypothesis, but no evidence for the demobilization hypothesis.

To explain the mixed results, several considerations come up. Most of these relate to the type of vote issue, the type of legislator, and the type of interest group under study. Firstly, PAC influence is likely to vary with the scope and visibility (salience) of the vote issue. Issues which are not of general interest might be more easily affected. For example, the significant effect found by Stratmann (1991, 1995a) is based on a study of votes on subsidies to the farming sector. The issue has a narrow focus, benefits are concentrated and costs are distributed widely. Relatedly, special interest groups can be expected to be more effective when their goals and actions receive less public attention. Jones and Keiser (1987) and Schroedel (1986) indeed find some evidence that contributions have a larger effect when there is low public visibility. Moreover, it can be argued that the analysis of a specific bill or a series of individual bills might be too narrow to get an accurate reflection of an interest group's overall influence. Interest groups are often interested in a variety of votes, and the single issue chosen might just be unrepresentative. Hence, a series of bills should be analyzed as a group, instead of analyzing bills as single issues. This does not necessarily lead to unambiguous conclusions, though, as the contradicting results of Bronars and Lott (1997) and Wilhite and Theilmann (1987) indicate.

A second explanation for the mixed results is that PAC influence might also be determined by the characteristics of legislators, like their need for funds. Tosini and Tower (1987) find that the percentage of campaign contributions donated by interest groups of the textile industry had a significant impact on representatives for a specific protectionist textile bill, but not on senators. An explanation could be that in the time period considered the entire House was up for reelection, whereas only one third of the Senate was in the process of campaigning. Frendreis and Waterman (1985) also provide some empirical support for the hypothesis that PAC contributions have more influence when an election draws nearer.

The third consideration concerns the contribution strategy of the interest group under study. The activities of interest groups do not only affect the positions of legislators, they are also affected by these positions. Both campaign contributions and policy decisions are endogenous variables and the relationship between them is an interdependent one. Single-equation estimation, employed in a number of (earlier) articles, may suffer from a simultaneity bias. This bias may lead to both over- and underestimation of interest group influence, depending on the strategy of the group (see Chappell, 1981, 1982, Snyder, 1991, Stratmann, 1991, 1995a). On the one hand, if an interest group mainly donates to those legislators who already have a congruent viewpoint, then the impact of money on the stand taken by legislators is likely to be overestimated. On the other hand, if a group mainly donates to legislators who do not take a favorable stand, in the hope of swaying their position, the impact of money is more likely to be underestimated. Hence, if different groups employ different strategies, single-equation estimates may lead to mixed results. Apart from being insightful in itself, an appraisal of the contribution strategies of interest groups adds to the understanding of the (mixed) results regarding the impact of contributions on legislative behavior. Therefore, we briefly review the empirical results obtained in respect to the contribution strategies of interest groups.

Contribution strategies
The theoretical literature provides two competing models for the contribution strategy of interest groups.[9] If an interest group takes the policy positions of candidates as given it will mainly try to get favored candidates elected and address its donations to "friends", especially in those races that are expected to be close. This is the so-called position-induced or support model. If, on the other hand, a group takes the election chances of candidates as given, it will mainly try to influence the policy position of candidates, especially of those candidates which are taking a stance which is not (yet) in line with the group's preferences. This so-called service-induced or exchange model leads to the prediction that an interest group will address its money to candidates from which they are likely to get favors in return, that is, the likely winners of an election, the to be persuaded candidates, and the candidates that are likely to be (or to become) powerful legislators.[10]

A number of studies directly test both competing models against each other for several types of interest groups.[11] Other studies provide information on the determinants of donations without explicitly putting the two models to the test.[12] For an assessment of the contribution strategy we will use information from both types of studies. As was argued above, the important relationships are between campaign contributions, on the one hand, and the expected election outcome, the policy position of candidates, and their power, on the other hand. We will now discuss the empirical results obtained in regard to each of these three relationships.

First, expectations regarding the election outcome are often made operational by the actual vote share in either the previous or the present election. Sometimes incumbency is taken as a proxy for electoral security. Welch (1980, 1981) is among the earlier studies. He finds that (economic) interest groups aim more money at likely winners, using a measure based on the actual vote share. There also seems to be a rather robust positive relationship between incumbency and contributions. Most of the PACs' contributions go to incumbents and not to challengers. These results provide evidence for the exchange model. More recently, the picture has become a little more subtle. For example, Snyder (1990, 1992, 1993) finds that the contributions from economic interest groups fit the "quid pro quo' model much better than those from ideological groups and individuals. Ideological groups are found to focus their attention on close races, hinting at a support strategy (see also Welch, 1979). What is more, a number of studies found this positive relationship between contributions and expected closeness to hold for economic interest groups as well. Jacobson (1980) provides an explanation for the latter result. Incumbents can usually get the amount of money they want for campaigning, and they simply want more when they are in a close race.[13] Hence, the conjunction of these results indicates that interest groups, especially economic groups, tend to address their money to likely winners (i.c. incumbents). More money, however, is donated in races which are expected to be close.

Second, to identify a candidate's policy position, by and large two kinds of measures are being used: first, a general measure, based on past voting records (for incumbents) and party affiliation, and, second, more specific measures, based on the actual vote issue(s) under study. The more specific measures obtain swing legislators on a specific issue by considering how the legislator would vote based on her or his constituency interests (Stratmann, 1992b, Welch, 1982), or the measures are based on the correlation between the estimated vote and contributions equation in a simultaneous equations model (Chappell, 1982, Stratmann, 1991). Studies that use a general measure obtain rather robust results. Interest groups mainly give along ideological lines, with corporate PACs donating to conservatives and labor PACs supporting liberals. Corporate PACs tend to be less partisan, though, giving to both Republicans and Democrats.[14] For the more specific measures, which try to identify a group's allies on particular pieces of legislation, the results are mixed. Some claim to find evidence that economic interest groups tend to give to congressmen initially predisposed to vote in their favor, others suggest that they contribute to legislators who are likely to be undecided.[15] For ideological groups in particular, the former result seems strongest. In the gun control issue considered by Langbein (1993), for instance, the NRA and Handgun Control donated only to like-minded legislators.

Third, the power of (prospective) legislators and their ability to supply political favors, is largely determined by their position in the legislature. Specifically, party affiliation, committee membership and seniority play a role here. Being a member of the majority party, for instance, might make it more likely that the preferred policy outcome can be provided. This may explain why corporate PACs give substantial amounts of money to (conservative) representatives from the Democratic party. (All the empirical studies reviewed here relate to a time period that the Democratic party controlled the U.S. House of Representatives.) Moreover, several studies indicate that interest groups, in particular economic interest groups, tend to give more money to representatives who join important committees[16], thus providing support for the exchange model. Committees largely control the agenda and hence their members possess political clout with respect to the specific jurisdiction of the committee. Interestingly, Grier et al. (1990) and Grier and Munger (1993) find that the effect of committee membership on contributions received is larger for the U.S. House of Representatives than for the Senate, where the power of committees in shaping bills is diminutive. Lindsay and Maloney (1988) and Regens et al. (1991, 1994) obtain the same dissimilarity between the House and the Senate with respect to seniority. However, even though seniority is correlated with committee membership, its effect on contributions received seems less clear cut. Evidence suggests that seniority is positively related to contributions from corporate PACs, but the results for labor PACs and ideological groups are more ambiguous.[17]

In summary, there is empirical evidence for elements of both the exchange model and the support model. On the one hand, interest groups appear to contribute to secure, strong, but undecided candidates. On the other hand, interest groups support like-minded candidates that are expected to be in a close race. However, corporate PACs appear to lean more heavily on an exchange strategy than ideological groups. Labor PACs seem somewhere in between.[18] Hence, the differences in contributing strategy can help explain the mixed results obtained concerning the impact of money on votes (in one-equation estimation). In particular, the impact of corporate contributions is more likely to be underestimated than those of ideological PACs, because one-equation estimation underestimates the impact of contributions on voting behavior when the politician is a priori inclined to vote against the group's wishes.

Lobbying
Studies that incorporate interest group activities other than donating to campaigns are rare. We have grouped these activities under the header "lobbying". Most studies that fit within this category report a significant influence of lobbying on policy.[19] For example, Kau and Rubin (1979) consider voting on five bills in the House on which two public interest lobbies explicitly endorsed a specific policy position (see also Cropper et al., 1992). They take the per capita membership of the interest groups in a state as a measure of the strength of the influence activity and show that these interest groups were effective in influencing votes of representatives. Segal et al. (1992) examine the voting behavior of senators on sixteen nominations for the U.S. Supreme Court. The number of organized interest groups presenting testimony for and against the nominee at the Judiciary Committee hearings appeared to have a profound effect. A similar conclusion is reached by Caldeira and Wright (1998). Using data on three nomination votes they found that the number of interest groups that had direct contact either with the senator or with the senator's staff had a significant impact on the senator's confirmation vote.

Other interesting studies that find a significant influence of lobbying include Wright (1990). He investigates the influence of the number of lobbying contacts on committee voting. Questionnaires are used to identify the number of groups lobbying on each side of the issue and substantial influence is reported. Sometimes, letters, visits and phone calls are also found to have a discernible influence (cf. Congleton and Bennett, 1995, Langbein, 1993). Schneider and Naumann (1982), investigating Switzerland's direct democracy, find that motions to the government and the parliament by small and medium sized business are successful in dampening the spending increase of six out of nine items. On the other side, and to a somewhat lesser extent, farmers and trade unions seem able to further government spending. Their results suggest that, in line with previously discussed results in regard to the influence of campaign contributions, lobbying influence is likely to be dependent on the issue under consideration and to vary over interest groups (see also Esty and Caves, 1983).

Some studies do not focus on a specific lobbying activity, but find a significant impact of measures of an interest group's overall lobbying activity on public policy. For instance, Hoyt and Toma (1993) find lobbying expenditures by the National Education Association to have a significant impact on the salary of teaching staff. Leigh (1994) reports that the number of auto club lobbyists per capita, which is used as a measure of the strength of the pro-safety lobby, had a positive effect on the number of required state-wide vehicle inspections per year. Lastly, Mixon (1995) reports that the (estimated) number of registered lobbyists in an urban center, who are assumed to be mainly concerned with business interests, significantly reduced the average number of days the center was registered to be in violation of carbon emissions standards.

Just like donors of campaign contributions, lobbyists do not only affect the positions of politicians, they are also affected by these positions. The studies mentioned above, however, typically employ a single-equation estimation technique. For a full evaluation of the results, it is important to look at the lobbying strategies of interest groups. As in case of campaign contributions, lobbying may be used either to support and encourage "friends", or to persuade legislators who are predisposed to take a stand which is against the group's interest. If the former lobbying strategy is being used, single-equation models are likely to overestimate the impact of lobbying. In case of the latter strategy, the effect of lobbying may well be underestimated.

The few studies that investigate the lobbying strategy of interest groups provide some tentative evidence that lobbying is mainly intended to alter policy positions.[20] (Interestingly, this is contrary to the conventional wisdom of the early sixties, see e.g. Bauer et al. (1963) and Milbrath (1963).) Austen-Smith and Wright (1994), for instance, test a game-theoretic model which may explain the number of groups lobbying senators predisposed to vote for or against the 1987 nomination of Bork for the U.S. Supreme Court (see also Caldeira and Wright, 1998). Their results indicate that groups mainly lobbied "unfriendly" legislators. Friendly senators were mainly lobbied to counteract lobbies from opposing groups.[21] This result shows that lobbying is intended to alter voting intentions of legislators. Similarly, Hoyt and Toma (1993) find that lobbying expenditures by the state affiliates of the National Education Association are higher in states where the income position of teachers is relatively bad. Unfortunately, too few studies exist to draw a firm conclusion. If anything, the one-equation results are more likely to underestimate than to overestimate the impact of lobbying, however.

A final issue we want to touch upon is the relation between the contribution strategy and the lobbying strategy. Wright (1990) argues that contributing and lobbying are complementary activities that have to be analyzed jointly. In his empirical analysis the number of groups lobbying appears to depend (weakly) on the number of groups contributing, suggesting that contributions may facilitate access and reinforce lobbying messages.[22] Humphries (1991) concludes in a similar vein that contributions are used to support lobbying activities. Apart from some evidence that a corporation is more likely to have a PAC if it is represented by lobbyists, he also finds a positive link between the number of corporate lobbyists and the amount of money contributed to the corporate PAC. Langbein (1993) investigates both contributions and lobbying of "membership" interest groups. Membership groups are defined as interest groups that heavily rely on individual persons, rather than on firms or other organizations, for financial and political support. She argues that, as opposed to contributions, lobbying is less visible and can be strategically directed to opponents or pivotal legislators without antagonizing the group's membership. Indeed, the legislators targeted by lobbying on average appear to be more moderate on the gun control issue than those addressed by money, suggesting that lobbying is more exchange and less support oriented than contributing.

2.1.2   Characteristics of interest groups and their environment

When no data on interest group activity are available, the only route open is to try and relate public policy variables to structural characteristics of the interest groups (cf. the structure-performance approach in industrial organization). An important issue in this respect is whether the unobserved political actions of a group have an independent role or a facilitating role in shaping public policy. In the former case both a group's political actions and its structural characteristics independently influence policy. In the latter case, political actions are a sort of transaction cost that has to be born by the interest group in order for its favorable structural attributes to have an impact. In this case, it may be expected that a group with favorable characteristics is more likely to engage in political activities. Therefore, under the facilitating role of political actions a group's characteristics can be expected to serve as reasonably good instrumental variables for the group's political activities. Of course, the relationship between public policy variables and interest group characteristics is also of interest in itself, for interest groups may exert influence without engaging in explicit political activities (cf. Lindblom, 1977).

Esty and Caves (1983) and Lopez and Pagoulatos (1996) find evidence for both a facilitating and an independent role of political expenditures.[23] On the one hand, an industry's political expenditures can be explained fairly well by its structural attributes.[24] On the other hand, political expenditures appear to have a significant impact on a group's political success even when various kinds of structural attributes of an industry are controlled for. Because of the first result it seems quite possible to find good proxies for an interest group's political activity, when direct data on these activities are unavailable. The latter result indicates that one must keep in mind that political activities may vary over interest groups - and have an independent impact on policy outcomes - in a way which is not completely parallel to their structural characteristics.

Most variables included as proxies for activity relate either to the stake of a group to influence policy or to the free-rider problem of collective action. The results of these two measures will be reviewed first. Then we look at variables that are not proxies for a group's activities, but independent determinants of its political success. These variables include attributes of the group under study, as well as characteristics of the environment of the interest group that strengthen or weaken its influence on the policy process.

Stakes and collective action

An interest group that has a large stake in influencing policymakers and regulatory agents is hypothesized to be more politically active and hence to have a larger impact on policy. Variables typically used to measure the stake of a group of producers or an industry in influencing government policies are the average size of the firms involved and the percentage of proprietorial income.[25] According to Gardner (1987, p. 305): "The size of an average production unit, given the number of producers, is expected to have a positive effect on protection because the size of the gains per interest group member will be greater, increasing incentives for participation in pressure group activities." The importance of proprietorial income as a measure of stake is illustrated by the following observation of Guttman (1978, p. 475): "..owner-farmers would tend to favor [government] investment in agricultural research more than tenant farmers, because owner-farmers would capture a larger share of the increase in the producers' surplus." Another variable used to measure stakes is the degree of government involvement in an industry (Esty and Caves, 1983). The defense industry, for instance, is often thought to have a large stake in influencing political decision-making. Also, strong competition or import penetration is expected to increase an industry's demand for government intervention (Jarrell, 1978, Trefler, 1993). If competitive pressures are high and prices are low, an industry is more likely to vie for protective regulation. The estimated influence of the aforementioned stake variables mentioned is often significant and usually of the predicted sign. These results suggest that this approach, to take measures of stakes as a proxy for a group's political activities, is a sensible one if direct data on political activities are unavailable.

A second proxy for a group's political activities is the degree to which it is able to solve the free-riding problem of collective action. If free riding is severe, political activity will be low. A distinction can be made here between direct and indirect measures of free riding. For ideological groups and labor unions, often a direct measure is available in the form of the number or the percentage of members.[26] Membership rates of organized interest groups are often found to have a significant positive impact, as predicted. For example, Meier and McFarlane (1992) find that the number of members in a certain pro-choice interest group positively affected state funding of abortions in the U.S., and Kirchgässner and Pommerehne (1988) find that union membership has a significant positive impact on government expenditure in Switzerland. Hence, when a direct measure of free riding is used often a significant effect on the possibility to achieve favorable public policy is obtained.

There are also many interests that are not represented by a formal organization, or membership data are unavailable. In that case indirect proxies for the level of collective action have to be used, like the mere number of producers in an industry or some measure of concentration. Such indirect measures give more ambiguous results, however. Geographical concentration and concentration of sales are often hypothesized to alleviate free riding due to an increased possibility of communication and orchestration of political action. Most scholars indeed find an increased level of political influence with higher degrees of concentration, but there are many that find no effect or even a negative effect.[27] Equally ambiguous are the results of the use of numbers as a proxy for the free rider effect. A large number of potential participants in collective action is usually hypothesized to increase the free riding problem. Sometimes indeed a negative effect of numbers on influence is reported. More often, however, a positive effect is found.[28] Hence, on a general level there seems to be relatively little direct empirical support for Olson's (1965) influential theoretical study of collective action.

For both concentration and numbers, one could also conclude that the relationship between these proxies and the independent variables is not only driven by free-riding effects but also by effects which relate to the political strength of an interest group.[29] For example, it is sometimes argued that a geographically dispersed industry not only has more problems of organization, but also has a leverage on a larger group of local representatives and, hence, a broader political base of support (Schonhardt-Bailey, 1991, Snyder, 1989). Similarly, large groups not only have a larger free-riding problem, they also have more (electoral) resources. Probably the relationship between numbers and influence is not a linear one, and perhaps not even a monotonic one. This conclusion is supported by Guttman (1978) who finds that the effect of the number of producers on influence is more likely to be positive when there is a relatively large number of producers in a state (he considers the U.S.). This would suggest a marginally *increasing* effect of numbers. This finding is not robust, however. Miller (1991) finds that the number of farmers has a positive impact on its influence in developed countries but a negative impact in less developed countries. With the relative number of farmers in developed countries being much smaller, this finding would suggest a marginally *decreasing* effect of numbers on influence.[30] Nevertheless, the conclusion remains that the relationship between numbers and influence is an intricate one. It may be non-monotonic and, in addition, dependent on the type of interest group (Van Velthoven and Van Winden, 1986).

Political success
Some interest groups receive a better hearing in politics than others. Moreover, a group may have more influence at one time than at another time, or book a success with one policy issue but not with another. A wide variety of variables is employed in the empirical literature to pick up such differences in political success. Sometimes these variables are attributes of the interest group or the issue under study, on other occasions these variables refer to the political system, the public at large, or even the state of the economy. Though it is not a trivial task to try and classify these different variables, by and large three main groups can be distinguished. The first group of variables refers to the political strength of a

group, the second to the presence of oppositional or coalitional (lobbying) groups, and the third to the electoral pressures on the polity.

The first group of variables can be thought of as referring to a group's political strength or leverage. Certain attributes of a group or its environment are found to affect its success in the political arena. For example, although many authors argue that the size of a group or its geographical dispersion make collective action more difficult, it was argued above that these same variables might very well be among the attributes which give a group greater leverage in politics. It is very hard to disentangle the two effects, and there seems to be no systematic evidence that one of them is generally stronger, or stronger in particular cases. Equally ambiguous results are obtained for the average or relative income position of the members of an interest group. Some find a positive, others a negative effect on a group's political success.[31] Again, a double-tiered effect may be at work here. On the one hand, if a group's average or relative income position is bad, it has a larger stake to do something about it and it may be easier to demonstrate that political support is "really" needed. On the other hand, if income is bad, a group may lack the resources to back up or start political activities. Also in this case there is no evidence that one effect dominates the other. We shall now concentrate on some less ambiguous determinants of political success.

For producer groups some factors have been found that seem supportive of favorable government intervention.[32] If an industry is struck by large unemployment then this unfortunate feature clearly promotes a plea for protectionist measures. Also industries that are hurt by large duties on inputs have an easier time to get "compensating" protection. Hence, being in need favors your case. What also favors the position of producers is that prices of substitutes are high. This appealing argument is brought to the fore by Beghin (1990) and can be derived from a game-theoretic model of bargaining. He argues that producers are in a better position to bargain for high regulated prices if the outside option for consumers and government (to buy substitutes) is unfavorable. Furthermore, for non-industrial groups Plotnick (1986) finds that the "quality" of the membership matters. Studying the inter-state differences in child-support, he finds that the level of transfers is negatively related to the percentage of non-white recipients and the percentage of mothers with illegitimate children. Having members with low social status seems to harm your case. Summarizing the findings with respect to political strength, there appears to be some indication that being in need, having a strong bargaining position, and being of high social status help to get favorable government intervention. The evidence here is rather scant, however.

The second set of variables identified above that relates to the political success of an interest group concerns the presence of friends and foes. A fairly robust finding is that a group has more political success if it is likely to encounter a (strong) coalitional force in the political arena, and, conversely, that the presence of an oppositional lobby is harmful to its case.[33] For instance, Teske (1991) finds that the presence of a government funded consumer advocacy harms the case of the telecommunications business. Others argue that the composition of government may provide an interest group with a coalitional force within government,

increasing the group's political success. An example in this respect is the empirical
evidence suggesting that the relative number of females in local councils has a
positive effect on the supply of day-care for children by municipalities, an effect
found for both the Netherlands (Van Dijk et al., 1993) and Sweden (Gustafsson
and Stafford, 1992). Often, however, less direct measures of opposition and
support are used. Esty and Caves (1983), for instance, count the number of
industries that are customers of the industry under study and find that the presence
of many customers is harmful for a group's case. Here the supposition is that
customers fear to lose from high prices created by an industry's political efforts.
Less ad hoc and more convincing perhaps is Gardner's (1987) hypothesis that
producers are less likely to obtain high regulated prices, the larger the burden that
such prices lay on consumers (cf. the theoretical model of Becker, 1983, 1985).
Gardner uses demand and supply elasticities to measure the deadweight losses of
redistribution and finds these to be significant. Hence, though sometimes measured
a little ad hoc, the presence of (potential) friends or foes appears to affect the
political success of an interest group in an predictable way.

Now we turn to a third and final group of variables that affect a group's
reception in politics. The variables in this group all somehow relate to the electoral
or democratic pressures on the polity. The general supposition is that self-
interested politicians have to make a trade-off between, on the one hand, the
benefits they receive from special interest groups in return for special favors, and,
on the other hand, the electoral damage that may result from supplying these
special favors. Variables that affect this trade-off will thus have an impact on the
amount of favors provided to these groups. In particular, if policymakers are in
severe electoral competition, there is less discretion to disregard voters' interests
and less room to give in to demands by special interest groups. Of course, if an
interest group can mobilize considerable electoral resources, either directly or
indirectly, then electoral pressure might favor a group's position (e.g., Plotnick,
1986, Stigler, 1971). We already saw, for instance, that campaign contributions
appear to have a stronger impact on legislative voting when an election draws near
and there is increased need for campaign funds. On balance the literature suggests,
however, that interest groups have an easier time influencing politicians when the
latter are under little democratic pressure and have considerable discretion.[34]

Several measures have been used to account for electoral pressure. One
of them is income. Many studies find that per capita income has a positive impact
on particular government spending categories.[35] Some argue that per capita
income represents a measure of the opportunity costs of voters to monitor the
political process, so that a higher income level makes it easier for interest groups
to vie for an increase of particular expenditures. Others, however, argue that a
demand side effect is at work, as many of these government expenditures are
luxury goods to the voters. A second measure of electoral pressure, also producing
equivocal results, is the degree of urbanization. McCormick and Tollison (1981)
report a negative impact on interest group influence. They conclude that
urbanization makes it easier for the public to engage in counterbalancing lobbying.
Contrarily, Stigler (1971) finds a positive impact and argues that urbanization

makes it easier for an interest group to solicit electoral support for its case. A third, perhaps less ambiguous, measure is implied by the suggestion that a well-informed constituency makes congress less susceptible to the demands of special interests and more prone to serve the public interest.[36] Becker (1986), for instance, finds that it is harder for the dentist profession to obtain entry restrictions if a state is characterized by a well-educated electorate. Graddy and Nichol (1989) obtain the same result for registered nurses. Relatedly, as we saw in the previous section, campaign contributions are less likely to buy a favorable legislative vote when there is high media coverage of the issues concerned. All in all, the picture is that electoral pressure decreases politicians' susceptibility to specific demands by interest groups.

2.1.3 Summary and evaluation

It is generally believed that interest groups carry much weight in the determination of public policy. The empirical results reviewed and discussed in the previous two subsections provide us with more tangible evidence for the political influence of these groups. Even though in a number of cases equivocal results are obtained, it seems justified to present the following "stylized facts":

1. Campaign contributions and lobbying alter a legislator's (voting) behavior, particularly with respect to bills with a narrow focus and low public visibility.
2. The strategy of ideological interest groups is oriented towards supporting like-minded legislators, that of corporate interest groups is more aimed at changing the political position of legislators. Labor interest groups employ an intermediate strategy.
3. The larger the organized membership of an interest group, the larger its political influence will be.
4. An interest group's stake in influencing public policy is a positive determinant of both its political activity and its success.
5. The relation between the number of potential participants of collective action and influence on policy outcomes is an intricate one, driven by both free-riding effects and effects on the group's (electoral) resources. The same holds for measures of geographical concentration.
6. The presence of an oppositional (coalitional) force in the political arena hurts (helps) a group's case in politics.
7. Strong electoral pressures on the polity and the presence of a well-informed electorate, lower the influence of special interest groups.

We end this overview by pointing at two shortcomings of the current empirical literature that future studies should take hold of (a more comprehensive discussion of the problems and perspectives of the empirical literature can be found in Potters and Sloof, 1996). Firstly, lack of data constrains the assessment of the influence of interest groups on public policy in important ways. In this respect two interrelated problems, in particular, are worth mentioning. The first relates to the frequent use of proxies (instrumental variables) for the relevant but unobservable variables. The use of proxies often leads to problems of interpretation. For instance, in regard to "stylized fact" 5 above, the number and geographical dispersion of potential participants may both represent free-riding effects and the political strength of an interest group. Likewise, the use of proxies may lead to serious mismeasurement of control variables (constituency interests, legislator's ideology), and hence to over- or underestimation of the (relative) influence of interest groups (cf. Smith, 1995). The second problem concerns the rather limited scope of the empirical analyses in a number of respects and, hence, the question of generalizability. The empirical studies are strongly biased towards one country (the U.S.), one specific means of influence (campaign contributions), the impact of interest group behavior on legislative voting, and are narrow policy areas. Besides, though many argue that dynamics are important, most studies investigate the influence of interest groups in a static framework.[37] In short, the empirical literature could benefit greatly from broadening its database, to include more countries, a larger variety in the means of influence employed by interest groups, and to account for the dynamic aspects in the relationship between interest groups and public policymakers.

   Second, many empirical studies lack sufficient theoretical underpinning. As a result, the empirical model becomes rather ad hoc and the estimated equations may fail to have a clear cut interpretation. Indeed, often empirical results can be interpreted in several ways and explained by different factors. This is illustrated by the frequent statement in the formulation of hypotheses that the sign of a particular coefficient can be positive or negative for different reasons. In order to avoid the multiplicity of possible interpretations, competing hypotheses should be derived from underlying theoretical models.[38] In that way theory can guide the analysis by structuring the search for new material and providing a base for coherence and embeddedness.

2.2   THE POLITICAL INFLUENCE OF INTEREST GROUPS: THEORETICAL MODELS

The observed importance of interest groups for the formation of public policy makes them an interesting object for theoretical inquiry. A substantial theoretical literature already exists (see Austen-Smith, 1997a, Potters and Van Winden, 1996, and Van Winden, 1997, for comprehensive overviews). A shortcoming of this literature is that many studies, predominantly the earlier ones, are not explicit about what kind of activities interest groups undertake to influence public policy. As a result, they do not explain why interest groups are able to affect policy. In these models it is just assumed that interest groups exert "pressure" on government through spending resources, i.e. they use a black box production process of political pressure. Moreover, government is modeled in reduced form because it is assumed to react mechanistically, i.e. in a predescribed, exogenously given way, to interest group pressure (cf. the influence function approach of Becker, 1983, 1985).[39] Because these models are not explicit about which activities are involved, pressure may very well represent any activity of the interest group, or the aggregate influence of all instruments used by the interest group together.

More recent studies are more explicit about the kind of activities interest groups use when modelling the impact of interest groups on public policy. Ainsworth and Sened (1993), Lohmann (1993a, 1993b, 1994a, 1995a) and Potters and Van Winden (1990), for instance, model "political pressure" as the strategic transmission of information through costly political action. In Potters and Van Winden's model costly political action refers to the carrying out of a sanction by an interest group when the government does not comply to the group's wishes. For instance, a public sector union organizes a strike when the government does not give in to its wage claim. In the models of Ainsworth and Sened and Lohmann costly political action refers to signing a petition, joining an interest group and/or taking part in protest demonstrations. [40] Besides being explicit about the kind of activities, in these models the influence of political pressure is explained rather than assumed from the outset. Government behaves strategically rather than mechanistically.

Just like the empirical literature, most theoretical papers that are explicit about the means of influence involved focus on the impact of campaign contributions on public policy. As mentioned when we discussed the empirical evidence on the contribution strategies of interest groups, theoretical models of campaign contributions can be divided, by and large, along a motivational line. Campaign contributions are either intended to obtain specific favors from politicians (service-induced model), or are used to get the favored candidate elected (position-induced model).[41] In the former type of models politicians value campaign contributions because they are assumed to buy votes. Likewise, in the latter type of models a positive relationship between campaign expenditures and votes is assumed rather than explained. In other words, in both types of models voters respond mechanistically to political campaign expenditures.

More recently, "lobbying" activities have also received considerable attention in the theoretical literature. They have been modeled as explicit bribes, as implicit payments through "wining-and-dining", and as a means of strategic information transmission.[42] In the latter case the information may refer to the electoral salience of a group's cause, or relate to how policy eventually maps into outcomes. Of course, when lobbying is modeled as money transfers to the politician, there is only a small difference between lobbying and campaign contributions. As a matter of fact, some authors use the terms lobbying and contributions interchangeably (e.g. Ball, 1995).

Finally, some studies analyze interest group influence by means of the voting power of their individual members. The probabilistic voting framework of Coughlin et al. (1990a, 1990b) predicts that the influence of an interest group is positively related to the size and the homogeneity of the group. In the rather different cooperative game setup of Dougan and Snyder (1996) individuals form coalitions of voters when voting on the redistribution of income. These authors conclude, among other things, that policies intended to reduce income inequality between groups are more likely to involve direct payments than policies that redistribute wealth across groups with similar incomes.

The theoretical work presented in this monograph first of all adds to the existing literature on interest groups by providing a (position-induced) model for the influence of campaign contributions, without relying on the assumption that voters behave mechanistically (cf. Chapter 4). More importantly, however, is that the endogenous choice an interest group may have between different means of influence is studied. Our survey of the empirical literature indicated that different means of influence may differ in their effectiveness. Also, different types of interest groups may have a different scope for using different means. Furthermore, the empirical evidence that contributing and lobbying are interrelated indicates that there may exist interaction (complementarity) effects between several means of influence. In all the theoretical models discussed above, however, interest groups have only one specific means of influence available. Therefore, the question why an interest group uses one rather than another — and, thus, why different groups employ different means — cannot be addressed. Likewise, issues such as whether different means of influence are complements or substitutes in the production of influence, remain underexposed. Only a very few existing models are concerned with the endogenous choice of a specific type of instrument, or with the combined influence of a package of instruments. The models presented in this monograph add to this small, but growing, literature. We will now briefly review the few existing studies in this field and informally introduce the models presented in the next chapters.

Moore and Suranovic (1992) analyze an industry's choice between lobbying the legislative branch and/or the executive branch in obtaining government protection. The choice of a specific option is simply determined by its exogenously given relative success. The model does not indicate what the determinants of this relative success are, because lobbying is modeled as a black box process.[43] The main point of the paper is that policies lowering the probability of success of (one or both of) the two means of influence may reduce national welfare. The reason is that the industry may switch to an alternative option of protection seeking which uses more resources, or starts using the alternative option after an unsuccessful first attempt using its most preferred option. A similar criticism as the one raised against the Moore and Suranovic model applies to Hoyt and Toma (1989). They adapt the model of Becker (1983) by allowing interest group lobbying both at the local and the state level. Their specific assumptions concerning the influence functions, namely that lobbying at the state level is (far) more effective, immediately yield the conclusion that interest groups mainly direct their lobbying efforts at the state level. It is not addressed what the determinants of these influence functions are.

Austen-Smith (1993) is more specific about the type of instruments interest groups use. Moreover, their influence is explained rather than assumed. He models lobbying as the strategic transmission of information. In his model only an interest group may acquire information about the final consequences of certain policies. The interest group may lobby the (closed rule) legislature both at the agenda setting stage (committee members) and at the vote stage (House members). In equilibrium the decision at which stage(s) to lobby depends on the relative preferences of the House, the committee and the lobbyist. It appears that there can coexist influential lobbying at both stages. But, lobbying the committee is likely to be influential, whereas lobbying the House is often only informative (that is, not affecting the voting behavior of its members). Note that this result is in line with the empirical results obtained by Hall and Wayman (1990) and Wright (1990) regarding the effectiveness of campaign contributions (cf. Section 2.1).

Lastly, in Austen-Smith (1995) and Lohmann (1995b) interaction effects between campaign contributions and informational lobbying are analyzed, where the former activities buys access for the latter (see also Austen-Smith, 1997b). In the first study lobbying is modeled as a cheap talk message concerning the group's private information on both its preferences and the consequences of a particular policy. The potential gain to the interest group from having the possibility to tell its story (access) is then increasing in the similarity with the legislator's preferences. Campaign contributions can signal the value of access to the interest group, and thereby indicate the informational value (credibility) of the lobbying message regarding the consequences of legislation. Interestingly, complementarity effects between campaign contributions and lobbying essentially disappear when the interest group is already informed on the consequences of policy prior to making contributions. In that case all information that can be credibly transmitted, can be transmitted through contributions alone. Like in Austen-Smith's study, in Lohmann (1995b) contributions are used to enhance the credibility of a cheap talk lobbying message in case the interests of an interest group conflict with those of the policymaker.

In this monograph three different models are presented in which an interest group has various means of influence at its disposal. Contrary to Moore and Suranovic (1992) and Hoyt and Toma (1989), in all these game theoretic models government is modeled as an active player in the game. The models differ from Austen-Smith (1995) and Lohmann (1995b) in the means of influence focused at. In Chapter 4 the choice between two different ways of supporting a candidate running for political office is analyzed. An interest group may either contribute money to the candidate's campaign, or reach out to the voters directly via direct endorsements. In this model the interest group in fact influences public policy indirectly by influencing the electorate. Chapter 5 studies the choice of an interest group between words (lobbying) and actions (pressure) in order to affect the decision of a single policymaker. In this model the interest group directly influences the policy decision of a policymaker. Besides incorporating into one model both lobbying and pressure, the model extends the existing literature by analyzing (repeated) lobbying in an intertemporal framework without relying on a black box process (as e.g. in Wirl, 1994). Finally, the third model presented in Chapter 6 looks at the interest group's choice of lobbying at different tiers of the government organization. Depending on how policy authority is divided within government, the interest group may either lobby (transmit information to) the legislative branch and/or the executive branch. An important characteristic of this model (and the one presented in Chapter 5) is that the group's choice of instrument is directly and actively influenced by government itself. By taking a specific course of action, government may exclude the choice of a specific type of instrument. Also, in this model the politician's choice to delegate policy authority to the bureaucracy in the presence of interest group influence is analyzed. These two features, among other things, distinguishes the third model from the Austen-Smith (1993) model discussed above.

2.3   SUMMARY

We started out in this chapter with an overview of the empirical evidence of the importance of interest groups in the formation of government policies. Apart from providing a justification for the subject of this monograph, this overview is of interest in itself because it provides a comprehensive assessment: activities and characteristics of interest groups appear to influence public policy significantly (see the seven "stylized facts" listed in Subsection 2.1.3). The discussion of the empirical results obtained in the literature provides a basic knowledge of the how and when of interest group influence. For instance, specific empirical evidence suggests that means of influence may interact, and that they may differ in their effectiveness. A strategically acting interest group will, therefore, look for an optimal choice of instruments. In particular this latter observation inspired the larger part of the theoretical work presented in the sequel of this monograph.

In Section 2.2 we directed our attention towards the theoretical literature. It was observed that many theoretical studies are not explicit about the kind of influence activities involved when modeling the influence of interest groups on public policy. Those studies that do make explicit the means of influence interest groups employ typically focus on one specific means in isolation. Activities like political pressure, lobbying, voting, and, in particular, contributing to political campaigns received attention. Only a few theoretical models study the endogenous choice an interest group has between several means of influence, and these models were reviewed in more detail. In that way, Section 2.2 provided the theoretical background for the game-theoretical material to be presented in later chapters.

In all three theoretical models that will be discussed in Chapters 4, 5 and 6, the fundamental characteristic of the influence of interest groups is the transmission of information. Moreover, government, interest groups and voters (when present) act strategically, and influence is explained rather than assumed. In order to get the reader acquainted with game-theoretic models of influence through the strategic transmission of information, the next chapter discusses the basic asymmetric information game that underlies all three models.

NOTES

1. The material presented here is a substantially shortened version of Potters and Sloof (1996).

2. Other existing surveys of the empirical literature have a more restrictive scope. For instance, Noll (1989) gives a detailed assessment of the 'interest group' versus the 'public interest' theory of regulation, Swinnen and Van der Zee (1993) review applications in the field of agricultural policies, and Smith (1995) focuses on interest group influence in the U.S. Congress.

3. Of course, interest groups may aim their influence activities at the electorate in trying to influence public policy indirectly. Only a few studies investigate the influence of interest groups on voter behavior, however. For instance, Schneider and Naumann (1982) find that the recommendations of interest groups sometimes had a significant influence on voters in Swiss referenda. Lupia (1994) and Rapoport et al. (1991) also find that interest group endorsements may affect individual voters.

4. In general, any factor that may affect policy choices should be controlled for. For instance, it seems rather likely that individual legislators are partly affected by the "party-line" when voting on a specific bill, that is, by the stance the party (s)he belongs to takes on the issue. When studying voting behavior of individual legislators it thus seem important to control for party affiliation.

5. Roughly, the distinction runs parallel to the difference between structural form and reduced form estimation. In the former case, endogenous political decision variables are related to endogenous (decision) variables of interest groups. In the latter case, endogenous political decision variables are related to exogenous (characteristics) variables of interest groups.

6. See Bental and Ben-Zion (1975), Ben-Zion and Eytan (1974) and Welch (1974). For a historical overview of the theoretical economics literature on interest groups, and a discussion of the theoretical origins of most of the empirical research reviewed here, see Mitchell (1990) and Mitchell and Munger (1991).

7. The investment approach assumes that the goal of campaign contributions is to alter the policy position of the politician, rather than to help the favored candidate in getting elected. In that way, the investment approach assumes that the *exchange* model rather than the *support* model applies for contribution strategy of interest groups. These two competing models are discussed in full detail below when we turn to the contribution strategy of interest groups.

8. Studies belonging to the first group are, among others, Chappell (1981), Coughlin (1985), Frendreis and Waterman (1985), Kau, Keenan and Rubin, (1982), Kau and Rubin (1982, 1993), McArthur and Marks (1988), Masters and Zardkoohi (1988), Peltzman (1984), Saltzman, (1987), Silberman and Durden (1976), Stratmann (1991, 1992a, 1995a), and Wilhite and Theilmann (1987). Studies that find support for only moderate influence include Abler (1991), Bronars and Lott (1997), Chappell (1982), Grenzke (1989a), Langbein (1993), Wayman (1985), Welch (1982), Wilhite (1988) and Wright (1985).

9. See e.g. Austen-Smith (1997a), Morton and Cameron (1992), Potters and Van Winden (1996), Van Winden (1997), and Welch (1974).

10.In passing, we note that the support model, contrary to the exchange model, excludes the occurrence of "split giving". Langbein (1993) and Poole and Romer (1985) indeed find that contributors usually do not give to both candidates in the same race or to both sides of an issue. Schlozman and Tierney (1986), however, find split giving to occur quite often.

11. Abler (1991), Chappell (1982), Herndon (1982), Keim and Zardkoohi (1988), Langbein (1993), Levitt (1998), Maitland (1985), Saltzman (1987), Snyder (1990, 1992, 1993), Stratmann (1991, 1995b), Welch (1979, 1980, 1981, 1982), Wright (1985).

12. Bennett and Loucks (1994), Endersby and Munger (1992), Gopoian (1984), Grenzke (1989a, 1989b), Grier and Munger (1986, 1991, 1993), Grier, Munger and Torrent (1990), Jacobson (1980), Kau and Rubin (1993), Kau, Keenan and Rubin (1982), Kroszner and Stratmann (1995), Lindsay and Maloney (1988), Munger (1989), Palda and Palda (1985), Poole and Romer (1985), Poole et al. (1987), Regens et al. (1991, 1994), Romer and Snyder (1994), Schlozman and Tierney (1986), Silberman and Yochum (1980), Stratmann (1992b, 1994, 1995a), Wilhite and Theilmann (1986a, 1986b, 1987).

13. Note that the two basic models of the contribution strategies of interest groups do not explicitly take account of the demand for contributions by legislators. However, demand side considerations are important, especially for the exchange model in which the legislator has to supply political favors in return for funds (cf. Giertz and Sullivan, 1977). The contributions actually observed are the outcome of both demand and supply considerations. When the former are accounted for, it becomes more difficult to distinguish the two competing contributions strategies empirically. For instance, when demand considerations are taken into account, the exchange model is compatible with both larger and smaller contributions to likely winners (cf. Welch, 1982). For a more detailed account of the issues raised here, see Potters and Sloof (1996).

14. See, e.g., Chappell (1982), Gopoian (1984), Grier and Munger (1991), Keim and Zardkoohi (1988), Poole and Romer (1985), Saltzman (1987), Welch (1982).

15. See, e.g., Abler (1991), Chappell (1982) and Welch (1982), and Stratmann (1991, 1992b), respectively. Welch (1982), however, does not take his results as evidence against the exchange model. He introduces the "ex post exchange model" in which contributions are rewards for candidates' favorable past behavior, rather than a means to ingratiate with candidates who are undecided ("ex ante exchange model"). Evidently, it is hard to distinguish the ex post exchange model empirically from the support model. Stratmann (1995b) explicitly incorporates timing considerations in the empirical analysis of PACs' contribution strategies. He argues that PACs use the timing of contributions as a tool to prevent legislators from reneging on an implicit vote-for-contributions contract, and finds that farm PACs in the 1985-1986 cycle contributed to undecided (based on constituency interests) legislators mainly *after* a favorable vote. He concludes that these PACs attempted to purchase votes rather than getting their preferred candidate elected.

16. Bennett and Loucks (1994), Endersby and Munger (1992), Grenzke (1989b), Grier and Munger (1986, 1991), Kroszner and Stratmann (1995), Munger (1989), Romer and Snyder (1994), Saltzman (1987), Stratmann (1991), Welch (1982) and Wilhite and Theilmann (1987). See Regens et al. (1991) and Wright (1985) for exceptions.

17. See, e.g., Grier and Munger (1986, 1991, 1993), Kau, Keenan and Rubin (1982), Kau and Rubin (1993), Keim and Zardkoohi (1988). There is also some evidence that seniority may have a non-monotonic effect on contributions received (cf. Grier and Munger, 1991, Poole and Romer, 1985, Silberman and Yochum, 1980, Wilhite and Theilmann, 1986a, 1986b).

18. Explanations given in the literature for the different strategies employed by different types of interest groups include, the different kind of goals interest groups try to pursue (e.g. Welch, 1979), the way in which they raise (financial) support from their membership (e.g. Wright, 1985, Langbein, 1993), and dissimilarities in market power among interest groups (e.g. Herndon, 1982, Keim and Zardkoohi, 1988).

19. We found only two studies that report no, or a very limited effect. Quinn and Shapiro (1991) find that the lobbying activity of business did not have a significant effect on tax policy, and Fowler and Shaiko (1987) report that grass roots lobbying by members of environmental interest groups only had a modest impact on a senator's vote.

20. By referring to the literature on subgovernments, i.e. the "iron triangle" between interest groups, (sub)committee members and bureaucrats, Kollman (1997) points at a potential problem when trying to determine the lobbying strategy of interest groups. He argues that the representation on the side of the government constrains the choice of interest groups who to lobby. He finds that the ideology of interest groups surrounding a particular committee largely corresponds with the ideology of the (members of the) committee itself, such that interest groups actually have no choice but to lobby "friends". Hence empirical findings that seem to support the support model (i.e. lobbying of friendly legislators) may in fact reveal the systematic bias in representation by interest groups and committees in the specific policy areas, and thereby actually provide support for the subgovernment model.

21. The empirical analysis of Austen-Smith and Wright is heavily criticized by Baumgartner and Leech (1996a, 1996b). In a reply, however, Austen-Smith and Wright (1996) convincingly refute this critique. Moreover, they present additional evidence supporting their model.

22. Direct quantitative evidence that money indeed buys access, measured as the number of minutes interest groups spend with a politician during the week, is provided by Langbein (1986). A telling example that politicians sell access to contributors is given in Lewis (1996, p. 29). He presents the secret menu of contracts available to donors of the Democratic party, which was first disclosed by the Chicago Sun Times in June 1995. The list indicates that a $100,000 contributor would get, among other things, two meals with President Clinton and two meals with Vice President Gore.

23. See also Wright (1989) who finds some evidence that lobbying and contributions are used to reinforce the appearance of the interest group's existing organizational strength in the geographic constituency the legislator represents. He argues that political activities mainly serve a facilitating role. Grenzke (1989a) raises a similar point.

24. This result is fairly robust for political activity in general (cf. Grier et al., 1991, 1994, Humphries, 1991, Kennelly and Murrell, 1991, McKeown, 1994, Pittman, 1976, 1977, 1988, and Zardkoohi, 1985).

25. See, e.g., Gardner (1987), Guttman (1978), Kamath (1989), Miller (1991), Pincus (1975), Salamon and Siegfried (1977).

26. E.g., Graddy (1991), Guttman (1978), Kahane (1996), McArthur and Marks (1988), Meier (1987), Meier and McFarlane (1992), Naert (1990), Nelson and Silberberg (1987), Pashigian (1985), Peltzman (1984) and Plotnick (1986).

27. A positive effect of concentration is found by Esty and Caves (1983), Gardner (1987), Guttman (1980), Kalt and Zupan (1984), and Trefler (1993). Negative, ambiguous or insignificant effects are reported in Becker (1986), Cahan and Kaempfer (1992), Pincus (1975), Quinn and Shapiro (1991), and Salamon and Siegfried (1977).

28. For a negative effect, see Miller (1991), Plotnick (1986), Trefler (1993), Young (1991). For a positive effect, see, e.g., Abler (1991), Becker (1986), Boucher (1991), Congleton and Shugart (1990), Kristov et al. (1992), Leigh (1994), Pincus (1975), Renaud and Van Winden (1991), Stigler (1971). For ambiguous or insignificant effects, see Graddy and Nichol (1989), Michaels (1992), Neck and Schneider (1988), Renaud and van Winden (1988), Salamon and Siegfried (1977).

29. In addition, concentration may also affect a group's stake in obtaining favorable policy. For instance, a highly concentrated industry may not need political assistance to secure high profits. In that case, concentration not only lowers the costs, but also the potential benefits (stakes) of political action. Whether concentrated industries are more or less politically active — and, thus, which of the two effects dominates — is debated in the literature (cf. Grier, Munger and Roberts, 1991, 1994, Munger, 1988, Pittman, 1976, 1977, 1988, Zardkoohi, 1985, 1988).

30. It is possible that in the case of international comparisons there are too many disturbing factors to warrant conclusions on the effect of numbers. For instance, the effect of numbers might be lower in some less developed countries because of the lower democratic quality of governments in these countries.

31. Positive effects are found by Kamath (1989), Trefler (1993), and Stigler (1971), negative effects by Bullock (1992), Congleton and Shugart (1990), Gardner (1987), and Salamon and Siegfried (1977). See Cahan and Kaempfer (1992) for an insignificant effect.

32. See, e.g., Beghin (1990), Boucher (1991), Cahan and Kaempfer (1992), Pincus (1975), Quinn and Shapiro (1991), Tigges and Clarke (1992), Trefler (1993).

33. See, e.g., besides the studies cited in the examples, Abrams and Settle (1993), Boucher (1991), Globerman and Kadonaga (1994), Graddy (1991), Graddy and Nichol (1989), Guttman (1978), Kamath (1989), Plotnick (1986), and Stigler (1971).

34. See, e.g., Abler (1991), Becker (1986), Boucher (1991), Graddy (1991), McCormick and Tollison (1981), Richardson and Munger (1990), Teske (1991), Young (1991).

35. See, e.g., Congleton and Shugart (1990), Guttman (1978), Hoyt and Toma (1993), McCormick and Tollison (1981), Kristov et al. (1992), Mueller and Murrell (1986), Naert (1990), Plotnick (1986).

36. Some indirect support for this suggestion is provided by studies showing that education increases an individual's willingness to participate in politics (Kau and Rubin, 1979, Snyder, 1993, Welch, 1981), and studies that report a positive relationship between the level of education and the number of interest groups in a state (Benson and Engen, 1988).

37. For an exception see Snyder (1992). He incorporates dynamic aspects when investigating the contribution strategies of interest groups and finds that long-term considerations are in fact important for (economic) interest groups. This conclusion is challenged by McCarty and Rothenberg (1996), however. They find some indirect empirical evidence that legislators typically do not punish groups that contributed to the opponent in previous elections, and thereby conclude that the evidence for commitment to long-term relationships is weak. Finally, based on a dynamic model of contribution strategies allowing for both a support (election-influencing) and an exchange (favors-buying) motive, Levitt (1998) argues that the relative importance of the support motive is easily underestimated in a static framework.

38. See, e.g, Austen-Smith and Wright (1994), Beghin (1990) and Snyder (1990) for nice illustrations of the assistance of a theoretical model in deriving (non-trivial) competing hypotheses.

39. Because interest group influence is assumed rather than explained, models based on the influence function approach also lack a well-defined benchmark to evaluate the impact of the influence of interest groups on the political economy (Austen-Smith, 1997a, Van Winden, 1997).

40. See Lohmann (1994b) for a nice interpretation of the "Monday mass demonstrations", held in Leipzig around the time of the fall of the Berlin Wall (November 9, 1989), in the context of her model.

41. See Austen-Smith (1997a) and Morton and Cameron (1992) for an overview of the theoretical models of campaign contributions, and Dixit et al. (1997) and Grossman and Helpman (1994, 1995, 1996) for one of the most recent models not included in these surveys. Grossman and Helpman (1996) include both motivations in their model.

42. Concerning explicit bribes see Shleifer and Vishny (1994), Snyder (1991) and Young (1978). Baye et al. (1993) model the "wining-and-dining" activities by lobbyists as (implicit) payments in an all-pay auction. In Ainsworth (1993), Ainsworth and Sened (1993), Austen-Smith (1993, 1995), Austen-Smith and Wright (1992), Lohmann (1995b), Milgrom and Roberts (1986), Potters and Van Winden (1992) and Rasmusen (1993) lobbying is modeled as a game of strategic information transmission. In Ball (1995) and Glazer and Konrad (1995) lobbying serves several functions. Besides being a transfer of cash payments lobbying has a signaling externality through the amount of money transferred.

43. Moore and Suranovic (1992) refer to lobbying the executive branch as seeking administrative protection, and to lobbying the legislative branch simply as lobbying. In their notation both options have a fixed resource cost wL_{ap} and wL_{lob} (with w the wage and L_i the labor input needed for the specific influence attempt), a fixed probability of success s_{ap} and s_{lob}, and both leading to exactly the same increase in profit ($\Delta\pi$) when successful. Now, when $(s_{ap}-s_{lob})\Delta\pi > w(L_{ap}-L_{lob})$ the industry chooses to pursue administrative protection, otherwise it lobbies the legislature.

REFERENCES

Abler, D.G., 1991, Campaign contributions and House voting on sugar and dairy Legislation, *American Journal of Agricultural Economics* 73, 11-17.

Abrams, B.A. and R.F. Settle, 1993, Pressure group influence and institutional change: Branch banking legislation during the Great Depression, *Public Choice* 77, 687-705.

Ainsworth, S., 1993, Regulating lobbyists and interest group influence, *Journal of Politics* 55, 41-56.

Ainsworth, S. and I. Sened, 1993, The role of lobbyists: Entrepeneurs with two audiences, *American Journal of Political Science* 37, 834-866.

Austen-Smith, D., 1993, Information and influence: Lobbying for agendas and votes, *American Journal of Political Science* 37, 799-833.

Austen-Smith, D., 1995, Campaign contributions and access, *American Political Science Review* 89, 566-581.

Austen-Smith, D., 1997a, Interest groups: Money, information and influence, in: D.C. Mueller, ed., Perspectives on public choice. A handbook (Cambridge University Press, Cambridge) 296-321.

Austen-Smith, D., 1997b, Endogenous informational lobbying, Mimeo (Northwestern University, Evanston).

Austen-Smith, D. and J.R. Wright, 1992, Competitive lobbying for a legislator's vote, *Social Choice and Welfare* 9, 229-57.

Austen-Smith, D. and J.R. Wright, 1994, Counteractive lobbying, *American Journal of Political Science* 38, 25-44.

Austen-Smith, D. and J.R. Wright, 1996, Theory and evidence for counteractive lobbying, *American Journal of Political Science* 40, 543-564.

Ball, R., 1995, Interest groups, influence and welfare, *Economics and Politics* 7, 119-146.

Bauer, R., De Sola Pool, I. and A. Dexter, 1963, American business & public policy. The politics of foreign trade (Aldine Atherton, Chicago).

Baumgartner, F.R. and B.L. Leech, 1996a, The multiple ambiguities of "counteractive lobbying", *American Journal of Political Science* 40, 521-542.

Baumgartner, F.R. and B.L. Leech, 1996b, Good theories deserve good data, *American Journal of Political Science* 40, 565-569.

Baye, M.R., Kovenock, D. and C.G. de Vries, 1993, Rigging the lobbying process: An application of the all-pay auction, *American Economic Review* 83, 289-294.

Becker, G., 1986, The public interest hypothesis revisited: A new test of Peltzman's theory of regulation, *Public Choice* 49, 223-234.

Becker, G.S., 1983, A theory of competition among pressure groups for political influence, *Quarterly Journal of Economics* 98, 371-400.

Becker, G.S., 1985, Public policies, pressure groups, and dead weight costs, *Journal of Public Economics* 28, 329-347.

Beghin, J.C., 1990, A game-theoretic model of endogenous public policies, *American Journal of Agricultural Economics* 72, 138-148.

Bennett, R.W. and C. Loucks, 1994, Savings and loan and finance industry PAC contributions to incumbent members of the House banking committee, *Public Choice* 79, 83-104.

Benson, B.L. and E.M. Engen, 1988, The market for laws: An economic analysis of legislation, *Southern Economic Journal* 54, 732-745.

Bental, B. and U. Ben-Zion, 1975, Political contribution and policy-some extensions, *Public Choice* 19, 1-12.

Ben-Zion, U. and Z. Eytan, 1974, On money, votes, and policy in a democratic society, *Public Choice* 17, 1-10.

Boucher, M., 1991, Rent-seeking and the behavior of regulators: An empirical analysis, *Public Choice* 69, 51-67.

Bronars, S.G. and J.R. Lott, 1997, Do campaign donations alter how a politician votes? Or, do donors support candidates who value the same thing that they do?, *Journal of Law and Economics* 40, 317-350.

Bullock, D., 1992, Objectives and constraints of government policy: The countercyclity of transfers to agriculture, *American Journal of Agricultural Economics* 74, 617-629.

Cahan, S.F. and W.H. Kaempfer, 1992, Industry income and congressional regulatory legislation: Interest groups vs. median voter, *Economic Inquiry* 30, 47-57.

Caldeira, G.A. and J.R. Wright, 1998, Lobbying for justice: Organized interests, Supreme Court nominations and the United States Senate, *American Journal of Political Science* 42, 499-523.

Chappell, H.W., 1981, Campaign contributions and voting on the Cargo Preference Bill: A comparison of simultaneous models, *Public Choice* 36, 301-312.

Chappell, H.W., 1982, Campaign contributions and Congressional voting: A simultaneous probit-tobit model, *Review of Economics and Statistics* 61, 77-83.

Congleton, R.D. and R.W. Bennet, 1995, On the political economy of state highway expenditures: Some evidence of the relative performance of alternative public choice models, *Public Choice* 84, 1-24.

Congleton, R.D. and W.F. Shugart I, 1990, The growth of social security: Electoral push or political pull, *Economic Inquiry* 28, 109-132.

Coughlin, C.C., 1985, Domestic content legislation: House voting and the economic theory of regulation, *Economic Inquiry* 23, 437-448.

Coughlin, P.J., Mueller, D.C. and P. Murrell, 1990a, A model of electoral competition with interest groups, *Economics Letters* 32, 307-311.

Coughlin, P.J., Mueller, D.C. and P. Murrell, 1990b, Electoral politics, interest groups, and the size of government, *Economic Inquiry* 29, 682-705.

Cropper, M.L., Evans, W.N., Berardi, S.J., Ducla-Soares, M.M. and P.R. Portney, 1992, The determinants of pesticide regulation: A statistical analysis of EPA decision making, *Journal of Political Economy* 100, 175-197.

Dixit, A., Grossman, G.M. and E. Helpman, 1997, Common agency and coordination: General theory and application to government policy making, *Journal of Political Economy* 105, 752-769.

Dougan, W.R. and J.M. Snyder, 1996, Interest-group politics under majority rule, *Journal of Public Economics* 61, 49-71.

Endersby, J.W. and M.C. Munger, 1992, The impact of legislator attributes on union PAC campaign contributions, *Journal of Labor Research* 13, 79-97.

Esty, D.C. and R.E. Caves, 1983, Market structure and political influence: New data on political expenditures, activity and success, *Economic Inquiry* 21, 24-38.

Fowler, L.L. and R.G. Shaiko, 1987, The grass roots connection: Environmental activists and Senate roll calls, *American Journal of Political Science* 31, 484-510.

Frendreis, J.P. and R.W. Waterman, 1985, PAC contributions and legislative behavior: Senate voting on trucking deregulation, *Social Science Quarterly* 66, 401-412.

Gardner, B.L., 1987, Causes of U.S. farm commodity programs, *Journal of Political Economy* 95, 290-310.

Giertz, J.F. and D.H. Sullivan, 1977, Campaign expenditures and election outcomes: A critical note, *Public Choice* 31, 157-162.

Glazer, A. and K.A. Konrad, 1995, Strategic lobbying by potential industry entrants, *Economics and Politics* 7, 167-179.

Globerman, S. and D. Kadonaga, 1994, International differences in telephone rate structures and the organization of business subscribers, *Public Choice* 80, 129-142.

Gopoian, J.D., 1984, What makes PACs tick? Analysis of the allocation patterns of economic interest groups, *American Journal of Political Science* 28, 259-281.

Graddy, E., 1991, Toward a general theory of occupational regulation, *Social Science Quarterly* 72, 676-695.

Graddy, E. and M.B. Nichol, 1989, Public members on occupational licensing boards: Effects on legislative regulatory reforms, *Southern Economic Journal* 55, 610-625.

Grenzke, J.M., 1989a, PACs and the Congressional supermarket: The currency is complex, *American Journal of Political Science* 33, 1-24.

Grenzke, J.M., 1989b, Candidate attributes and PAC contributions, *Western Political Quarterly* 42, 245-264.

Grier, K.B. and M.C. Munger, 1986, The impact of legislator attributes on interest-group campaign contributions, *Journal of Labor Research* 7, 349-361.

Grier, K.B. and M.C. Munger, 1991, Committee assignments, constituent preferences and campaign contributions, *Economic Inquiry* 29, 24-43.

Grier, K.B. and M.C. Munger, 1993, Comparing interest group PAC contributions to House and Senate incumbents, 1980-1986, *Journal of Politics* 55, 615-643.

Grier, K.B., Munger, M.C. and B.E. Roberts, 1991, The industrial organization of corporate political participation, *Southern Economic Journal* 57, 727-738.

Grier, K.B., Munger, M.C. and B.E. Roberts, 1994, The determinants of industry political activity, 1978-1986, *American Political Science Review* 88, 911-926.

Grier, K.B., Munger, M.C. and G.M. Torrent, 1990, Allocation patterns of PAC monies: The U.S. Senate, *Public Choice* 67, 111-128.

Grossman, G.M. and E. Helpman, 1994, Protection for sale, *American Economic Review* 84, 833-850.

Grossman, G.M. and E. Helpman, 1995, Trade wars and trade talks, *Journal of political Economy* 103, 675-708.

Grossman, G.M. and E. Helpman, 1996, Electoral competition and special interest politics, Review of Economic Studies 63, 265-286.

Gustafsson, S. and F. Stafford, 1992, Child care subsidies and labor supply in Sweden, *Journal of Human Resources* 27, 204-230.

Guttman, J.M., 1978, Interest groups and the demand for agricultural research, *Journal of Political Economy* 86, 467-484.

Guttman, J.M., 1980, Villages as interest groups: The demand for agricultural extension services in India, *Kyklos* 33, 122-141.

Hall, R.L. and F.W. Wayman, 1990, Buying time: Moneyed interests and the mobilization of bias in Congressional committees, *American Political Science Review* 84, 797-820.

Herndon, J.F., 1982, Access, record, and competition as influences on interest group contributions to Congressional campaigns, *Journal of Politics* 44, 996-1019.

Hoyt, W.H. and E.F. Toma, 1989, State mandates and interest group lobbying, *Journal of Public Economics* 38, 199-213.

Hoyt, W.H. and E.F. Toma, 1993, Lobbying expenditures and government output: The NEA and public education, *Southern Economic Journal* 60, 405-417.

Humphries, C, 1991, Corporations, PACs and the strategic link between contributions and lobbying activities, *Western Political Quarterly* 44, 353-372.

Jacobson, G.C., 1980, Money in Congressional Elections (Yale University Press, New Haven).

Jarrell, G.A., 1978, The demand for state regulation of the electric utility industry, *Journal of Law and Economics* 21, 269-296.

Jones, W. and K.R. Keiser, 1987, Issue visibility and the effects of PAC money, *Social Science Quarterly* 68, 170-176.

Kahane, L.H., 1996, Senate voting patterns on the 1991 extension of the fast track procedure: prelude to NAFTA, *Public Choice* 87, 35-53.

Kalt, J.P. and M.A. Zupan, 1984, Capture and ideology in the economic theory of politics, *American Economic Review* 74, 279-300.

Kamath, S.J., 1989, Concealed takings: capture and rent-seeking in the Indian sugar industry, *Public Choice* 62, 119-138.

Kau, J.B., Keenan, D. and P.H. Rubin, 1982, A general equilibrium model of Congressional voting, *Quarterly Journal of Economics* 97, 271-293.

Kau, J.B. and P.H. Rubin, 1979, Public interest lobbies: Membership and influence, *Public Choice* 34, 45-54.

Kau, J.B. and P.H. Rubin, 1982, Congressmen, constituents and contributors: Determinants of roll call voting in the House of Representatives (Martinus Nijhoff, Boston).

Kau, J.B. and P.H. Rubin, 1993, Ideology, voting and shirking, *Public Choice* 76, 151-172.

Keim, G. and A. Zardkoohi, 1988, Looking for leverage in PAC markets: Corporate and labor contributions considered, *Public Choice* 58, 21-34.

Kennelly, B. and P. Murrell, 1991, Industry characteristics and interest group formation: An empirical study, *Public Choice* 70, 21-40.

Kirchgässner, G. and W. Pommerehne, 1988, Government spending in federal systems: A comparison between Switzerland and Germany, in: J.A. Lybeck and M. Henrekson, eds., Explaining the growth of government (North Holland, Amsterdam) 327-356.

Kollman, K., 1997, Inviting friends to lobby: Interest groups, ideological bias, and Congressional committees, *American Journal of Political Science* 41, 519-544.

Kristov, L., Lindert P. and R. McClelland, 1992, Pressure groups and redistribution, *Journal of Public Economics* 48, 135-163.

Kroszner, R.S. and T. Stratmann, 1995, Interest group competition and the organization of Congress: Theory and evidence from financial services political action committees, Mimeo (University of Chicago, Chicago).

Langbein, L.I., 1986, Money and access: Some empirical evidence, *Journal of Politics* 48, 1052-1062.

Langbein, L.I., 1993, PACs, lobbies and political conflict: The case of gun control, *Public Choice* 77, 551-572.

Leigh, J.P., 1994, Non-random assignment, vehicle safety inspection laws and high way fatalities, *Public Choice* 78, 373-387.

Levitt, S.D., 1998, Are PACs trying to influence politicians or voters?, *Economics and Politics* 10, 19-35.

Lewis, C., 1996, The buying of the President (Avon Books, New York).

Lindblom, C.E., 1977, Politics and markets: The world's political-economic system (Basic Books, New York).

Lindsay, C.M. and M.T. Maloney, 1988, Party politics and the price of payola, *Economic Inquiry* 26, 203-221.

Lohmann, S., 1993a, A signaling model of informative and manipulative political action, *American Political Science Review* 87, 319-333.

Lohmann, S., 1993b, A welfare analysis of political action, in: W.A. Barnett, M.J. Hinich and N.J. Schofield, eds., Political economy: Institutions, competition, and representation (Cambridge university press, Cambridge) 437-461.

Lohmann, S., 1994a, Information aggregation through costly political action, *American Economic Review* 84, 518-530.

Lohmann, S., 1994b, The dynamics of informational cascades: The Monday demonstrations in Leipzig, East Germany, 1989-1991, *World Politics* 47, 42-101.

Lohmann, S., 1995a, A signaling model of competitive political pressures, *Economics and Politics* 7, 181-206.

Lohmann, S., 1995b, Information, access, and contributions: A signaling model of lobbying, *Public Choice* 85, 267-284.

Lopcz, R.A. and E. Pagoulatos, 1996, Trade protection and the role of campaign contributions in U.S. food and tobacco industries, *Economic Inquiry* 34, 237-248.

Lupia, A., 1994, Shortcuts versus encyclopedias: information and voting behavior in California insurance reform elections, *American Political Science Review* 88, 63-76.

Maitland, I., 1985, Interest groups and economic growth rates, *Journal of Politics* 47, 44-58.

Masters, M.F. and A. Zardkoohi, 1988, Congressional support for unions' positions across diverse legislation, *Journal of Labor Research* 9, 149-165.

McArthur, J. and S.V. Marks, 1988, Constituent interest vs. legislator ideology: The role of political opportunity cost, *Economic Inquiry* 26, 461-470.

McCarty, N. and L.S. Rothenberg, 1996, Commitment and the campaign contribution contract, *American Journal of Political Science* 40, 872-904.

McCormick, R. and R.D. Tollison, 1981, Politicians, legislators and the economy. An inquiry into the interest group theory of government (Kluwer, Boston).

McKeown, T.J., 1994, The epidemiology of corporate PAC formation, 1975-84, *Journal of Economic Behavior and Organization* 24, 153-168.

Meier, K.J., 1987, The political economy of consumer protection: An examination of state legislation, *Western Political Quarterly* 40, 343-359.

Meier, K.J. and D.R. McFarlane, 1992, State policies on funding of abortions: A pooled time series analysis, *Social Science Quarterly* 73, 690-698.

Michaels, R.J., 1992, What's legal and what's not: The regulation of opiates in 1912, *Economic Inquiry* 30, 696-713.

Milbrath, L., 1963, The Washington Lobbyists (Rand McNally, Chicago).

Milgrom, P. and J. Roberts, 1986, Relying on the information of interested parties, *Rand Journal of Economics* 17, 18-32.

Miller, T.C, 1991, Agricultural price policies and political interest group competition, *Journal of Policy Modelling* 13, 489-513.

Mitchell, W.C., 1990, Interest groups: Economic perspectives and contributions, *Journal of Theoretical Politics* 2, 85-108.

Mitchell, W.C. and M.C. Munger, 1991, Economic models of interest groups: An introductory survey, *American Journal of Political Science* 35, 512-546.

Mixon, F.G., 1995, Public choice and the EPA: Empirical evidence on carbon emissions violations, *Public Choice* 83, 127-137.

Moore, M.O. and S.M. Suranovic, 1992, Lobbying vs. administered protection: Endogenous industry choice and national welfare, *Journal of International Economics* 32, 289-303.

Morton, R. and C. Cameron, 1992, Elections and the theory of campaign contributions: A survey and critical analysis, *Economics and Politics* 4, 79-108.

Mueller, D.C. and P. Murrell, 1986, Interest groups and the size of government, *Public Choice* 48, 125-145.

Munger, M.C., 1988, On the political participation of the firm in the electoral process: An update, *Public Choice* 56, 295-298.

Munger, M.C., 1989, A simple test of the thesis that committee jurisdictions shape corporate PAC contributions, *Public Choice* 62, 181-186.

Naert, F., 1990, Pressure politics and government spending in Belgium, *Public Choice* 67, 49-63.

Neck, R. and F. Schneider, 1988, The growth of the public sector in Austria: An explanatory analysis, in: J.A. Lybeck and M. Henrekson, eds., Explaining the growth of government (North Holland, Amsterdam) 231-263.

Nelson, D. and E. Silberg, 1987, Ideology and legislator shirking, *Economic Inquiry* 25, 15-25.

Noll, R.G., 1989, Economic perspectives on the politics of regulation, in: R. Schmalensee and R.D. Willig, eds., Handbook of industrial organization, Vol. 2 (North-Holland, Amsterdam) 1253-1287.

Olson, M., 1965, The logic of collective action (Harvard University Press, Cambridge).

Palda, K.F. and K.S. Palda, 1985, Ceilings on campaign spending: Hypothesis and partial test with Canadian data, *Public Choice* 45, 313-331.

Pashigian, B.P., 1985, Environmental regulation: Whose self-interests are being protected?, *Economic Inquiry* 23, 551-584.

Peltzman, S., 1984, Constituent interest and Congressional voting, *Journal of Law and Economics* 27, 181-210.

Pincus, J.J., 1975, Pressure groups and the pattern of tariffs, *Journal of Political Economy* 83, 757-777.

Pittman, R., 1976, The effects of industry concentration and regulation on contributions in three 1972 U.S. Senate campaigns, *Public Choice* 23, 71-80.

Pittman, R., 1977, Market structure and campaign contributions, *Public Choice* 31, 37-52.

Pittman, R., 1988, Rent-seeking and market structure: Comment, *Public Choice* 58, 173-185.

Plotnick, R.D., 1986, An interest group model of direct income redistribution, *Review of Economics and Statistics* 68, 594-602.

Poole, K.T. and T. Romer, 1985, Patterns of PAC contributions to the 1980 campaigns for the U.S House of Representatives, *Public Choice* 47, 63-111.

Poole, K.T., Romer, T. and H. Rosenthal, 1987, The revealed preferences of political action committees, *American Economic Review* 77, 298-302.

Potters, J. and R. Sloof, 1996, Interest groups. A survey of empirical models that try to assess their influence, *European Journal of Political Economy* 12, 403-442.

Potters, J. and F. van Winden, 1990, Modelling political pressure as transmission of information, *European Journal of Political Economy* 6, 61-88.

Potters, J. and F. van Winden, 1992, Lobbying and asymmetric information, *Public Choice* 74, 269-92.

Potters, J. and F. van Winden, 1996, Models of interest groups: Four different approaches, in: N. Schofield, ed., Collective decision-making: Social choice and political economy (Kluwer, Boston) 337-362.

Quinn, D.P. and R.Y. Shapiro, 1991, Business political power: The case of taxation, *American Political Science Review* 85, 851-874.

Rapoport, R.B., Stone, W.J. and A.I. Abramowitz, 1991, Do endorsements matter? Group influence in the 1984 Democratic caucuses, *American Political Science Review* 85, 193-203.

Rasmusen, E., 1993, Lobbying when the decisionmaker can acquire independent information, *Public Choice* 77, 899-913.

Regens, J.L., Elliott, E. and R.K. Gaddie, 1991, Regulatory costs, committee jurisdictions, and corporate PAC contributions, *Social Science Quarterly* 72, 751-760.

Regens, J.L., Gaddie, R.K. and E. Elliott, 1994, Corporate PAC contributions and rent provision in Senate elections, *Social Science Quarterly* 75, 152-165.

Renaud, P.S.A. and F. van Winden 1988, Fiscal behaviour and the growth of government in the Netherlands, in: J.A. Lybeck and M. Henrekson, eds., *Explaining the growth of government* (North Holland, Amsterdam) 133-156.

Renaud, P.S.A. and F. van Winden, 1991, Behavior and budgetary autonomy of local governments, *European Journal of Political Economy* 7, 547-577.

Richardson, L.E. and M.C. Munger, 1990, Shirking, representation, and Congressional behavior: Voting on the 1983 amendments to the Social Security Act, *Public Choice* 67, 11-33.

Romer, T. and J.M. Snyder, 1994, An empirical investigation of the dynamics of PAC contributions, *American Journal of Political Science* 38, 745-769.

Salamon, L.M. and J.J. Siegfried, 1977, Economic power and political influence: The impact of industry structure on public policy, *American Political Science Review* 71, 1026-1043.

Saltzman, G.M., 1987, Congressional voting on labor issues: The role of PACs, *Industrial and Labor Relations Review* 40, 163-179.

Schlozman, K. and J. Tierney, 1986, Organized interests and American democracy (Harper and Row, New York).

Schneider, F. and J. Naumann, 1982, Interest groups in democracies - how influential are they? An empirical examination for Switzerland, *Public Choice* 38, 281-303.

Schonhardt-Bailey, C., 1991, Lessons in lobbying for free trade in 19th-century Britain: To concentrate or not, *American Political Science Review* 85, 37-58.

Schroedel, J.R., 1986, Campaign contributions and legislative outcomes, *Western Political Quarterly* 39, 371-389.

Segal, J.A., C.M. Cameron and A.D. Cover, 1992, A spatial model of roll call voting: Senators, constituents, presidents, and interest groups in Supreme Court confirmations, *American Journal of Political Science* 36, 96-121.

Shleifer, A. and R.W. Vishny, 1994, Politicians and firms, *Quarterly Journal of Economics* 109, 995-1025.

Silberman, J. and G.C. Durden, 1976, Determining legislative preferences on the minimum wage: An economic approach, *Journal of Political Economy* 84, 317-329.

Silberman, J. and G. Yochum, 1980, The market for special interest campaign funds: An exploratory research, *Public Choice* 35, 75-83.

Smith, R.A., 1995, Interest group influence in the U.S. Congress, *Legislative Studies Quarterly* 20, 89-139.

Snyder, J.M., 1989, Political geography and interest-group power, *Social Choice and Welfare* 6, 103-125.

Snyder, J.M., 1990, Campaign contributions as investments: The U.S. House of Representatives, 1980-
    1986, *Journal of Political Economy* 98, 1195-1227.
Snyder, J.M., 1991, On buying legislatures, *Economics and Politics* 3, 93-109.
Snyder, J.M., 1992, Long-term investing in politicians: Or, give early, give often, *Journal of Law and
    Economics* 35, 15-43.
Snyder, J.M., 1993, The market for campaign contributions: Evidence for the U.S. Senate 1980-1986,
    Economics and Politics 5, 219-240.
Stigler, G., 1971, The theory of economic regulation, *Bell Journal of Economics and Management
    Science* 2, 3-21.
Stratmann, T., 1991, What do campaign contributions buy? Deciphering causal effects of money and
    votes, *Southern Economic Journal* 57, 606-620.
Stratmann, T., 1992a, The effects of logrolling on Congressional voting, *American Economic Review*
    82, 1162-1176.
Stratmann, T., 1992b, Are contributions rational? Untangling strategies of political action committees,
    Journal of Political Economy 100, 647-664.
Stratmann, T., 1995a, Campaign contributions and Congressional voting: Does the timing of contribu-
    tions matter?, *Review of Economics and Statistics* 77, 127-136.
Stratmann, T., 1995b, The market for Congressional votes: Is Timing of contributions everything?,
    paper presented at the 1995 joint meeting of the Public Choice Society and the Economic
    Science Association, Long Beach, California.
Stratmann, T., 1996, How reelection constituencies matter: Evidence from political action committees'
    contributions and Congressional voting, *Journal of Law and Economics* 39, 603-660.
Swinnen, J. and F.A. van der Zee, 1993, The political economy of agricultural policies: A survey,
    European Review of Agricultural Economics 20, 261-290.
Teske, P., 1991, Interests and institutions in state regulation, *American Journal of Political Science* 35,
    139-154.
Tigges, L.M. and M.J. Clarke, 1992, Community, class and cohesion in the passage of corporate
    takeover legislation, *Social Science Quarterly* 73, 798-814.
Tosini, S.C. and E. Tower, 1987, The Textile Bill of 1985: The determinants of Congressional voting
    patterns, *Public Choice* 54, 19-25.
Trefler, D., 1993, Trade liberalization and the theory of endogenous protection: An econometric study
    of U.S. import policy, *Journal of Political Economy* 101, 138-160.
Van Dijk, L, Koot-du Buy, A.H.E.B. and J.J. Siegers, 1993, Day-care supply by Dutch municipalities,
    European Journal of Population 9, 315-330.
Van Velthoven, B. and F. van Winden, 1986, Social classes and state behavior, *Journal of Institutional
    and Theoretical Economics* 142, 542-570.
Van Winden, F., 1997, On the economic theory of interest groups: Towards a group frame of reference
    in political economics, forthcoming in: *Public Choice*.
Wayman, F.W., 1985, Arms control and strategic arms voting in the U.S. Senate, *Journal of Conflict
    Resolution* 29, 225-251.
Welch, W.P., 1974, The economics of campaign funds, *Public Choice* 20, 83-97.
Welch, W.P., 1979, Patterns of contributions: Economic interest and ideological groups, in: H.E.
    Alexander, ed., Political finance (Sage, Beverly Hills).
Welch, W.P., 1980, Allocation of political monies: Economic interest groups, *Public Choice* 35, 97-
    120.
Welch, W.P., 1981, Money and votes: A simultaneous equation model, *Public Choice* 36, 209-234.
Welch, W.P., 1982, Campaign contributions and legislative voting: Milk money and dairy price
    supports, *Western Political Quarterly* 35, 478-495.
Wilhite, A., 1988, Union PAC contributions and legislative voting, *Journal of Labor Research* 9, 79-
    90.
Wilhite, A. and J. Theilmann, 1986a, Unions, corporations, and political campaign contributions: The
    1982 House Elections, *Journal of Labor Research* 7, 175-185.
Wilhite, A. and J. Theilmann, 1986b, Women, blacks, and PAC discrimination, *Social Science
    Quarterly* 67, 283-298.

Wilhite, A. and J. Theilmann, 1987, Labor PAC contributions and labor legislation: A simultaneous logit approach, *Public Choice* 53, 267-276.

Wirl, F., 1994, The dynamics of lobbying — A differential game, *Public Choice* 80, 307-323.

Wright, J.R., 1985, PACs, contributions, and roll calls: An organizational perspective, *American Journal of Political Science* 79, 400-414.

Wright, J.R., 1989, PAC contributions, lobbying, and representation, *Journal of Politics* 51, 713-729.

Wright, J.R., 1990, Contributions, lobbying, and committee voting in the U.S. House of Representatives, *American Political Science Review* 84, 417-438.

Young, H.P., 1978, The allocation of funds in lobbying and campaigning, *Behavioral Science* 23, 21-31.

Young, S.D., 1991, Interest group politics and the licensing of public accountants, *The Accounting Review* 66, 809-817.

Zardkoohi, A., 1985, On the political participation of the firm in the electoral process, *Southern Economic Journal* 51, 804-817.

Zardkoohi, A., 1988, Market structure and campaign contributions: Does concentration matter?, *Public Choice* 58, 187-191.

3 GAME-THEORETIC PRELIMINARIES

In their survey of theoretical models of interest groups Potters and Van Winden (1996) distinguish four different approaches. One of these concerns a strand of literature that focuses on the influence of interest groups through the transmission of information (Austen-Smith, 1997, separates out the same class). The models discussed in subsequent chapters all follow this approach. The purpose of this chapter is threefold. First, it intends to give the reader some feeling and intuition for models of strategic information transmission. Second, in this chapter a basic model is discussed that forms one of the building blocks for the extended models analyzed in subsequent chapters. Third, this chapter is meant to give the reader somewhat more detailed insight in the existing literature in this area.

Much of the material presented in this chapter is not original, and draws heavily on Potters and Van Winden (1992, 1996). This material is included, however, to make this monograph self-contained. We start by describing the (informational lobbying) model studied by Potters and Van Winden (1992) in an abstract way without reference to a particular application. In the second section the three specific applications of this basic model used in subsequent chapters are introduced. Section 3.3 elaborates on some extensions of the basic model. The final section summarizes.

3.1   A BASIC SIGNALING GAME

Description of the game. Consider the following two player incomplete information game between a male sender and a female decisionmaker (the receiver). The sender possesses private information concerning the actual state of the world t. The action x the decisionmaker prefers to take depends on the value of t, that is, information concerning the state of world is relevant for the decisionmaker. Before the decisionmaker has to decide on her action, the sender may send a (costly) signal concerning the actual state of the world. The order of play in this costly signaling game is as follows:

(i)   Nature draws the state of the world $t \in \{t_1, t_2\}$, with $P(t=t_2)=p$.
(ii)   Only the sender observes the actual state of the world (his type).
(iii)   The sender chooses signal $s \in S \equiv \{n\} \cup M$. A message $m \in M$ bears cost $c \geq 0$
         ($c(s)=c$ if $s=m \in M$), whereas $s=n$, interpreted as no message, bears no
         cost(i.e. $c(n)=0$).
(iv)   The decisionmaker receives signal s (but cannot verify its accuracy).
(v)   The decisionmaker chooses between action x_1 and x_2.
(vi)   The actual state of the world is revealed and payoffs are obtained.

The setup of the game is known to both players. Note that the costs c of sending
a message $m \in M$ are assumed to be independent of both the content of the message
(the particular element m of M) and the actual state of the world. The normalized
payoffs over action-state pairs are given by the following table:

Table 3.1 Gross payoffs in
the basic signaling game

	$t=t_1$		$t=t_2$	
$x=x_1$	d_1	0	0	0
$x=x_2$	0	e_1	d_2	e_2

Additional assumptions: $d_i>0$
for $i=1,2$

The first entry in each of the four lower-right cells of Table 3.1 refers to the
decisionmaker's payoff. From the assumption that $d_i>0$ for $i=1,2$ it follows that
when $t=t_1$ ($t=t_2$) the decisionmaker prefers to take action $x=x_1$ ($x=x_2$).[1] Regarding
these preferences two cases can be distinguished: the case in which the
decisionmaker, on the basis of the prior, would choose $x=x_1$ (requiring
$p<d \equiv d_1/(d_1+d_2)$), and the opposite case where she is a priori inclined to choose
$x=x_2$ ($p>d$). The second entry in each cell denotes the sender's gross payoff. His
net payoff follows by subtracting the costs (either 0 or c) of sending signal s.
Concerning the preferences of the sender three basic situations can be distinguished (knife-edge cases are ignored):

(I)   No conflict with the decisionmaker's interests: $e_1<0<e_2$.
(II)   Full conflict with the decisionmaker's interests: $e_2<0<e_1$.
(III)   Partial conflict with the decisionmaker's interests: $e_1,e_2>0$.

Equilibrium analysis. The theoretical prediction of what will be the outcome of the
basic signaling game when players are fully rational is given by the set of
equilibria. The standard equilibrium notion used for the type of game under

consideration is perfect Bayesian equilibrium (PBE).[2] In equilibrium each player behaves optimally and chooses a strategy such as to maximize own expected payoffs, given the (equilibrium) strategies of the other players. For a formal definition of a PBE of the game considered here, some additional notation is needed. Let $\rho(s)$ denote the probability that the decisionmaker chooses $x=x_2$ after receiving signal $s \in S \in S$, and let $\sigma(s|t_i)$ refer to the probability that the sender sends signal s when he has private information that $t=t_i$ ($i=1,2$). Finally, $q(s)$ is used to denote the decisionmaker's posterior belief that $t=t_2$ after having received signal s. A PBE of this game is now defined as a set of strategies $(\rho(\cdot),\sigma(\cdot))$ and a belief $q(\cdot)$ that satisfy the following three conditions:

(e1)   if $\sigma(s|t_i)>0$ for some $s \in S$, then s maximizes $e_i\rho(s)-c(s)$. In addition, $\sum_{s \in S}$ $\sigma(s|t_i)=1$ for $i=1,2$.

(e2)   if $\rho(s)>0$ (<1) for some $s \in S$, then $q(s) \geq (\leq)$ d.

(e3)   for all $s \in S$, $q(s)=p\sigma(s|t_2)/[p\sigma(s|t_2)+(1-p)\sigma(s|t_1)]$ whenever the denominator is positive.

The first condition requires that the sender only chooses those signals that maximize his expected payoff, taking the strategy of the decisionmaker as given. The second condition states that the decisionmaker maximizes expected payoffs, given her posterior beliefs. The last condition requires these posterior beliefs to be consistent with Bayes' rule whenever possible.

Because the sender's signal itself does not affect the decisionmaker's payoffs and does not affect her choice set, (costly) signals influence the action taken by the decisionmaker only via their impact on the information she obtains from them. Therefore, an interesting question is how much information can be revealed in equilibrium. Of course, when the costs c of sending messages $m \in M$ are too high, neither type of sender wants to send a costly message and no information is revealed in equilibrium. Hence, when $c>|e_1|$ and $c>|e_2|$ information revelation cannot occur in equilibrium. Both sender types choose $s=n$ and the decisionmaker bases her decision completely on her prior information. So, $c<\max\{|e_1|,|e_2|\}$ is assumed to hold in the sequel. The scope for information revelation is now analyzed for each of the three basic situations distinguished above with respect to the sender's preferences (cases I, II, and III, respectively).

In case of no conflict of interest between the sender and the decisionmaker (case I), the sender has no incentive at all to report untruthfully and to mislead the decisionmaker. Therefore, there is no reason for the decisionmaker to mistrust a message coming from the sender. Using the assumption $c<\max\{-e_1,e_2\}$, it follows easily that full information revelation can occur in equilibrium. For instance, the sender type that gains most from the decisionmaker taking his preferred decision chooses $m \in M$ with probability one, the other type chooses $s=n$ for sure.[3] There is no problem regarding the scope for information transfer in this case. When there is full conflict of interests between the sender and the decisionmaker (case II), the sender always has an incentive to misinform the decisionmaker about his private information. Due to the rational expectations character of a PBE, then, the decisionmaker will always interpret a (costly) message in a way which is unfavorable for the sender. This induces both sender types to choose s=n for sure in equilibrium, and information transfer is not possible. Lastly, when $0 < e_1, e_2$ (or, similarly, $e_1, e_2 < 0$) the scope for information transfer is not immediately clear. The remaining part of this section analyzes this case of partial conflict of interests (we assume $0 < e_1, e_2$).

When $e_i > 0$ for $i = 1, 2$ the sender's preference ordering over the decisionmaker's actions x_1 and x_2 is independent of t; each type strictly prefers x_2. As a result, both types of sender want the decisionmaker to believe that $t = t_2$, in order to induce a choice of $x = x_2$. Only the type t_2 sender possesses the "right" information from this perspective, that is, information such that the claim the sender wants to make indeed holds. The sender of type t_2 is therefore referred to as the "good" type, the sender of type t_1 as the "bad" type.

Since both sender types strictly prefer action $x = x_2$ to be chosen, messages which are equally costly cannot induce different reactions in equilibrium. All messages $m \in M$ should induce the same equilibrium reaction strategy by the decisionmaker. This follows because after deciding to send a costly message $m \in M$, the content of the message — the choice of a specific element from M — is essentially cheap talk. Since the sender's preferences are independent of his private information, cheap talk cannot affect the decisionmaker's decision.[4] In short, in a setting of partially conflicting interests the informational value of a message $m \in M$ merely lies in its costs, and not in its content.

From the previous discussion it follows that nothing is lost, in terms of equilibrium outcomes, when M is reduced to a singleton set and a costly message is just identified with its costs. Therefore, in the sequel of this section let $M = \{c\}$. Message c can then be interpreted as saying that "t equals t_2". Likewise, below we will identify signal s=n with its costs as well, and write s=0 instead. Incorporating the reduction of S to $S = \{0, c\}$, the following standard terminology is used in order to compare and classify different (perfect Bayesian) equilibria by the amount of information which is revealed:

(i)   In a *pooling* equilibrium both sender types choose to send the costly message s=c with exactly the same probability.

(ii)   In a *hybrid* equilibrium the two sender types choose to send the costly message s=c with positive, but nonequal, probabilities, or the two types choose to send no message (s=0) with positive, but nonequal, probabilities (or both).

(iii)   In a *separating* equilibrium one sender type chooses s=c with certainty, whereas the other type always chooses signal s=0.

In a pooling equilibrium each sender type employs exactly the same strategy and the decisionmaker does not get additional information from observing (the absence of) a costly message s=c. That is, in such an uninformative equilibrium the decisionmaker does not get useful information to update her prior belief. In a hybrid equilibrium some, but not all information is revealed. Since in this kind of equilibrium there is a signal (s=0 or s=c) which both types choose with positive probability, the observation of this signal does not lead to a decisive answer concerning the sender's type. However, due to the fact that both types choose this signal with different probabilities, some inferences can be made to update the prior belief. In a separating equilibrium all information is revealed because the strategies of the sender types are completely opposed. After every signal $s \in \{0,c\}$ the decisionmaker knows the type of the sender for sure and she takes the same decision as in case she knew the sender's type beforehand.

Using the classification of equilibria above, we now turn to the conditions under which an informative — that is, hybrid or separating — equilibrium exists. It appears that information revelation is only possible when two conditions are satisfied. Firstly, and rather trivially, information revelation requires that the costs of sending message c are not prohibitive for the good type sender ($c \le e_2$). Intuitively, when it is always too expensive for the good type sender to spend c on a costly message, the observation of such a message is a clear signal of the sender being of the bad type. But, the bad type wants to conceal his identity (information), and thus certainly does not send costly message s–c when this leads to all information being revealed. So, when $e_2 < c$ only pooling on s=0 can occur in equilibrium. A second condition follows from Proposition 3.1 below, which, in turn, directly follows from (the proof of) Proposition 1 in Potters and Van Winden (1992).[5]

Proposition 3.1
When $e_1 > e_2 > 0$ only pooling PBE exist.   □

From Proposition 3.1 a necessary condition follows for information transfer to occur in equilibrium, to wit the condition $e_1 \le e_2$ on the sender's preferences. In other words, for a costly message s=c to potentially possess informational value the sender should value the decision he prefers ($x=x_2$) weakly more highly when the decisionmaker is satisfied with her decision ex post. Such a "sorting" condition, requiring a degree of consonance between the objectives of the decisionmaker and the agent affected by the decision (the sender), is characteristic for models concerning the strategic transmission of information (cf. Cho and Sobel, 1990). The intuition for this condition is as follows. In case the condition does not hold the bad type sender always has an incentive to exactly mimic the behavior of the good type sender. When $e_1 > e_2$ the bad type has a larger stake in persuading the decisionmaker to choose $x=x_2$ than the good type sender. As a consequence, when the good type is motivated to send a costly message s=c in order to get the decisionmaker to choose $x=x_2$, the bad type is motivated to do this as well. Under these circumstances, costly messages cannot be informative.

Taking both conditions together, it follows that $\max\{c, e_1\} \le e_2$ is a necessary condition for information transfer to be possible in equilibrium. The following proposition — based on Propositions 2 and 3 in Potters and Van Winden (1992) — leads to the conclusion that this condition is also sufficient for information revelation to be possible in equilibrium. In this proposition the simplified notation $\sigma(t_i) \equiv \sigma(c \mid t_i)$ and $1 - \sigma(t_i) \equiv \sigma(0 \mid t_i)$ for $i=1,2$ is used.

Proposition 3.2[6]
Using $d \equiv d_1/(d_1 + d_2)$ and $r \equiv p(1-d)/d(1-p)$, the equilibrium paths of the PBE of the model with partial conflict of interests $(e_1, e_2 > 0)$ are given by:[7]

$\underline{p < d:}$
(i) $c < e_1 < e_2$
    E1: $\sigma(t_1) = r$, $\sigma(t_2) = 1$, $\rho(0) = 0$, $\rho(c) = c/e_1$; $q(0) = 0$, $q(c) = d$.
    E2: $\sigma(t_1) = \sigma(t_2) = 0$, $\rho(0) = 0$; $q(0) = p$.

(ii) $e_1 < c < e_2$
    E2.
    E3: $\sigma(t_1) = 0$, $\sigma(t_2) = 1$, $\rho(0) = 0$, $\rho(c) = 1$; $q(0) = 0$, $q(c) = 1$.

$\underline{p > d:}$
(i) $c < e_1 < e_2$
    E4: $\sigma(t_1) = 0$, $\sigma(t_2) = 1 - 1/r$, $\rho(0) = 1 - c/e_2$, $\rho(c) = 1$; $q(0) = d$, $q(c) = 1$.
    E5: $\sigma(t_1) = \sigma(t_2) = 0$, $\rho(0) = 1$; $q(0) = p$.
    E6: $\sigma(t_1) = \sigma(t_2) = 1$, $\rho(c) = 1$; $q(c) = p$.

(ii) $e_1 < c < e_2$
    E3, E4 and E5. $\qquad\qquad\qquad\qquad\qquad\qquad\qquad\qquad$ □

Proposition 3.2 shows that a separating equilibrium only exists when the costs of sending s=c are prohibitive for the bad type, but not for the good type sender. In the unique separating equilibrium E3 the good type sends a costly message with certainty, whereas the bad type rationally refrains from sending s=c. In the hybrid equilibria the message strategies of the two sender types are partially, but not completely, distinct. When the decisionmaker is a priori inclined to choose x=x₁ (p<d), the good type always sends costly message s=c, and the bad type only now and then sends s=c (i.e., plays a mixed strategy, cf. E1). The absence of a costly message provides the decisionmaker conclusive evidence that the sender is of the bad type, which induces her to choose x=x₁. After observing a costly message, however, the decisionmaker will take either action with positive probability. In the opposite case (p>d), the bad type never sends a costly message s=c, whereas the good type only sometimes does so (cf. E4). A costly message induces the decisionmaker to take action x=x₂ with certainty, whereas silence is followed by taking either action with positive probability. Lastly, in a pooling equilibrium either (i) the sender never sends a costly message (E2 and E5), or (ii) the sender always sends s=c (E6). The first kind of pooling equilibrium always exists, the second only when the decisionmaker is a priori inclined to choose x=x₂ (p>d). Note that when this latter case applies (p>d), all three types of equilibria — pooling, hybrid and separating — may exist at the same time (viz. if $e_1 < c < e_2$).
    Although not explicitly stated in Proposition 3.2, the equilibria also exist on the edges. Rather than $c < e_1 < e_2$ and $e_1 < c < e_2$, already $c \leq e_1 \leq e_2$ and $e_1 \leq c \leq e_2$, respectively, are sufficient conditions for the equilibria to exist. These conditions

are not incorporated in Proposition 3.2, though, because on the edges typically a continuum of — from an information revelation perspective essentially the same — equilibria exist. In the sequel such knife-edge cases will be ignored. The results of Propositions 3.1 and 3.2 and the discussion above yield the following corollary for the situation with partial conflict of interests ($e_1, e_2 > 0$):

Corollary 3.1
A necessary and sufficient condition for information revelation to be possible in equilibrium is given by the "Information Revelation" (IR) condition $\max\{c, e_1\} < e_2$. $\square$

When information is transferred in equilibrium the decisionmaker modifies her decision as compared with her decision solely based on prior information.[8] In equilibrium the decisionmaker responds positively to costly messages. That is, the decisionmaker will never adjust her prior belief p downwards when she observes a costly message s=c. This adjustment is rational given that in equilibrium the good type sender is at least as likely to send a costly message as the bad type sender is. In short, in equilibrium a costly message by the sender increases — or better, does not decrease — the probability that the alternative he prefers is chosen by the decisionmaker.

Comparative statics. Due to the existence of multiple equilibria (cf. Proposition 3.2), general comparative statics results are difficult to obtain from the model. Some results are worth mentioning, however. We focus on the informative equilibria.[9] Firstly, consider the *impact* of a costly message. When p<d the probability that the decisionmaker chooses x=x2 after observing s=c is weakly increasing in c and weakly decreasing in e_1. In a sense, the decisionmaker discounts the informational value of a costly message, depending on the stake (e_1) the bad type has in misinforming her relative to the cost c of sending the message. A similar result holds when p>d since in this case the gain from sending s=c, measured by the increase in the probability $\rho(c)-\rho(0)$ that x=x2 is chosen, is weakly increasing in c and weakly decreasing in e_2.[10] The (additional) informational value of a costly message to the decisionmaker is discounted using the stake (e_2) the good type has in informing her relative to the cost of sending s=c. Besides, sending s=c always results in having the preferred action chosen when p>d, but not necessarily so when p<d. In other words, the decisionmaker responds more favorable to a costly message when she is already a priori inclined to take the action preferred by the sender. Secondly, the *expected occurrence* of a costly message is weakly increasing in e_1 and weakly decreasing in c when p<d. Together with the fact that only pooling on no-message occurs when e_2 drops below c, this result suggest that costly messages are more likely when the cost decreases or when the stake increases. Interestingly, the frequency of observing a costly message equals p/d in the hybrid equilibrium E1, and thus is highest when p is close to d. This suggests that a costly message is more likely to occur when the decisionmaker only needs a small push to take the preferred action.

Welfare implications. In order to obtain the welfare implications of the model, some payoff comparisons are made. First, consider the decisionmaker. Due to the rational expectations character of an equilibrium, and the fact that receiving a message bears no costs to the decisionmaker, the possibility of information transmission never harms her. However, only in the separating equilibrium E3 the decisionmaker is better off — in terms of expected payoffs — than in case of a decision based on prior information.[11] Rather surprisingly, the decisionmaker is not worse off in the pooling equilibria compared to the more informative hybrid equilibria. In other words, the decisionmaker does not benefit in terms of expected payoffs from additional, but incomplete information. In this context, it should be realized that the strategic transmission of information involves that the decisionmaker is sometimes misled. In the hybrid equilibrium when $p<d$ the decisionmaker is sometimes misled by the costly message of the bad type sender. In the other hybrid equilibrium E4 (only possible when $p>d$) she is now and then misled by the absence of a costly message from the good type sender. On the other hand, in these equilibria the decisionmaker is sometimes completely informed; namely when she does not observe a costly message in case $p<d$, or does observe message $s=c$ when $p>d$. Overall, in a hybrid equilibrium the expected loss from sometimes being misled and the expected gain from now and then being fully informed cancel out, and the decisionmaker receives the same expected payoff as in case costly messages — and, thus, information transmission — were not possible.

Next we turn to the welfare implications for the sender.[12] We compare the expected payoff the sender obtains in the informative equilibria with his expected payoff in case the decisionmaker's action is based on a priori information (i.e., when costly messages are not possible). Rather intuitively, both sender types appear to be better (not worse) off having the option to send $s=c$ when the decisionmaker is a priori inclined to take the unfavorable action $x=x_1$ ($p<d$). However, in the opposite case ($p>d$) the sender would rather see that costly messages were not possible. So, from an ex ante perspective, the sender only values the option to send costly messages in case the decisionmaker is not a priori inclined to choose the preferred action.

In the pooling or non-informative equilibria the behavior of the decisionmaker is unaffected by the sender's expenditures. Hence, when expenditures are made in these types of equilibria they constitute a pure social waste. The sender would be better off and the decisionmaker would not be worse off if costly messages were not possible. Note, however, that such a wasteful pooling equilibrium only exists in case the decisionmaker is already a priori inclined to take the preferred action. From a similar point of view, the expenditures made on costly messages in the hybrid equilibrium E4 can also be seen as a pure social waste.

Equilibrium refinements. We end the description of the basic model with a brief elaboration on the use of equilibrium refinements. The PBE concept does not restrict out-of-equilibrium beliefs, that is, beliefs in case PBE condition (e3) does not apply. Hence these beliefs may take any form, even unreasonable ones.

Equilibrium refinements typically require the out-of-equilibrium beliefs to satisfy certain plausibility conditions. In the simple game considered here, out-of-equilibrium beliefs are only invoked in the pooling equilibria E2, E5 and E6, since only in these equilibria either signal s=0 or signal s=c is not sent in equilibrium. Therefore, refinements based on restricting these beliefs will only question the plausibility of these three equilibria. Two of such refinements proposed in the literature, belonging to different families of forward induction criteria (cf. Mailath et al., 1993, Umbhauer, 1994), are considered here.

The first, universal divinity (cf. Banks and Sobel, 1987), is also used by Potters and Van Winden (1992). Universal divinity requires that beliefs are concentrated on the type that has the weakest disincentive to deviate from the prescribed equilibrium path.[13] Potters and Van Winden (1992) show that E2 is not universally divine. In addition, by using this refinement they are able to sharpen Proposition 3.1 above in the sense that, when $e_1 > e_2 > 0$, in all universally divine equilibria the sender types pool on no-message. That is, through universal divinity the possibility of pooling on sending s=c is excluded when $e_1 > e_2 > c$.

The second refinement is the Consistent Forward Induction Equilibrium Path (CFIEP) concept of Umbhauer (1994). In a loose way, this refinement envisages an out-of-equilibrium message as a signal by some set of types that the beliefs at that message should be determined by an alternative PBE path. Hence an equilibrium path can only be removed by another equilibrium path.[14] In our basic signaling game CFIEP deletes E2 and E6 as plausible equilibria. In fact, pooling on s=c is always excluded by CFIEP, and by using this refinement Proposition 3.1 can be strengthened in the same way as when using universal divinity. In short, by deleting the pooling on s=c equilibrium (in some instances), equilibrium refinements weaken the prediction that outlays on costly messages may constitute a pure social waste. On the other hand, by deleting some non-informative (pooling) equilibria the likeliness of information revelation is emphasized when refinements are used.

Summary. The results obtained from analyzing the basic signaling game can be summarized as follows. In case the sender and the decisionmaker agree upon what the best action is in every state of the world, full information revelation is an equilibrium phenomenon whenever the costs of sending costly messages are not prohibitive. On the other hand, when the interests of the sender fully conflict with those of the decisionmaker, information transfer cannot occur in equilibrium. With respect to the most interesting case of partially conflicting interests the following results are obtained:

Result 3.1

(a)   The informational value of a costly message merely lies in its costs, and not in its content.

(b)   Information transfer only occurs in equilibrium when the preferences of the sender and the decisionmaker are sufficiently aligned (sorting condition), and the costs of sending a costly message are not prohibitive for the sender with the "right" information.

(c)   In equilibrium sending a costly message increases the probability that the sender's favored action is actually chosen.

(d)   The impact of a costly message (relative to no message) is increasing in its costs and decreasing in the sender's stake in persuading the decisionmaker. The decisionmaker responds more favorable to a costly message when she is already a priori inclined to take the sender's preferred action. Costly messages are more likely to occur when the costs are lower, the sender's stakes are higher, and the decisionmaker only needs a small push.

(e)   The decisionmaker only strictly benefits from information revelation when separation occurs. The sender only values the possibilities for costly information transfer when the decisionmaker is predisposed towards the not preferred action. Expenditures on costly messages may constitute a pure social waste.   □

In the next section these results are interpreted in light of the specific applications of the basic signaling game used in subsequent chapters.

3.2   THE THREE WAYS IN WHICH THE BASIC SIGNALING GAME WILL BE APPLIED

In this section the three applications of the basic signaling game are discussed that will be used as building blocks for the models analyzed in later chapters. These three applications concern political campaigns, direct endorsements, and informational lobbying. By discussing the basic ingredients of the models that follow already in this chapter, we can direct our attention in subsequent chapters more easily towards the main questions posed in this monograph.

For expositional reasons, and to enable a full appreciation of the basic model, we first describe how the basic signaling game can be applied to advertising in general. Then we direct our attention towards a special form of advertising, namely political advertising, and the applications of political campaigns and direct endorsements are discussed. Lastly, the topic of informational lobbying is addressed in the context of the basic signaling game.

Advertising. In the industrial organization literature a distinction is made between, on the one hand, experience goods, and, on the other hand, search goods (cf. Tirole, 1988). The former are characterized by the fact that important aspects of the quality of the good are impossible to establish except through the use of the product. The quality of the latter type of goods can be determined ex ante (and, thus, prepurchase) through inspection, though. As a result of the product's characteristics, advertisements for experience goods cannot, in principle, credibly contain much direct information. The clearest message such an advertisement then carries is that the producer is spending a lot of money on the advertisement campaign. The basic signaling game of the previous section can be interpreted as a simple model of such directly uninformative advertising for the quality of an experience good.[15]

A brief illustration of applying the basic signaling game in this context may be instructive.[16] A consumer (decisionmaker) has to choose between buying coffee from an established brand with a known quality ($x=x_1$), or trying out a new brand ($x=x_2$). The producer (sender) of the new brand prefers the consumer to buy his product, irrespective of whether this new brand is better ($t=t_2$) or not ($t=t_1$) for the consumer. The consumer, of course, wants to buy the best brand available, which is the established one when $t=t_1$, and the new one when $t=t_2$. The producer and the consumer thus have a partial conflict of interests. The producer knows the quality of his own product, and may advertise his product to the consumer at a fixed cost. The equilibrium analysis presented in the previous section shows that expenditures on coffee ads are only informative when high quality producers gain more from the consumer buying the good than low quality producers do. Kihlstrom and Riordan (1984) and Milgrom and Roberts (1986b) argue that this sorting condition is likely to be met by referring to repeated purchases by consumers. Only high quality producers will induce repeated purchases and, thus, have higher expected sales. In summary, the basic signaling game provides a rationale for why spending money on directly uninformative advertising leads to higher sales.[17]

Repeated purchases may cause advertisements for experience goods to have some indirect informational content through their costs. In a somewhat different (equilibrium) setting, such repeated purchases may cause advertisements for, in principle, experience goods to have some direct informational content as well. For instance, in a model where lying bears a reputation cost through the loss of future sales, producers may have an incentive to adhere to "truth-in-advertising" as an equilibrium result. In other words, market forces may blur the distinction between experience goods and search goods, and largely determine whether a good can be seen as either more like a search good, or more like an experience good. This problem becomes more apparent when we turn to the topic of political advertising.

Political campaigns and direct endorsements. Campaigning by political candidates is just a special form of advertising.[18] On the one hand, political candidates seem to have some search good characteristics. Political campaigns seem to have some direct informational content through clarifying and supplying useful information about the policy position ("quality") of the candidate. On the other hand, political candidates also seem to have some experience goods characteristics. Firstly, voters may think of a candidate's campaign as just being non-informative, hollow campaign rhetoric. Secondly, the non-transparency of many policy proposals may cause voters not to understand the impact of these policies.[19] In that case, the actual quality of a candidate can only be assessed through post election experience once the candidate is in office.

Economists tend to start from the assumption that, in principle, politicians are like experience goods (cf. Lott, 1987). However, due to the reelection goal politicians are assumed to have and the selection forces in the political market, politicians become, to some extent, like search goods. The reputational approach of Nelson (1976) emphasizes the long term effects of lying; the loss of reputation as the costs of not keeping your campaign promises induces, to a certain extent, "truth-in-political-advertising".[20] According to Lott (1987) and Lott and Reed (1989) the past voting record of an incumbent indicates his (her) policy stance and credibility. Voters naturally select legislators over time on the basis of this past voting record. True ideologues, those who voted in accordance with their own preferences, can provide this past voting record signal at a much lower cost. Competition between legislators, then, will efficiently sort the political market.[21] As a result, the politician seeking reelection becomes more like a search good. Under the latter approach that emphasizes the signaling value of a past voting record, challengers seeking office for the first time are still more like experience goods.

In the model presented in the next chapter we apply the basic signaling game to political campaigning and direct endorsements. Acknowledging the informational value of past experience with political candidates, we assume that incumbent politicians are more like search goods, and challengers more like experience goods. Under this assumption, the basic signaling game can be used to describe an election with an incumbent and a challenger (sender) competing for the vote of a representative voter (decisionmaker). The voter must either vote for the incumbent ($x=x_1$), or for the challenger ($x=x_2$). Using the incumbent's past voting record she is informed about what kind of policy is to be expected once he is reelected. On the other hand, she is uncertain about the policy position of the challenger.[22] A priori there are two possibilities (states of the world): either the future policy position of the challenger fits her preferences better than the incumbent's position ($t=t_2$), or it fits her preferences worse ($t=t_1$). The goal of the challenger is to get elected, which gives him an incentive to make the voter believe that his policy is better, irrespective of whether in fact this is true or not. Hence the case of partial conflict of interests between the challenger and the voter applies. In the political campaigning model, the challenger may engage in costly campaigning in order to persuade the voter to elect him. To run an effective campaign the challenger has to spend a fixed amount of money (c), independent of the content of the campaign. Of course, the challenger may also abstain from campaigning.

The results obtained in Section 3.1 can now be interpreted in the context of this particular application. Result 3.1(a) shows that the informational value of a political campaign may lie solely in its costs.[23] As already hinted at by Austen-

Smith (1991) and Calvert (1986, pp. 53-54), the amount of campaign expenditures provides information about the preferences of the candidate and, hence, her or his policy position. A recent experimental study by Lupia (1992, 1994) provides empirical support for this contention. In his model voters use the observation that an agenda setter has paid a certain fixed amount of money to propose an alternative to the commonly known status quo, as a cue for the content of this unobserved alternative. In case the agenda setter makes no expenditures he cannot propose an alternative and the status quo prevails. The Lupia model is linked to the basic signaling game for the case p<d (the voter a priori prefers the incumbent), where the agenda setter replaces the sender and the status quo replaces action x=x_1 (the incumbent's policy position). The experiments showed that the observation that the setter paid a substantial amount of money in order to propose an alternative to the status quo was an effective substitute for complete information about the content of this alternative.

Result 3.1(b) suggests that for campaign expenditures and information transmission to occur in equilibrium it is required that the candidate's preferences are not too divergent from the voter's. When candidates are mainly driven by reelection considerations, this condition is likely to hold (cf. the repeated purchases argument of Milgrom and Roberts, 1986b). By Result 3.1(c) the campaigning model explains why voters respond positively to campaign spending, as is found by and large in the empirical literature (cf. Morton and Cameron, 1992). Rather than assuming that "campaign spending buys votes" from the outset, the model yields this relationship as an equilibrium result. In that way, the model provides a micro-foundation for the vote production function used in many theoretical models that relate voters' decisions to campaign spending.[24] The comparative statics results and welfare implications of the model also have nice interpretations. For instance, in line with empirical evidence (e.g., Kau and Rubin, 1982, Poole and Romer, 1985), one of the comparative statics results predicts that campaign expenditures are highest when the election is expected to be close. Furthermore, the welfare implications indicate that expenditures on political campaigns may benefit the challenger and/or the electorate at some instances, but constitute a pure social waste at others.[25]

Instead of — or better, in addition to — the political advertising done by the candidate himself, interested third parties may use advertisements to recommend voters to elect a specific candidate. Interest groups, for instance, follow politics much more closely and are therefore better informed and better able to assess the policy positions of candidates. An interest group's endorsement, then, may reveal information about the candidates' policies to less informed voters because the latter are aware of the interests of the interest group. In this respect, the basic signaling game can be used to obtain a formal model for direct endorsements.[26] The sender is replaced by an informed interest group, and the decisionmaker again by a representative voter. The interest group knows the policy positions of both the incumbent and the challenger (whereas the voter only knows the first), and may try to persuade the voter by means of a costly endorsement to elect either one of them. The analysis presented in the previous section then, among other things, predicts that only when the preferences of the interest group and the voter are sufficiently aligned, direct endorsements can be informative. To the extent that endorsements are indeed meant to inform voters, this result is in line with the theoretical result obtained by Snyder (1991, p. 105) who finds that "..moderate interest groups should be more likely than extremist interest groups to try to bring their issue to the attention of the general public."

Informational lobbying. In the Chapters 5 and 6 of this monograph the basic signaling game is applied in a rather different context, viz. informational lobbying. According to the more descriptive literature, one of the key characteristics of lobbying is the transmission of information (Birnbaum, 1992). Interest groups typically possess information relevant to policymakers which the latter do not have themselves. The information may for instance refer to the (interpretation of) the consequences of a certain policy (cf. Smith, 1984), to the true demand for certain public goods (Ainsworth and Sened, 1993), or relate to the electoral salience of a group's cause (Ainsworth, 1993). Although policymakers realize that interest groups have such valuable information, they are also aware of the strategic incentives interest groups have in presenting this information.[27] A number of recent theoretical studies, therefore, focus on the informational role of lobbying in a signaling game context. Two of them, Ainsworth (1993) and Potters and Van Winden (1992), make use of the exact setup of the basic signaling game of the previous section, where the interest group/lobbyist represents the sender and a policymaker represents the decisionmaker.[28] Important issues that can be tackled with the use of the basic signaling game include questions like: When is lobbying influential?; How much information will be revealed through lobbying?; Does lobbying improve the decision of the policymaker? The answers to these questions correspond with the results discussed at length in Section 3.1 and will not be repeated here, except for two observations. First, note that by the equilibrium result that the policymaker will positively respond to outlays on lobbying, an informational micro-foundation is provided for the use, and possibly the specifica-tion, of an influence function (cf. Lohmann, 1995a, 1995b). Second, recall that the policymaker only benefits from lobbying (signaling) when full information revelation occurs. Ainsworth (1993) uses this observation to argue that the policymaker should create an institutional structure such that a separating equilibri-um can occur. In anticipation with their interaction with lobbyists, policymakers can structure the institutional features under which they will interact. Lobbying regulations and costs of access, like registration requirements, can be used to enable complete revelation of information.

3.3   EXTENSIONS OF THE BASIC SIGNALING GAME

The basic signaling game presented in Section 3.1 can be extended in several interesting ways. In this section a number of such extensions are discussed, and related to the three applications of campaigning, endorsing and lobbying. At the same time, this section relates the basic signaling game to other existing models of interest group influence through the transmission of information. In the next three chapters we will frequently return to the extensions discussed here.

Endogenous costs of sending a message. Starting from the result that in a situation with partial conflict of interests the informational value of a message lies solely in its costs, a straightforward extension is to increase the informativeness of costly messages by allowing their costs to be endogenously determined. Allowing only for cost levels between 0 and c would not change much, however. As long as the upper bound c satisfies the conditions that are given in Proposition 3.2, the existence of the various types of equilibria remains essentially unchanged. Only if no upper bound is put on the level of expenditures, the scope for information transmission is enlarged. Specifically, c drops out of the IR condition of Corollary 3.1. Interpreting this result in the context of campaign expenditures, it follows that the existence of upper bounds on campaign expenditures may in effect reduce the scope for information transfer and an informed decision by the voter (cf. Palda and Palda, 1985).[29] Since voters typically a priori prefer incumbents ($p<d$), such upper bounds may also limit the electoral opportunities of challengers.[30]

Multiple senders. A second extension of the basic signaling game is to allow for multiple senders. Rather intuitively, the presence of two equally informed senders with opposing interests increases the scope for information revelation (cf. Gilligan and Krehbiel, 1989, Potters, 1992).[31] Full separation by both senders may occur in equilibrium in cases where separation by only one sender does not constitute equilibrium behavior. The sender with the "bad" information can be deterred from sending a (false) message by the separating message strategy of the sender with the "good" information. In such an equilibrium there is an oversupply of information (cf. Lohmann, 1993, 1994), and part of the expenditures on costly messages constitute a social waste. Similar results are obtained in the slightly different setup (see below) of Austen-Smith and Wright (1992). The main result they obtain from their two senders model, however, is that (costly) messages are often sent just by one sender, typically the sender that disagrees with the predisposition of the decisionmaker. The a priori favored sender usually only sends costly messages to counteract the influence attempts by the opposing sender. The intuition behind this result is that a sender has nothing to gain from information transmission if he is the only one who transmits information to a decisionmaker whose a priori inclination is to support his case. A second reported result is that, the more important the issue to the senders, the more likely is the decisionmaker to decide under complete information. This prediction, however, is qualified in Sloof (1997a), where it is argued that this comparative statics result is not general.

In fact, Austen-Smith and Wright (1992) use their two-sender model to analyze competitive lobbying, and they conclude that lobbying is mainly intended to alter policy positions, rather than supporting friends (cf. the exchange model discussed in Chapter 2). Friends are only lobbied to counteract the influence of opposing groups. However, the model also has a nice interpretation in the context of political campaigns. Compared with the basic signaling game, a two-sender model allows the incumbent to engage in campaigning as well. It follows from the Austen-Smith and Wright study, then, that the a priori favored candidate, typically the incumbent, will particularly focus on "counter-campaigning". If, a priori, the vote is likely to be in his favor, the incumbent will mainly spend money to counteract the campaign of the challenger.[32] Hence, the expenditures of the challenger are a major determinant of the expenditures of the incumbent (cf. the empirical evidence in Jacobson, 1980, Kau, Keenan and Rubin, 1982, and Wilhite and Theilmann, 1987). Incidentally, these results provide a rationale for the focus in our basic political campaigning model on the explanation of the challenger's expenditures.

Verification of messages. Besides incorporating multiple senders the setup of Austen-Smith and Wright (1992) differs from the basic signaling game in another notable way; the decisionmaker can verify at a certain cost the information provided by the senders, and senders are penalized exogenously when they are caught lying.[33] In fact, the counteractive behavior of the a priori favored sender serves as a substitute for the decisionmaker's checking on messages from the opposing sender. When the costs of verifying messages are high, verifying occurs less often, and the a priori favored sender is more likely to engage in counteractive information transmission. In a similar vein, Rasmusen (1993) investigates the decisionmaker's choice whether to obtain completely reliable information. She may either acquire her own information, or rely on (and possibly verify) the information that may be transmitted by a single sender. In his model there is no exogenous penalty for lying. The sender's incentive for truth telling lies in the possibility that the decisionmaker checks the information before making his decision, so that expenditures on false statements are wasted. The model of Rasmusen (1993) is a straightforward extension of the basic signaling game, and comes down to assuming that the decisionmaker can check the accuracy of a costly message $s=m$ (with $m \in M$) at cost k_1, and that she may also investigate the actual state of the world herself at cost k_2 when no such message is received ($k_2 \geq k_1 > 0$ by assumption). In comparison with the basic signaling game, Rasmusen replaces stage (iv) by stage (iv)^R:

(iv)^R The decisionmaker receives signal $s \in S \equiv \{n\} \cup M$ and chooses k, the expenditure of discovering t. After costly message $s \in M$ $k \in \{0, k_1\}$, and after $s = n$ $k \in \{0, k_2\}$. k=0 implies that the decisionmaker gets no additional information about t, $k = k_i$ (i=1,2) that she learns the true value of t.

It is easily verified that when $k_1 > d_1 d_2 / (d_1 + d_2)$ verification and independent investigation never occur in equilibrium, and the game essentially reduces to the basic signaling game. A comprehensive equilibrium analysis for the game when $k_1 \leq d_1 d_2 / (d_1 + d_2)$ is more involved, and therefore omitted here.[34] The main conclusions Rasmusen (1993) draws from his model are immediately recognized without the full formal analysis, though. Compared with a situation where the decisionmaker can only rely on his own (independent) investigation, allowing information transfer by the sender never hurts the decisionmaker (in expected payoff terms), and is beneficial to her when such information transfer indeed occurs. The information coming from the sender de facto serves as a substitute for the decisionmaker's own investigation.[35] The first type of information is costless, though not completely reliable, whereas the second source of information is fully reliable, but costly. In contrast to the decisionmaker, allowing costly messages may either weakly hurt or weakly help the sender, depending on the sender's preferences over x_1 and x_2, the prior belief of the decisionmaker, and whether the latter would investigate herself when costly messages were excluded.

A nice application of the Rasmusen model concerns the topic of oversight. Following McCubbins and Schwartz (1984) a number of scholars argue that interest groups may help legislators to (partly) overcome the agency problem in controlling the bureaucracy. By acting as "fire alarms" interest groups may signal information about the performance of the bureaucracy. On the other hand, through "police patrols" the legislature may also monitor bureaucrats herself. The results discussed above then suggest that the monitoring role of "fire alarms" may substitute for the politicians' own monitoring, and that politicians may benefit from the watchdog role of interest groups, even when different interests induce these groups to strategically transmit information.[36] Similarly, interest group endorsements may help voters to control politicians, and serve as a substitute for voters' own acquisition of information about the performance of politicians.

Instead of verification by the decisionmaker herself, it is possible that a costly message from the sender may be screened by an intermediary agent. For instance, a lobbying report of an interest group may be first evaluated by a bureaucratic agency before it reaches a politician. Potters and Van Winden (1992) analyze such an extension (for the case of partial conflict of interests between the sender and the decisionmaker, cf. case III in Section 3.1). Rather than spending c on a message, the sender may now hire at this cost an outside consultant who evaluates the message. With probability α_i this consultant reports "$t = t_i$" when actually $t = t_i$. Due to the assumption $0 \leq \alpha_1 \leq \alpha_2 \leq 1$ there is an increased scope for information transfer compared with the basic model. When $\alpha_2 / \alpha_1 \geq 1/d$, even costless, evaluated messages can be informative.

Persuasion games. The last extension discussed above takes us from assumptions made in signaling games to assumptions characteristic of so-called "persuasion games" (cf. Grossman, 1981, Milgrom, 1981). Contrary to the basic signaling game, in a basic persuasion game (i) senders have to select a truthful message from the set of available messages, and (ii) all messages that are truthful are available to them. Loosely put, in a persuasion game senders always have to report truthfully, though they need not be very precise, and senders are always able to prove any true claim they make. The main idea is that a sender tries to influence a decisionmaker by selectively providing relevant data. The driving force behind information revelation is the different potential for sending certain messages; sender types with "good" information are allowed to send certain messages which sender types with "bad" information cannot select. In comparison, the driving force behind information transfer in signaling games is the sorting condition; types with "good" information have a stronger propensity to send certain messages than those with "bad" information.

The basic signaling game of Section 3.1 is most easily transformed into a basic persuasion game in the following way. Let $M=\{m_1,m_2\}$, where m_i represents the message "$t=t_i$" (i=1,2). Let the cost of sending a message from M equal zero, that is let $c(m)=c=0$. The sender, knowing that $t=t_i$, can only choose between signals s=n and $s=m_i$, with signal n having the interpretation of saying "$t=t_1$ or $t=t_2$". Although the sender cannot lie, he need not be very precise and may essentially say nothing (as signal n entails). It is easily seen that full disclosure of information — in the literature typically referred to as "unravelling" — certainly occurs when there is no conflict of interests (cf. case I in Section 3.1), and does not occur in case of full conflict of interests (case II in Section 3.1). In the remaining case of partial conflict of interests unravelling necessarily occurs when p<d, but not so when p>d. When p<d the decisionmaker employs a skeptical strategy; she expects that any information withheld by the sender is unfavorable. Specifically, signal n is always interpreted as stating that $t=t_1$, i.e. in the least favorable way for the sender. This is not the case when p>d. Then, based on her prior information, the decisionmaker already chooses the alternative that is most preferred by the sender ($x=x_2$), and full disclosure need not occur in equilibrium. A pooling equilibrium exists in which both sender types send signal s=n, and the decisionmaker chooses $x=x_2$. Of course, an equilibrium in which unravelling occurs also exists (let type t_2 choose m_2 with probability 1). In short, separation (unravelling) can always occur in equilibrium, and sometimes necessarily occurs in equilibrium. The scope for information transfer is larger than in the basic signaling game.

Okuno-Fujiwara et al. (1990) and Lipman and Seppi (1995) argue that assuming that the sender can always prove any true claim is too strong. In reality, definitive proofs of some true facts may simply not exist. An interesting situation to consider, therefore, is the case of partial provability in which not all true messages are available to the sender. Such a case for instance applies in the setup described in the previous paragraph when message m_2 is not possible. That is, when it is not possible to unrefutable prove that $t=t_2$. When the sender, referred to as S2, always prefers $x=x_2$ to be chosen, unravelling does not occur. (For a less trivial example that partial provability may upset the unravelling result, see the Cournot duopoly example of Okuno-Fujiwara et al. (1990, example 3).) The main point of Lipman and Seppi (1995) is that conflicting interests among multiple senders may restore the transmission of information in equilibrium. For instance, suppose there is another sender S1 who strictly prefers action $x=x_1$ to be chosen, irrespective of the actual state of the world. In that case the decisionmaker receives a pair of messages. In case $p>d$ sender S1 will always completely reveal his information when he knows that $t=t_1$, in order to get his preferred alternative chosen. Full disclosure is the unique equilibrium outcome in this case ($p>d$). The decisionmaker even does not need to know "who is who", that is, which of the two senders is actually S1. Of course, also under complete provability (senders can prove any true claim) the scope for unravelling increases when there are competing senders (cf. Milgrom and Roberts, 1986a).

A number of authors take as a natural application of (extended versions of) a persuasion game the setting of informational lobbying (see e.g. Lagerlöf, 1997, and Lipman and Seppi, 1995). The strong assumption that lobbyists cannot lie can be justified by repeated interaction considerations; lying leads to loss of reputation and is therefore (prohibitively) costly. The costs of lying in the lobbying model of Austen-Smith and Wright (1992) have a similar justification. From this perspective the signaling game setup (lying bears no cost) and the persuasion game setup (lying cannot occur, i.e. bears infinite cost)[37] can be seen as benchmark cases of a more general static model with exogenous costs of lying. In a repeated setting the costs of lying could even come about endogenously. For instance, in a dynamic version of the model of Section 3.1 the sender may initially want to report truthfully to gain credibility and to build up or maintain a reputation (cf. Sobel, 1985). Extending the basic signaling game to incorporate dynamic features introduces a lot of new issues. For instance, rather than costly messages, then also sanctions — that is, the actual enforcement of a threat — can be used to influence the decisionmaker (cf. Potters and Van Winden, 1990). The model presented and discussed in Chapter 5 extends the basic signaling game in such a way.

Receiver uncertainty. A final extension we want to address is the topic of receiver uncertainty. Under receiver uncertainty the decisionmaker is uncertain about whether the sender is informed, or about how much information he possesses. In that case the informed types with unfavorable information can pool with the uninformed. This may increase the scope for information transfer in a setting where lying bears no exogenous cost (cf. Austen-Smith, 1993a, 1994). When lying is prohibited, and thus unravelling is the rule rather than the exception (Milgrom, 1981, Seidmann and Winter, 1997), such uncertainty may actually reduce the scope for information transmission, though. A nice result Shin (1994) obtains in respect to receiver uncertainty when lying is prohibited is that reports from a sender with an a priori superior information position may be treated more skeptical by the decisionmaker. Suppose, for instance, that in the two-sender persuasion game setup discussed previously sender S1 (S2) is informed about t with probability $\beta_{S1}>0$ ($\beta_{S2}>0$). The two senders simultaneously send truthful signals on t, with necessarily $s_i=n$ if Si is uninformed for $i=1,2$ (it is again assumed that senders can prove any true claim they make). Now, consider the following strategies, labelled "sanitization" strategies by Shin (1994); when t is favorable for a sender, and the sender is informed, he reveals this information, otherwise he sends signal s=n. Under these equilibrium strategies we get $q((n,n))>p$ iff $\beta_{S1}>\beta_{S2}$. So, when S1 has a superior information position ex ante ($\beta_{S1}>\beta_{S2}$), the decisionmaker is more skeptical about his (uninformative) reports. In general, the effect of the increased potential for the better informed sender to engage in "news management" may outweigh the effect of his larger expertise. This result corresponds, at least superficially, with a result obtained by Austen-Smith (1993c) in a signaling game setting with no receiver uncertainty. He finds that the relationship between influence and the level of expertise may be non-monotonic.

3.4 SUMMARY

In this chapter we started out in Section 3.1 with a description of a basic model of strategic information transmission. A decisionmaker was able to extract information from the messages coming from a sender, even in cases where the interests of the sender were somewhat (but not too much) opposed. Section 3.2 introduced the three applications of this basic signaling game that will be used in later chapters, viz. political campaigning, direct endorsements, and informational lobbying. A number of extensions of the basic signaling game were discussed in Section 3.3, and related to the three applications of interest. In this way, a number of related models could be given attention.

The next chapter builds on the political campaigning and direct endorsements applications of the basic signaling game. Chapters 5 and 6, on the other hand, start from the application to informational lobbying.

NOTES

1. When $d_1 d_2 < 0$ information about t is of no importance to the decisionmaker and, therefore, not relevant. The decisionmaker always takes the same action. In case $d_1 < 0$ and $d_2 < 0$ reversing the labels of x_1 and x_2 leads to the same setup as in the main text. In this way, the assumption $d_i > 0$ i=1,2 is without loss of interesting generality.

2. See Gibbons (1992). In the game considered here, perfect Bayesian equilibrium is equivalent to the slightly different notion of sequential equilibrium (cf. Fudenberg and Tirole, 1991).

3. Information transfer is not inevitable, though, for both sender types choosing s=n can also occur in a PBE. To see this, take out-of-equilibrium beliefs after receiving message $m \in M$ equal to the prior. In addition, when M is not a singleton set or both $-e_1 > c$ and $e_2 > c$, several fully revealing equilibria exist and there may be a problem of coordination. Finally, besides the fully revealing equilibria, also partial revealing equilibria (the hybrid equilibria defined below) may exist at the same time.

4. In other words, cheap talk cannot be *influential* (cf. Austen-Smith 1993a). It can still be informative in equilibrium, though, as the following example of a PBE illustrates. Let $d_i = e_i = i$ for i=1,2, $p = \frac{1}{2}$ and $c = \frac{1}{4}$. Moreover, let M be restricted to only two elements, $M = \{m_1, m_2\}$ say. It is easily verified that the following strategies and beliefs describe a PBE whenever $0 \le \varepsilon \le 1/6$: $\sigma(m_1 | t_1) = \sigma(m_2 | t_2) = \frac{1}{2} + \varepsilon$, $\sigma(m_2 | t_1) = \sigma(m_1 | t_2) = \frac{1}{2} - \varepsilon$, $\rho(n) = 0$, $\rho(m_1) = \rho(m_2) = 1$; $q(n) = 0$, $q(m_1) = \frac{1}{2} - \varepsilon$, $q(m_2) = \frac{1}{2} + \varepsilon$. Both sender types send a costly message for sure, but when $\varepsilon > 0$ type t_1 (t_2) sends message m_1 (m_2) somewhat more often than type t_2 (t_1) does. Receiving message m_1 instead of m_2 in equilibrium then reveals some information ($q(m_1) < p < q(m_2)$), but not "enough" to alter the action taken by the decisionmaker ($\rho(m_1) = \rho(m_2)$). In other words, although cheap talk may be informative, it has no informational value to the decisionmaker.

5. Potters and Van Winden (1992) focus on the PBEs of the game that survive the universal divinity refinement on out-of-equilibrium beliefs (cf. Banks and Sobel, 1987). Here we do not immediately restrict our attention, and look at all PBEs first.

6. See Potters and Van Winden (1992) for a proof of the proposition. Erroneously, equilibrium E4 (their LE4) is not mentioned in their Proposition 3 for the case p>d and $e_1 < c < e_2$.

7. Only the equilibrium strategies and beliefs along the equilibrium paths are specified. When Bayes' rule does not apply for signal s there is freedom in choosing out-of-equilibrium posterior belief q(s), and hence the out-of-equilibrium strategy $\rho(s)$. Due to this freedom, a specific equilibrium path may be compatible with various out-of-equilibrium strategies and beliefs. We do not list all the possible out-of-equilibrium beliefs and strategies that sustain a specific path, but only note that every equilibrium path is sustainable with $q(s) = \rho(s) = 0$ for each out-of-equilibrium signal s.

8. This observation is a consequence of our reduction of M to $M = \{c\}$. In that case all informative equilibria are *influential* (cf. Austen-Smith, 1993a). However, as already pointed out in note 4, when M contains more than one element informative equilibria exist which are not influential. These *essentially uninformative* equilibria are equivalent — in terms of equilibrium outcomes — to the uninformative equilibria in the model in which M is restricted to {c}. When not explicitly stated otherwise in the main text, the informative equilibria are also influential.

9. When p<d the informative equilibrium is unique, and either hybrid (c<e_1<e_2) or separating (e_1<c<e_2). In that case the comparative statics results follow from changes in the equilibrium strategies of the unique equilibrium, or from a switch from the hybrid (separating) to the separating (hybrid) equilibrium. When p>d a hybrid and a separating equilibrium may exist side by side. In that case we focus on the comparative statics of the hybrid equilibrium as its existence is implied by the existence of the separating equilibrium, whereas the reverse does not hold. Focusing on the separating equilibrium does not give substantially different results, however.

10. The results with respect to the stake parameter run counter to the second part of "stylized fact" 4 presented in Subsection 2.1.3, reflecting the empirical observation that an interest group's stake is a positive determinant of its political success. It must be realized, however, that this "stylized fact" is obtained from an overall assessment of empirical models in which means of influence are not restricted to just the strategic transmission of information (whereas they are in the basic signaling game). Besides, even in the context of the basic signaling game the result is not general. For instance, when 0<e_1<e_2<c and p<d only E2 is an equilibrium and always x=x_1 is chosen. Increasing the stake parameters e_i sufficiently, equilibrium E1 or E3 becomes possible. In these two equilibria the preferred action x=x_2 is chosen with positive probability, displaying larger political success than in E2. Also when case I of no conflict of interests is taken into account it becomes clear that the reported comparative statics results do not apply for all possible settings.

The particular comparative statics results reported illustrate the idea that messages of a disinterested sender may have a larger impact than those of a biased source. Some empirical support for this contention can be found, for instance, in Page et al. (1987). They find that reported statements of news commentators have a stronger impact on public opinion than those of special interest groups, allegedly because the former have a higher level of credibility.

11. In the separating equilibrium E3 the expected equilibrium payoff to the decisionmaker equals (1-p)d_1+pd_2. In all other equilibria her expected equilibrium payoff equals max{(1-p)d_1, pd_2}. In hybrid equilibrium E1 the decisionmaker is either indifferent between her two actions (this is the case after message c), or strictly prefers action x_1 (this occurs after message 0). Expected payoffs are thus most easily calculated by assuming that she always chooses x_1, directly yielding the expected payoff (1-p)d_1. From this perspective it is easily understood that E1 and pooling equilibrium E2 are payoff equivalent to the policymaker. A similar reasoning applies for hybrid equilibrium E4.

12. Denoting the expected payoff the bad and the good type sender get in equilibrium Ej as $U^*(t_1|Ej)$ and $U^*(t_2|Ej)$, respectively, we get: $U^*(t_1|E1)=U^*(t_1|E2)=U^*(t_1|E3)=0$, $U^*(t_1|E4)=e_1(1-c/e_2)$, $U^*(t_1|E5)=e_1$, $U^*(t_1|E6)=e_1-c$, and $U^*(t_2|E1)=c(e_2/e_1-1)$, $U^*(t_2|E2)=0$, $U^*(t_2|E3)=U^*(t_2|E4)=U^*(t_2|E6)=e_2-c$, $U^*(t_2|E5)=e_2$.

13. Formally the restriction is given as follows. Let E be an equilibrium yielding expected payoffs $U^*(t_1|E)$ and $U^*(t_2|E)$ to the bad and the good type sender, respectively. Let s be an out-of-equilibrium signal and BR(s) the set of potentially best responses ρ(s) to s. With equilibrium condition (e2) and any inference q(s)∈[0,1] possible after receiving signal s, we have BR(s)=[0,1]. Now define R_i(s) as the set of best responses to s for which the type t_i sender has a (weak) incentive to deviate from its equilibrium strategy; R_i(s)={ρ(s)∈BR(s)|e_i ρ(s)-c(s)≥$U^*(t_i|E)$}. If R_1(s)⊂R_2(s) the good type sender has an incentive to deviate when the bad type has, but not vice versa; universal divinity requires q(s)=1 in this case. Similarly, when R_2(s)⊂R_1(s) universal divinity requires q(s)=0. If neither case applies, the refinement does not put restrictions on q(s).

14. A formal description of the CFIEP refinement concept is somewhat involved and runs as follows. A PBE path E is a CFIEP iff it is not removed by any other PBE path. Now, a PBE path E is removed by another PBE path $E^{\#}$ when there is a signal s which is not sent in E but is sent in $E^{\#}$, such that for the two sets DEV={i∈{1,2}|$\sigma^{\#}(s|t_i)$>0 and $U(t_i|E^{\#},s)$>$U^*(t_i|E)$} and BR={ρ(s)∈argmax $\sum_{i\in DEV}$ u(ρ(s)|t_i)$\sigma^{\#}(s|t_i)$P(t=t_i)/($\sum_{i\in DEV}$ $\sigma^{\#}(s|t_i)$P(t=t_i))} conditions (i), (ii) and (iii) are satisfied: (i) DEV is non-

empty, (ii) for all $i \in \{1,2\} \backslash DEV$ holds that $U(t_i | E^\#, s) \leq U^*(t_i | E)$, and (iii) there is a $\rho(s)$ in BR such that for all $i \in DEV$ holds that $U(t_i | \rho(s)) \geq U^*(t_i | E)$. Here $\sigma^\#(s | t_i)$ denotes type t_i probability of sending signal s in equilibrium $E^\#$. $U(t_i | E^\#, s)$ denotes the expected payoff type t_i gets were he to send signal s (not necessarily his equilibrium message) in path $E^\#$, $U^*(t_i | E)$ his equilibrium expected payoffs in equilibrium E, and $U(t_i | \rho(s))$ his expected payoff of sending s when the decisionmaker reacts with strategy $\rho(s)$. Finally, $u(\rho(s) | t_i)$ denotes the decisionmaker's expected payoff when using strategy $\rho(s)$, given that $t = t_i$.

The undefeated equilibrium notion of Mailath et al. (1993) is motivated by exactly the same global consistency ideas as the CFIEP refinement concept, and is thus closely related. However, the undefeated equilibrium concept is only developed for pure strategy equilibria, and is somewhat less powerful than CFIEP. (With respect to the three pooling equilibria, E2 and E5 are undefeated, whereas E6 is defeated by E5.) Therefore, we focus on CFIEP.

15. The original idea of portraying seemingly wasteful advertising as signals to consumers about the quality of an experience good dates back to Nelson (1974). See also The Economist (1998) for a popular discussion of these ideas.

16. For the particular application of product advertising more elaborate models than the basic signaling game exist, see e.g. Hertzendorf (1993), Kihlstrom and Riordan (1984) and Milgrom and Roberts (1986b).

17. In this model expenditures on advertisements play an informational role by means of putting the consumer in a position to identify a high quality producer. As argued by Bagwell and Ramey (1994a, 1994b), expenditures on directly uninformative advertising may also serve a coordinating role. By directing consumers towards the same firm or a small number of firms, advertising enables consumers and firms to take advantage of mutually beneficial scale economies ("The more we sell, the lower the price, and the lower the price, the more we sell!").

18. Here we focus on the informative role of political advertising; voters' induced preferences are affected by campaigning because information is transmitted about the candidate's policy stance. However, campaigning may also be used to convince voters to elect a candidate regardless of her (his) position. This is defined persuasive campaigning (cf. Mueller and Stratmann, 1994).

19. The Dutch Parliamentary Election Study 1994 reports that about 72% of the 1527 people polled thinks that politics is sometimes too complicated (cf. Anker and Oppenhuis, 1995).

20. See also Banks (1990), Davis and Ferrantino (1996), and Harrington (1993). Some evidence suggests that according to the general voting public the incentive for truth is rather low in reality. For instance, Harrington (1992) reports that 68% of the people polled did not believe George Bush's 1988 campaign promise "Read my lips, no new taxes!". The aforementioned Dutch Parliamentary Election Study 1994 reports that about 90% of the sample agreed with the statement "Although they know better, politicians promise more than they can deliver."

21. Some empirical evidence supporting the sorting model shows that legislators do not alter their voting behavior in their last term. See Bender and Lott (1996) for an overview of the relevant empirical literature.

22. An indirect justification for these assumptions is provided by some empirical evidence that voting is more responsive to challenger than to incumbent spending (e.g. Jacobson, 1985, Kenny and McBurnett, 1992, Palda and Palda, 1985, 1998). It seems likely that this asymmetry is a consequence of the policy position of the incumbent being better known to the electorate than the position of the challenger is.

23. Besides in the model presented in Potters et al. (1997) that is based on the basic signaling game of Section 3.1, also in the models of Gerber (1996) and Prat (1997) political campaigns are directly uninformative. In this respect these models differ from most earlier theoretical models of political campaigning. Chappell (1994), for instance, assumes that campaign advertising is completely informative about the candidate's policy position and, moreover, truthful. Harrington (1992) assumes that political campaigns are costless and explores under what conditions these cheap talk messages are informative. Finally, in Banks (1990) campaigns are only costly to the winning candidate. These costs are assumed to represent the candidate's loss of reputation due to lying and are increasing in the difference between the policy proposed in the campaign and the policy implemented. In our model the costs of a campaign are independent of its content.

24. See e.g. Baron (1994), Magee et al. (1989), Grossman and Helpman (1996), Helsley and O'Sullivan (1994), and Snyder (1989). Some other theoretical studies do not use a vote production function directly, but assume that political campaigns convey useful information to the voter by decreasing the variance of perceived policy positions in a probabilistic voting model (see Austen-Smith, 1987, Gersbach, 1995, Hinich and Munger, 1989, Mayer and Li, 1994). In these models, however, the technical relationship between expenditures and the variance of perceived policies is assumed and not explained.

25. Here the possibility of costly campaigns benefits the voter because they enable the voter to make a more informed decision. In models where the policies of candidates are endogenous, campaigns may be beneficial to voters because the candidate's position in an equilibrium with campaigns may be closer to the median than without campaigns (cf. Gersbach, 1997, 1998).

26. Endorsements by informed parties directed at the general voting public are incorporated into models of elections by Grossman and Helpman (1997), McKelvey and Ordeshook (1985) and Grofman and Norrander (1990), and in models of direct legislation by Cameron and Jung (1992) and Lupia (1992). Only in Grossman and Helpman (1997) and Cameron and Jung (1992) the endorser acts strategically, but contrary to the basic signaling game considered here, these endorsements are costless for the endorser.

27. This is not just an academic assessment, as the following passage taken from *The Economist*'s analysis of the contemporary status of pressure groups in Britain nicely illustrates (The Economist, 1994). In addition, this passage also addresses one of the central issues of this monograph, viz. the strategy of an interest group in trying to influence public policy.

> "When it comes to getting things done, I would go for private contact with civil servants," a campaigner for the disabled declared in *Political Studies*, a journal. Civil servants rely much more on interest groups for information than is generally realized. A former permanent secretary at the Treasury, Sir Peter Middleton, has admitted: "if you want information, you've got to allow people to give you opinions too."
> The Child Poverty Action Group (CPAG), on the other hand, is an unabashed user of publicity as a battering ram. "Our main aim is to shift government, not chat with civil servants - coverage in the media is our strategy," one of its early directors told the *Journal of Social Policy*.

28. Ainsworth (1993) restricts the analysis to the case $d_1=d_2=1$, and assumes the sorting condition $e_1<e_2$ to hold from the outset. In a related model (cf. Ainsworth and Sened, 1993), the lobbyist provides information to both a decisionmaker and individuals potentially taking part in mass political action.

29. In the theoretical models of Gerber (1996) and Prat (1997) political campaigns are also solely informative through their costs. In contrast to the model based on (an extension of) the basic signaling game, however, in their model there is also an indirect cost of campaigning; the campaign is financed by interest groups that expect and indeed get favors in return, at the expense of the general voting public (cf. the model presented in Chapter 4). In such a setting, in case the indirect costs of granting favors exceed the direct benefits from a more informed vote decision, the voters may in fact gain from imposing spending limits or a ban on political advertising (cf. The Economist, 1997).

30. For a claim that this is indeed the case see Palda (1994a, 1994b) and Palda and Palda (1998). They use the aforementioned empirical evidence that voting is more responsive to challenger than to incumbent spending to suggest that spending limits are more harmful to challengers. Upper bounds on campaign expenditures currently exist in e.g. Canada and France.

31. Also in a situation where senders are not perfectly informed (but still have an informational advantage) and in which they obtain different signals (information) about the actual state of the world, the scope for information transfer increases (cf. Austen-Smith, 1993b). Interestingly, in the latter model the scope for information transfer is larger when the two senders move sequentially rather than simultaneously, and depends on the order in which the senders send their signals.
    Similarly, as Farrell and Gibbons (1989) show, the scope for information transfer may also increase in the presence of multiple "audiences" (decisionmakers).

32. Gerber and Lupia (1995) extend the Lupia (1992) agenda setter model briefly discussed in the previous section by adding an opponent (incumbent) to the agenda setter (challenger). Their assumption that the incumbent can only engage in counter-campaigning is justifiable in the light of this result.

33. A third difference is that in their model senders are not informed from the outset, but can acquire specialized information at a certain cost. Since information acquisition decisions are observable, such an assumption does not lead to substantively different results.

34. Rasmusen (1993) just considers the case $d_1=d_2$, $c<e_1=e_2$, and $p<d=\frac{1}{2}$. However, his equilibrium analysis is incomplete. Sloof (1997b) supplements his analysis and, thereby, qualifies part of his argument. The welfare results to be discussed in the main text are only one part of these qualifications, and they are presented here in the way as proposed by Sloof (1997b). Rasmusen (1997) responds to the comment that in his opinion the welfare results can still be presented in a stronger way, such that they really predict positive (negative) rather than non-negative (non-positive) changes in welfare after allowing for lobbying.

35. As shown by Austen-Smith and Wright (1992, Proposition 5), this result does not hinge on the assumption $k_2 \geq k_1$. They obtain that even when $k_2 < k_1$ the decisionmaker may prefer the indirect information acquisition through the messages of the sender rather than direct information acquisition at cost k_2.

36. Models explicitly concerned with the "fire alarms" role of interest groups obtain these conclusions as well; see Epstein and O'Halloran (1995), Hopenhayn and Lohmann (1996), Laffont and Tirole (1993, Chapter 15), Lohmann and Hopenhayn (1998), Lupia and McCubbins (1994), and Milner and Rosendorff (1996). In Banks and Weingast (1992) it is assumed, but not explained, that the presence of interest groups reduce the costs to the legislature of auditing the bureaucracy.

37. Leland (1981), in a comment on Grossman (1981), interprets the persuasion game as a signaling game with the cost of verifying messages going to zero and the cost of being caught lying going to infinity.

REFERENCES

Ainsworth, S., 1993, Regulating lobbyists and interest group influence, *Journal of Politics* 55, 41-56.

Ainsworth, S. and I. Sened, 1993, The role of lobbyists: Entrepreneurs with two audiences, *American Journal of Political Science* 37, 834-866.

Anker, H. and E.V. Oppenhuis, 1995, Dutch Parliamentary Election Study 1994 (Steinmetz Archive/SWIDOC, Amsterdam).

Austen-Smith, D., 1987, Interest groups, campaign contributions, and probabilistic voting, *Public Choice* 54, 123-139.

Austen-Smith, D., 1991, Rational consumers and irrational voters: A review essay on black hole tariffs and endogenous policy theory, *Economics and Politics* 3, 73-92.

Austen-Smith, D., 1993a, Information and influence: Lobbying for agendas and votes, *American Journal of Political Science* 37, 799-833.

Austen-Smith, D., 1993b, Interested experts and policy advice: Multiple referrals under open rule, *Games and Economic Behavior* 5, 3-43.

Austen-Smith, D., 1993c, Information acquisition and orthogonal argument, in: W.A. Barnett, M.J. Hinich, and N.J. Schofield, eds., Political economy: Institutions, competition, and representation (Cambridge University Press, Cambridge) 407-436.

Austen-Smith, D., 1994, Strategic transmission of costly information, *Econometrica* 62, 955-963.

Austen-Smith, D., 1997, Interest groups: Money, information and influence, in: D.C. Mueller, ed., Perspectives on public choice. A handbook (Cambridge University Press, Cambridge) 296-321.

Austen-Smith, D. and J.R. Wright, 1992, Competitive lobbying for a legislator's vote, *Social Choice and Welfare* 9, 229-57.

Bagwell, K. and G. Ramey, 1994a, Advertising and coordination, *Review of Economic Studies* 61, 153-172.

Bagwell, K. and G. Ramey, 1994b, Coordination economies, advertising, and search behavior in retail markets, *American Economic Review* 84, 498-517.

Banks, J.S., 1990, A model of electoral competition with incomplete information, *Journal of Economic Theory* 50, 309-325.

Banks, J.S. and J. Sobel, 1987, Equilibrium selection in signaling games, *Econometrica* 55, 647-661.

Banks, J.S. and B.R. Weingast, 1992, The political control of bureaucracies under asymmetric information, *American Journal of Political Science* 36, 509-524.

Baron, D.P., 1994, Electoral competition with informed and uninformed voters, *American Political Science Review* 88, 33-47.

Bender, B. and J.R. Lott, 1996, Legislator voting and shirking: A critical review of the literature, *Public Choice* 87, 67-100.

Birnbaum, J., 1992, The lobbyists (Times Books, New York).

Calvert, R.L., 1986, Models of imperfect information in politics (Harwood, Chur).

Cameron, C.M. and J.P. Jung, 1992, Strategic endorsements, Mimeo (Columbia University, New York).

Chappell, H.W., 1994, Campaign advertising and political ambiguity, *Public Choice* 79, 281-303.

Cho, I.K. and J. Sobel, 1990, Strategic stability and uniqueness in signaling games, *Journal of Economic Theory* 50, 381-413.

Davis, M.L. and M. Ferrantino, 1996, Towards a positive theory of political rhetoric: Why do politicians lie?, *Public Choice* 88, 1-13.

Epstein, D. and S. O'Halloran, 1995, A theory of strategic oversight: Congress, lobbyists, and the bureaucracy, *Journal of Law, Economics and Organization* 11, 227-255.

Farrell, J. and R. Gibbons, 1989, Cheap talk with two audiences, *American Economic Review* 79, 1214-1223.

Fudenberg, D. and J. Tirole, 1991, Perfect Bayesian equilibrium and sequential equilibrium, *Journal of Economic Theory* 53, 236-260.

Gerber, A., 1996, Rational voters, candidate spending, and incomplete information: A theoretical analysis with implication for campaign finance reform, Working Paper.

Gerber, E.R. and A. Lupia, 1995, Campaign competition and policy responsiveness in direct legislation elections, *Political Behavior* 17, 287-306.

Gersbach, H., 1997, Elections and campain contributions, divided support and regulation, Mimeo (Alfred-Weber-Institut, Heidelberg).

Gersbach, H., 1998, Communication skills and competing for donors, *European Journal of Political Economy* 14, 3-18.

Gibbons, R., 1992, A primer in game theory (Harvester Wheatsheaf, New York).

Gilligan, T.W. and K. Krehbiel, 1989, Asymmetric information and legislative rules with a heterogeneous committee, *American Journal of Political Science* 33, 459-490.

Grofman, B. and B. Norrander, 1990, Efficient use of reference group cues in a single dimension, *Public Choice* 64, 213-227.

Grossman, S.J., 1981, The informational role of warranties and private disclosure about product quality, *Journal of Law and Economics* 24, 461-483.

Grossman, G.M. and E. Helpman, 1996, Electoral competition and special interest politics, Review of Economic Studies 63, 265-286.

Grossman, G.M. and E. Helpman, 1997, Competing for endorsements, Working Paper (Princeton University, Princeton).

Harrington, J.E., 1992, The revelation of information through the electoral process: An exploratory analysis, *Economics and Politics* 4, 255-275.

Harrington, J.E., 1993, The impact of reelection pressures on the fulfillment of campaign promises, *Games and Economic Behavior* 5, 71-97.

Helsley, R.W. and A. O'Sullivan, 1994, Altruistic voting and campaign contributions, *Journal of Public Economics* 55, 107-119.

Hertzendorf, M.N., 1993, I'm not a high-quality firm - but I play one on TV, *Rand Journal of Economics* 24, 236-247.

Hinich, M.J. and M.C. Munger, 1989, Political investment, voter perceptions, and candidate strategy: an equilibrium spatial analysis, in: P.C. Ordeshook, ed., Models of strategic choice in politics (The University of Michigan Press, Ann Arbor) 49-67.

Hopenhayn, H. and S. Lohmann, 1996, Fire-alarm signals and the political oversight of regulatory agencies, *Journal of Law, Economics and Organization* 12, 196-213.

Jacobson, G.C., 1980, Money in Congressional Elections (Yale University Press, New Haven).

Jacobson, G.C., 1985, Money and votes reconsidered: Congressional elections, 1972-1982, *Public Choice* 47, 7-62.

Kau, J.B., Keenan, D. and P.H. Rubin, 1982, A general equilibrium model of Congressional voting, *Quarterly Journal of Economics* 97, 271-293.

Kau, J.B. and P.H. Rubin, 1982, Congressmen, constituents and contributors: Determinants of roll call voting in the House of Representatives (Martinus Nijhoff, Boston).

Kenny, C. and M. McBurnett, 1992, A dynamic model of the effect of campaign spending on congressional vote choice, *American Journal of Political Science* 36, 923-937.

Kihlstrom, R.E. and M.H. Riordan, 1984, Advertising as a signal, *Journal of Political Economy* 92, 427-450.

Laffont, J.-J. and J. Tirole, 1993, A theory of incentives in procurement and regulation (MIT press, Cambridge).

Lagerlöf, J., 1997, Lobbying, information, and private and social welfare, *European Journal of Political Economy* 13, 615-637.

Leland, H.E., 1981, Comments on Grossman, *Journal of Law and Economics* 24, 485-489.

Lipman, B.L. and D.J. Seppi, 1995, Robust inference in communication games with partial provability, *Journal of Economic Theory* 66, 370-405.

Lohmann, S., 1993, A welfare analysis of political action, in: W.A. Barnett, M.J. Hinich and N.J. Schofield, eds., Political economy: Institutions, competition, and representation (Cambridge university press, Cambridge) 437-461.

Lohmann, S., 1994, Information aggregation through costly political action, *American Economic Review* 84, 518-530.

Lohmann, S., 1995a, A signaling model of competitive political pressures, *Economics and Politics* 7, 181-206.

Lohmann, S., 1995b, Information, access, and contributions: A signaling model of lobbying, *Public Choice* 85, 267-284.

Lohmann, S. and H. Hopenhayn, 1998, Delegation and the regulation of risk, forthcoming in: *Games and Economic Behavior*.

Lott, J.R., 1987, Political cheating, *Public Choice* 52, 169-186.

Lott, J.R. and W.R. Reed, 1989, Shirking and sorting in a political market with finite-lived politicians, *Public Choice* 61, 75-96.

Lupia, A., 1992, Busy voters, agenda control and the power of information, *American Political Science Review* 86, 390-403.

Lupia, A., 1994, The effect of information on voting behavior and electoral outcomes: An experimental study of direct legislation, *Public Choice* 78, 65-86.

Lupia, A. and M.D. McCubbins, 1994, Learning from oversight: Fire alarms and police patrols reconstructed, *Journal of Law, Economics and Organization* 10, 96-125.

Magee, S.P., Brock, W.A. and L. Young, 1989, Black hole tariffs and endogenous policy theory (Cambridge University Press, Cambridge).

Mailath, G.J., Okuno-Fujiwara, M. and A. Postlewaite, 1993, Belief-based refinements in signaling games, *Journal of Economic Theory* 60, 241-276.

Mayer, W. and J. Li, 1994, Interest groups, electoral competition, and probabilistic voting for trade policies, *Economics and Politics* 6, 59-77.

McCubbins, M.D. and T. Schwartz, 1984, Congressional oversight overlooked: Police patrols versus fire alarms, *American Journal of Political Science* 28, 165-179.

McKelvey, R.D. and P.C. Ordeshook, 1985, Elections with limited information: a fulfilled expectations model using contemporaneous poll and endorsement data as information sources, *Journal of Economic Theory* 36, 55-85.

Milgrom, P., 1981, Good news and bad news: Representation theorems and applications, *Bell Journal of Economics* 12, 380-391.

Milgrom, P. and J. Roberts, 1986a, Relying on the information of interested parties, *Rand Journal of Economics* 17, 18-32.

Milgrom, P. and J. Roberts, 1986b, Price and advertising signals of product quality, *Journal of Political Economy* 94, 796-821.

Milner, H.V. and B.P. Rosendorff, 1996, Trade negotiations, information and domestic politics: The role of domestic groups, *Economics and Politics* 8, 145-189.

Morton, R. and C. Cameron, 1992, Elections and the theory of campaign contributions: A survey and critical analysis, *Economics and Politics* 4, 79-108.

Mueller, D.C. and T. Stratmann, 1994, Informative and persuasive campaigning, *Public Choice* 81, 55-77.

Nelson, P., 1974, Advertising as information, *Journal of Political Economy* 82, 729-754.

Nelson, P., 1976, Political information, *Journal of law and Economics* 19, 315-336.

Okuno-Fujiwara, M., Postlewaite, A., and K. Suzumura, 1990, Strategic information revelation, *Review of Economic Studies* 57, 25-47.

Page, B.I., Shapiro, R.Y. and G.R. Dempsey, 1987, What moves public opinion?, *American Political Science Review* 81, 23-43.

Palda, K.F., 1994a, How much is your vote worth? The unfairness of campaign spending limits (ICS Press, San Francisco).

Palda, K.F., 1994b, Desirability and effects of campaign spending limits, *Crime, Law and Social Change* 21, 295-317.

Palda, K.F. and K.S. Palda, 1985, Ceilings on campaign spending: Hypothesis and partial test with Canadian data, *Public Choice* 45, 313-331.

Palda, K.F. and K.S. Palda, 1998, The impact of campaign expenditures on political competition in the French legislative elections of 1993, *Public Choice* 94, 157-174.

Poole, K.T. and T. Romer, 1985, Patterns of PAC contributions to the 1980 campaigns for the U.S House of Representatives, *Public Choice* 47, 63-111.

Potters, J. and F. van Winden, 1990, Modelling political pressure as transmission of information, *European Journal of Political Economy* 6, 61-88.

Potters, J. and F. van Winden, 1992, Lobbying and asymmetric information, *Public Choice* 74, 269-92.

Potters, J. and F. van Winden, 1996, Models of interest groups: Four different approaches, in: N. Schofield, ed., Collective decision-making: Social choice and political economy (Kluwer, Boston) 337-362.

Prat, A., 1997, Campaign advertising and voter welfare, Discussion Paper No. 97118 (Center for Economic Research, Tilburg).

Rasmusen, E., 1993, Lobbying when the decisionmaker can acquire independent information, *Public Choice* 77, 899-913.

Rasmusen, E., 1997, Choosing among signalling equilibria in lobbying games: A reply to Sloof, *Public Choice* 91, 209-214.

Seidmann, D. and E. Winter, 1997, Strategic information transmission with verifiable messages, *Econometrica* 65, 163-169.

Shin, H.S., 1994, The burden of proof in a game of persuasion, *Journal of Economic Theory* 64, 253-264.

Sloof, R., 1997a, Competitive lobbying for a legislator's vote: A comment, *Social Choice and Welfare* 14, 449-464.

Sloof, R., 1997b, Lobbying when the decisionmaker can acquire independent information: A comment, *Public Choice* 91, 199-207.

Smith, R.A., 1984, Advocacy, interpretation and influence in the U.S. Congress, *American Political Science Review* 78, 44-63.

Snyder, J.M., 1989, Election goals and the allocation of campaign resources, *Econometrica* 57, 637-60.

Snyder, J.M., 1991, On buying legislatures, *Economics and Politics* 3, 93-109.

Sobel, J., 1985, A theory of credibility, *Review of Economic Studies* 52, 557-573.

The Economist, 1994, A nation of groupies, August 13.

The Economist, 1997, How to cut the costs of politics, February 8.

The Economist, 1998, Economics focus: The money in the message, February 14.

Tirole, J., 1988, The Theory of industrial organization (The MIT Press, Cambridge).

Umbhauer, G., 1994, Information transmission in signaling games: Confrontation of different forward induction criteria, in: B. Munier and M.J. Machina, eds., Models and experiments in risk and rationality (Kluwer, Dordrecht) 413-438.

Wilhite, A. and J. Theilmann, 1987, Labor PAC contributions and labor legislation: A simultaneous logit approach, *Public Choice* 53, 267-276.

4 CAMPAIGN CONTRIBUTIONS OR DIRECT ENDORSEMENTS?[1]

The basic signaling game presented in the previous chapter provides a simple explanation for the empirical observation that campaign expenditures buy votes (cf. Morton and Cameron, 1992). The mere fact that a candidate spends a lot of money on political advertisements may indicate that he will implement a policy platform the electorate likes. In the basic political campaigning model introduced in Section 3.2 it is assumed that the candidate has to rely on his own funds. In reality, however, the campaign chest of a candidate is often larded by interest groups. Hence, when a candidate is able to spend a lot of money on a campaign, this typically indicates that he received a large amount of contributions. In this respect Grossman and Helpman (1994, p. 836) suggest that the size of contributions received shows the candidate's ability as a fundraiser, and thus serves as a kind of indirect endorsement. This suggestion in turn raises two interesting questions, one with respect to the explanation of the impact of costly campaigns, and another with respect to an interest group's choice between direct and indirect endorsements. Firstly, are there circumstances under which political campaigns need contributions from interest groups in order to be effective? Secondly, why would interest groups contribute to the campaign of a candidate or party when they can also reach out to the voters directly via direct endorsements?

In this chapter these two questions are addressed. To that purpose we start from the model already introduced in Chapter 3 of an election with an incumbent and a challenger competing for office. Section 4.1 briefly describes this model and quickly repeats the relevant results obtained when campaigns and endorsements are considered in isolation. In Section 4.2 the basic political campaigning model is extended to account for the fact that political campaigns are typically funded by interest groups. The first question raised above is answered at the end of this section where we compare the extended model in which the political campaign is financed by an interest group with the basic model where the campaign is financed by the candidate himself. Section 4.3 studies the choice of an interest group between campaign contributions and direct endorsements, and thereby provides an answer to the second question posed. The analyses in Section 4.2 and 4.3 are based on the assumption that contributions are only observed when they are actually spent on the campaign. Section 4.4 elaborates on the consequences of relaxing this assumption. A concluding discussion is presented in Section 4.5.

4.1   CAMPAIGN EXPENDITURES, DIRECT ENDORSEMENTS AND VOTING

Recall the model described in Section 3.2 of an election with an incumbent and a challenger (both males) competing for office. A (female) representative voter must either vote for the incumbent ($x=x_1\equiv$incumbent), or for the challenger ($x=x_2\equiv$challenger). The incumbent's policy position is known to the voter, the position of the challenger is not. Compared with the incumbent's position, the policy position of the challenger may either fit the preferences of the voter better ($t=t_2\equiv$"better"), or fit her preferences worse ($t=t_1\equiv$"worse"). The challenger of type t_2 is referred to as the "good" type challenger, the challenger of type t_1 as the "bad" type challenger. The prior probability that the challenger is of the good type equals $p=P(t=$better). Of course, when the challenger is of the good type the voter wants to elect him, whereas when he is of the bad type she prefers the incumbent. The goal of the challenger is to get elected, irrespective of whether he is indeed "better" or not. In short, the challenger and the voter have partially conflicting interests.

Besides the challenger and the voter there is now also an interest group in the model. This interest group is also affected by governmental policy, but not necessarily in the same way as the voter is. The interest group may either completely agree with the voter on who the right candidate is, fully disagree, or just partially disagree on the best candidate available. Besides, contrary to the voter, the interest group follows politics rather closely and is therefore able to judge the policy platform of the challenger. In other words, the interest group is informed about the type of the challenger. Since the interest group possesses the same information as the challenger, it is said to be of the same type, for convenience; the group who knows that the challenger is of the good (bad) type is referred to as the good (bad) type interest group. However, it must be recognized that, although the interest group is always of the same type as the challenger, it may prefer the incumbent. Possessing the same information does not imply having the same interests. Without loss of generality we can normalize the payoffs of the three players over action-state pairs such that they are given by the following table:

Table 4.1 Gross payoffs in the election model

	$t=t_1\equiv$worse	$t=t_2\equiv$better
$x=x_1\equiv$incumbent	v_1 0 0	0 0 0
$x=x_2\equiv$challenger	0 u_1 g_1	v_2 u_2 g_2

Additional assumptions: $v_i, u_i>0$ for $i=1,2$

The first entry in each cell refers to the voter's payoff, the second to the challenger's payoff, and the third entry refers to the interest group's payoff. Since the voter in each state of the world should actually prefer the candidate she is hypothesized to prefer, v_1 and v_2 are taken to be strictly positive. Recall that two cases can be distinguished (neglecting ties): the case in which the voter, on the basis of the prior, would elect the incumbent (requiring that $p < v \equiv v_1/[v_1 + v_2]$), and the opposite case where she is a priori inclined to choose the challenger ($p > v$). The assumption that the challenger wants to get elected is reflected by assuming both u_1 and u_2 to be positive. A priori we do not put any restrictions on g_1 and g_2, and four different cases can be distinguished in regard to the interest group's preferences (as before, knife-edge cases are ignored):

(I)   No conflict with the voter's interests: $g_1 < 0 < g_2$.
(II)   Full conflict with the voter's interests: $g_2 < 0 < g_1$.
(III) Partial conflict with the voter's interests, and challenger preferred: $g_1, g_2 > 0$.
(IV)   Partial conflict with the voter's interests, and incumbent preferred: $g_1, g_2 < 0$.

The two questions posed in the introduction to this chapter in fact ask for a comparison between three different ways in which the challenger can be advertized and recommended to the voter. These three different ways are labeled (i) CAmpaigning, (ii) Direct Endorsements, and (iii) COntributions, respectively. The remainder of this section reviews the results obtained when studying the first two possibilities in isolation. The next section analyses the single case of COntributions, and compares the results with those obtained for the CAmpaigning model. In that way it provides an answer to the first question posed. The third section compares Direct Endorsements with COntributions, and thereby studies the interest group's choice between these two distinct ways of supporting the challenger.

The first way the challenger can be advertized to the voter is through a costly campaign financed by himself. In the campaigning model the challenger may either spend a fixed amount of money c (c>0) on campaigning, or spend no money at all. The voter observes the level of campaign expenditures (either 0 or c), as well as the content of the campaign ($m \in M$) when indeed a costly campaign is run. Thereupon she chooses between the incumbent and the challenger. Of course, the challenger must subtract any costs of campaigning from his gross payoff given in Table 4.1 in order to get his net payoff. From Result 3.1(a) follows that the informative value of a campaign lies solely in its costs. The specification of M is thus inessential. In addition, information revelation is only possible when a certain sorting condition on the preferences of the challenger is satisfied. Specifically, for the campaigning model the following corollary holds (cf. Corollary 3.1):

Corollary 4.1
A necessary and sufficient condition for information revelation through campaigning to be possible in equilibrium is given by the 'Information Revelation by CAmpaigning' (IRCA) condition: $\max\{c,u_1\}<u_2$.   □

For the amount of campaign expenditures to have informational value the challenger must value his election more highly when the voter is satisfied with her decision ex post. There are several reasons why this sorting condition $u_1<u_2$ might hold. It seems reasonable to assume that the challenger has an easier time during his term of office when his policy position is in harmony with the voter's preferences. The "power" of elected office (i.e. the ability to effectively determine and implement policies) may be positively related to the politician's popularity (cf. Harrington, 1992). Another justification follows from reelection considerations (cf. Chapter 3). When the voter is satisfied with her decision ex post the winner of the current election fits the preferences of the electorate rather well and has a higher chance of reelection in the next election.

   Instead of the challenger himself doing the advertising, the interest group may try to persuade the voter to elect the challenger. The direct endorsements model is, again, a straightforward application of the basic signaling game, with the interest group now being the sender. The group may either spend a fixed amount of money c on an endorsement, or spend no money at all. The voter observes the interest group's endorsement expenditures (either 0 or c), and the content of the endorsement $m\in M$ when indeed a costly endorsement is made. Then she elects one of the two candidates. The net payoffs of the interest group follow from its gross payoff presented in Table 4.1 by subtracting any expenditures it makes on endorsements. Lastly, the direct endorsements model restricts the setup of the basic signaling game in one, minor aspect. Because we are interested in the comparison of using different ways to promote the same candidate, viz. the challenger, we exclude the possibility that the interest group directly endorses the incumbent.[2] As a result, endorsements do not occur in case IV. In that case the interest group prefers the incumbent, thus never wants to recommend the voter to elect the challenger. Direct endorsements would only be used to support the incumbent, but these are excluded by assumption. Of course, excluding endorsements in support of the incumbent does not affect the choice of a challenger preferring interest group between endorsing the challenger or contributing to his campaign.

   In case of no conflict of interests between the interest group and the voter (case I), the interest group may either want to endorse the incumbent (bad type interest group) or the challenger (good type interest group). As a consequence of our restriction, however, in this case information transfer is only possible if the costs of endorsing are not prohibitive for the good type interest group ($c<g_2$). The analysis of the two remaining cases II and III is not affected by the exclusion of endorsing the incumbent, and the results of Section 3.1 immediately apply. Again, the informative value of endorsements for the challenger lie in their costs and not in their content. The condition under which costly endorsements indeed reveal information is repeated in the following corollary:

Corollary 4.2
A necessary and sufficient condition for information revelation through direct endorsements to be possible in equilibrium is given by the 'Information Revelation by Direct Endorsements' (IRDE) condition: $\max\{c,g_1\}<g_2$. □

Once more, costly endorsements are informative only when there is a sufficient degree of consonance between the interests of the interest group and those of the voter. On the basis of this result one would expect to see groups with moderate positions devoting more of their resources to direct endorsements than extremist groups do (cf. the theoretical results obtained by Snyder, 1991). The case $g_1<g_2$ could represent a situation in which an on-going relationship with a politician is valued more highly when the electorate would be satisfied with him ex post. A popular politician may have more discretion in providing specific benefits to interest groups. Another justification could be given by assuming the interest group to be "encompassing", that is, more aligned with the electorate's interests (cf. Olson, 1982).

The third way in which the challenger can be recommended to the voter is through indirect endorsements. In that case the challenger's campaign is financed by contributions from the interest group. Contrary to the campaigning model and the direct endorsements model, the COntributions model is not an immediate application of the basic signaling game. For that reason, the contributions model is discussed in more detail in a separate section.

4.2 CAMPAIGN CONTRIBUTIONS AND VOTING

Electoral campaigns are often financed by the contributions of interest groups. In the U.S., for instance, candidates running for a seat in Congress mainly rely on funds provided to them by donors. As was already pointed out in Chapter 2, the literature provides two, not mutually exclusive, motives for why interest groups contribute to a candidate's campaign. Interest groups may either try to get their favored candidates elected (position induced model), or they may try to influence the policy positions of the candidates (service induced model). Theoretical models of campaign contributions can be divided, by and large, along this motivational dimension. The model presented in this section focuses on the electoral motive of contributions, and thus belongs to the class of position induced models. Campaign contributions are used by the interest group to enable the challenger to transmit information to the voter through costly campaigning. Because the voter is aware of the fact that the interest group supplies the campaign funds, contributions serve as indirect endorsements.[3] In this way, the interest group tries to get the challenger elected.[4]

The contributions model has the following characteristics. The interest group has to decide whether to contribute a fixed amount of money to the campaign of the challenger or not, before the challenger takes his campaign decision. It is assumed that the interest group knows the (fixed) policy position of the challenger when it decides on its level of contributions. The challenger is entirely dependent on the contributions of the interest group to cover his campaign cost. Hence, when the challenger does not receive any contributions, he is restricted to not campaigning. The voter only observes campaign expenditures and the content of the campaign; contributions are not directly observed. However, she realizes that all campaign money spent by the challenger is provided by the group, and that the group knows the type of the challenger. Since the voter knows the preferences of the group and the challenger in each possible state of the world, she can use campaign expenditures as a signal of the challenger's type (policy position). To give the challenger a substantive role it is assumed that he can use contributions for other purposes than his campaign, and that these alternative uses yield direct utility. In this context one may think of expensive dinners, luxurious trips, and so on.[5] Another justification might be that the challenger wants to save for future elections or has to retire campaign debts from previous elections (cf. Grossman and Helpman, 1994). As a consequence, when the voter does not observe campaign efforts, she cannot be sure whether this is due to the fact that the challenger did not receive contributions, or decided to use them for other purposes. Lastly, for simplicity it is assumed that the challenger either has to spend the contributions received on his campaign or on other purposes; he cannot split contributions between these two alternatives.

The order of moves in the contributions model is as follows. Nature picks the type of challenger $t \in \{$worse, better$\}$ with $p=P(t=$better$)$, and this type is revealed to both the challenger and the interest group. Next the interest group decides whether to contribute a fixed amount c to the challenger or not and, subsequently, the challenger decides whether to spend any contributions received on a campaign or not. In case the challenger runs a costly campaign, he also chooses the content of the campaign $m \in M$. Finally, the voter observes the level of campaign expenditures, and when a costly campaign is run its content, and elects either the challenger or the incumbent. Gross payoffs are given in Table 4.1. To get the challenger's net payoff one must add the contributions that are not used for campaigning. The interest group's net payoff follows by subtracting its donation from the gross payoff.

Although the contributions model differs from the basic signaling game of Section 3.1, it bears a strong resemblance. To see this, let the voter represent the decisionmaker, and let the interest group and the challenger together represent the sender. The combined behavior of the interest group and the challenger determines whether the voter observes a costly campaign or not. It immediately follows from the analysis in the previous chapter that the content of the campaign, i.e. the specific element $m \in M$, is immaterial to the voter. Because the challenger decides on the content of the campaign completely by himself and, by assumption, always wants to get elected, the content of the campaign cannot be informative (influential). Result 3.1(a) thus trivially applies for the contributions model as well. Therefore, in the sequel we ignore the content of the campaign and focus on the informational value of its costs.

Different types of equilibria of the contributions model can again be classified according to the amount of information revealed. The classification of equilibria is now based on the differences in the campaigning strategies of the two types of interest group-challenger combinations. Note that the probability that the good (bad) type interest group-challenger combination runs a costly campaign is given by the probability that the good (bad) type interest group contributes times the probability that the good (bad) type challenger spends this money on his campaign. In a pooling equilibrium the resultant campaigning strategies do not differ between the two types of combinations, in a hybrid (separating) equilibrium these campaigning strategies are partially (completely) opposed.

The question of interest here is under what circumstances the three types of equilibria exist. A first, rather trivial, observation in this respect is that when the interest group prefers the incumbent, it will never contribute to the challenger's campaign. This immediately follows because in equilibrium the challenger would only spend these funds — that is, make them observable to the voter — when campaign spending enhances his probability of election. Hence in both cases of spending or no-spending of the contributions received, the interest group is worse off as compared to the situation of giving no contributions. As a result, when $g_2 < 0$ a costly campaign can never come from a good type combination, and the observation of such a campaign is a clear signal of the combination being of the bad type. Since the bad type challenger does not want to reveal his identity, he will never run a costly campaign in this case. So, when $g_2 < 0$ no information can be revealed in equilibrium, and thus, in the two cases II and IV only pooling equilibria (on no-campaigning) exist. The complete list of conditions for the existence of the various types of equilibria is given in Proposition 4.1 below (this proposition follows from Proposition 4.5 in Appendix 4.A).

Proposition 4.1
(a)   A *pooling* equilibrium always exists.
(b)   A necessary and sufficient condition for a *hybrid* equilibrium to exist is:
         if (i) $p < v$: $c < \min\{u_1, g_1\} < \min\{u_2, g_2\}$.
         if (ii) $p > v$: $\max\{c, \min\{u_1, g_1\}\} < \min\{u_2, g_2\}$.
(c)   A necessary and sufficient condition for a *separating* equilibrium to exist
         is: $\min\{u_1, g_1\} < c < \min\{u_2, g_2\}$.   □

Contrary to Proposition 3.2 for the basic signaling game, Proposition 4.1 does not specify the equilibrium paths of the contributions model in full detail. A precise description of all the equilibrium paths is given in Appendix 4.A. For our discussion, however, it is sufficient to point out that the message sending (campaigning) behavior of the interest group-challenger combination in the present contributions model is similar to the behavior of the sender in the basic signaling game. In fact, we can identify the stake for the interest group-challenger combination in getting the challenger elected with $\min\{u_i, g_i\}$, and use Proposition 3.2 to give us directly the equilibrium behavior of the interest group-challenger combination and the voter for both the cases I and III.[6]

In the non-informative pooling equilibria either (i) the interest group never contributes, or (ii) the interest group always contributes and the challenger always spends the contributions received on his campaign. Contrary to the first pooling equilibrium, the second only exists when the voter is a priori inclined to vote for the challenger $(p>v)$.[7] In the hybrid equilibria when $p<v$ the good type combination campaigns with certainty, the bad type combination with some fixed positive probability. Either the bad type interest group always donates and the challenger sometimes keeps the money, or the bad type interest group does not always provide funds and the challenger always uses the funds for his campaign, or both use a mixed strategy. In any case, the resultant campaigning strategy for the bad type combination is the same. When $p>v$ the bad type interest group-challenger combination never campaigns in the hybrid equilibria. Either the good type interest group or the good type challenger (or each of them) mixes between their pure strategies, but a campaign from the good type combination always occurs with the same positive probability. Lastly, in the single separating equilibrium the bad type interest group never donates, whereas the good type interest group always supplies funds and the good type challenger always uses these funds for his campaign.

Due to its close connection to the basic signaling game, the results obtained in Section 3.1 (cf. Result 3.1) almost directly apply to the contributions model. The most important one of them, at least for answering the two questions addressed in this chapter, is summarized in the following corollary of Proposition 4.1.

Corollary 4.3
A necessary and sufficient condition for information revelation through contributions to be possible in equilibrium is given by the 'Information Revelation by COntributions' (IRCO) condition: $\max\{c, \min\{u_1,g_1\}\}<\min\{u_2,g_2\}$.   □

By realizing that the stake for the interest group-challenger combination is given by $\min\{u_1,g_1\}$, i.e. by the stake of the agent in the combination who has the weakest incentive to get the challenger elected, it can be seen that Corollary 4.3 is in fact equivalent to Corollary 3.1. The IRCO restriction incorporates two separate conditions. The first condition $c<\min\{u_2,g_2\}$ entails that campaigning is not too costly for the good type combination. When this condition does not hold, the observation of an interest group funded campaign is a clear signal of the combination being of the bad type. As explained above, only pooling on no-campaigning can occur in that case. The second condition $\min\{u_1,g_1\}<\min\{u_2,g_2\}$ concerns the sorting condition for the interest group-challenger combination, in the sequel referred to as the "generalized" sorting condition. Necessarily, either the preferences of the challenger $(u_1<u_2)$ or the interest group $(g_1<g_2)$ must satisfy the (standard) sorting condition. Perhaps surprisingly, this latter condition is not sufficient. For instance, when $c<g_2<u_1<g_1<u_2$ the sorting condition for the challenger is satisfied $(u_1<u_2)$, but the generalized sorting condition is not met and information transfer is not possible. When the sorting condition is satisfied for

both the interest group and the challenger, however, the generalized sorting condition is satisfied as well.

Although already given for the sorting condition in the basic signaling game, it is instructive to briefly discuss the intuition behind the generalized sorting condition. In an informative equilibrium the good type interest group-challenger combination necessarily campaigns with a strictly larger probability than the bad type combination. When the generalized sorting condition is not met, the bad type combination always wants to campaign with certainty when the good type combination campaigns with positive probability. In the example with $c<g_2<u_1<g_1<u_2$, if the gains g_2 and the probability that the challenger spends the funds received are large enough for the good type interest group to contribute to the challenger, the stake for the bad type challenger u_1 is certainly large enough to justify the spending of contributions on a campaign (as $g_2<u_1$). Since $g_2<g_1$, however, this induces the bad type interest group always to contribute to the challenger and, thus, the bad type combination to campaign with certainty. This cannot lead to an informative equilibrium, but only to a pooling (on campaigning) equilibrium.

Corollary 4.3 illustrates that Result 3.1(b) also holds for the contributions model. As noted, the remaining results presented in Result 3.1 almost directly apply as well. First, in equilibrium the voter will never respond negatively to campaign spending by electing the challenger with a lower probability. Second, the comparative statics results for the contributions model predict that the impact of a campaign is weakly increasing in c and weakly decreasing in $\min\{u_1,g_1\}$ ($\min\{u_2,g_2\}$) when $p<v$ ($p>v$), and that $p>v$ is required to be assured of election after campaigning. On the other hand, a campaign, and thus a donation, is more likely when the costs are lower, the stakes of both the interest group and the challenger are higher, and when the electoral contest is expected to be close.

Third, with respect to the welfare implications it invariably holds that the voter will not lose from the possibility of contributions,[8] but only gains from an interest group funded campaign when separation is possible. Both the interest group and the challenger (weakly) value the possibilities for information revelation when $p<v$. So, from an ex ante perspective the interest group and the challenger value the possibility of making contributions in this case. When $p>v$ the interest group may either like the possibilities for information transfer (bad type interest group when $g_1<0$) or dislike them. In the latter case the voter is already inclined to take the decision preferred by the interest group, and information transfer is only costly. As the prior belief of the voter favors the challenger when $p>v$, the challenger (weakly) dislikes the possibility of information transfer and would rather see that campaigning, even with the funds of others, were not possible. In the hybrid equilibria, for instance, the gain from sometimes pocketing the contributions received does not offset the loss due to the voter sometimes electing the incumbent. Finally, in the pooling on campaigning equilibria political expenditures are a pure social waste.

At the end of this section we turn to answering the first of the two main questions addressed in this chapter: Do campaigns need contributions from interest groups in order to be effective? Put differently, does the scope for information transfer through political campaigns critically depend on the supplier of campaign funds? In order to tackle this question we compare the situation that the challenger has resources of his own (CAmpaigning model) with the situation that the funds are supplied by an interest group (COntributions model). Specifically, we compare the IRCA condition given in Corollary 4.1 with the IRCO condition given in Corollary 4.3.

It is easily seen that the two conditions are not overlapping; the fulfillment of IRCA does not entail the satisfaction of IRCO, nor does the reverse hold. On the one hand, the dependence on the interest group's money may reduce the possibilities for information transfer. When $c < g_2 < g_1 < u_1 < u_2$, for instance, the IRCA condition holds but the IRCO condition does not. The only way for campaigning to be informative is that the challenger has his own resources, because the interest group's interests are too opposed to the voter's interests to make information transfer through contributions possible. On the other hand, the fact that the challenger is dependent on the funds of the interest group sometimes enlarges the scope for information transfer. For example, when $c < g_1 < u_2 < u_1 < g_2$ the challenger's sorting condition is not met and campaigning by using own resources cannot reveal useful information in equilibrium. Information transfer is only possible in equilibrium when the challenger campaigns with the money supplied by the interest group. In this case the challenger exploits the fact that the preferences of the supplier of funds do satisfy the sorting condition. Thus, there are indeed circumstances under which political campaigns need contributions (indirect endorsements) from interest groups in order to be effective. More generally, it can be concluded that the challenger sometimes needs the fact that funds are supplied by an interest group in order that his campaign is informative in equilibrium, and in other cases needs money of his own.

Given that the same amount of information is being revealed, it is immaterial to the voter whether the campaign is funded by either the interest group or by the challenger himself. In case the two types of financing the campaign lead to different possibilities for information revelation, the voter prefers that type of funding that is the most informative (note, however, that the voter receives exactly the same expected payoff in the pooling and the hybrid equilibria). Of course, given that the two ways of financing a campaign lead to the same amount of information being revealed, the challenger would rather campaign with the money of others. However, since the type of funding may render different opportunities for information revelation, the challenger has no unambiguous preference for one type of funding over the other. (In this respect it must also be realized that when $p > v$ the challenger, and in case III also the interest group, dislikes the possibilities for information transfer.) For the interest group a similar preference ordering holds. Ceteris paribus it prefers the situation in which the challenger campaigns with his own money, but under different possibilities for information transmission it may prefer the contributions type of funding after all. In the next section we investigate under what circumstances the interest group prefers to give contributions rather than using direct endorsements in order to support the challenger.

4.3 CAMPAIGN CONTRIBUTIONS VERSUS DIRECT ENDORSEMENTS

In this section we are concerned with answering the second question of why interest groups would contribute to the campaign of a candidate when they can also reach out to the voters directly via direct endorsements. To that purpose we extend the analysis of the previous section by considering the possibility that the interest group can either contribute to the challenger's campaign, or endorse the challenger by sending a costly signal directly to the voter. It is assumed that donating and endorsing are equally costly. Though rather restrictive, this assumption allows us to focus on the differences in information revelation opportunities between two equally costly options for the interest group. For ease of exposition, we will draw our conclusions from comparing the results of the two separate models, viz. the contributions model of Section 4.2 and the direct endorsements model of Section 4.1. However, as follows from the general model considered in Appendix 4.A, exactly the same results are obtained when the interest group's choice between endorsing and contributing is actually made endogenous.

Because the interest group's preference for either direct or indirect endorsements relates to the amount of information that is being revealed, we first compare the possibilities for information transfer by using either channel. From Corollaries 4.2 and 4.3 it immediately follows that under full conflict of interests between the voter and the interest group (case II), and in case the interest group prefers the incumbent (case IV), neither contributions nor direct endorsements can be used to transmit information. Moreover, in these cases neither option will be used to support the challenger. Hence it suffices to explore the remaining cases I and III, which is done in the following proposition (the proposition follows from Proposition 4.5 presented in Appendix 4.A).

Proposition 4.2

(a) *Case I.* In the case of no conflict of interests between the interest group and the voter it holds that: when (full) information transfer is possible by funding the campaign of the challenger, full information revelation is possible by using direct endorsements. The reverse does not hold.

(b) *Case III.* When the interest group prefers the challenger it holds that:

if (i) $g_1 < g_2$: when information transfer is possible through campaign contributions, information revelation is also possible through direct endorsements. The reverse does not hold.

if (ii) $g_1 > g_2$: information transfer is certainly not possible through direct endorsements, but may be possible by contributing to the challenger's campaign. □

Proposition 4.2(a) shows that when the interests of the voter and the interest group are fully aligned, direct endorsements have a larger scope for information transfer than indirect endorsements. In this case the interest group does not need the challenger as an intermediary to transfer information. The second part of the proposition indicates that when the interest group prefers the challenger, contributions may be more informative for the voter than endorsements. When the sorting condition $g_1 < g_2$ holds contributions are only more informative when separation is possible by using contributions, but not by using direct endorsements. This is the case when only the stake of the bad type challenger u_1 is below c. Otherwise, endorsements are at least as informative as contributions and subsequent campaigning. In case the sorting condition does not hold for the preferences of the interest group ($g_1 > g_2$), the indirect route is more informative whenever the IRCO condition holds. These considerations indicate that under a range of circumstances interest group influence on voter behavior is possible only by using contributions, though both direct endorsements and contributions are options open to the interest group. Although in this case the preferences of the interest group carry information in the wrong direction, it can exploit — by using the indirect route — the fact that the challenger's preferences do satisfy the sorting condition. In short, the interest group may need the challenger as an intermediary to filter its partially opposing interests.

Proposition 4.2 does not yet establish when the interest group prefers to use contributions rather than direct endorsements to support the challenger. It is to this topic that we turn next. To that purpose we compare the expected utility obtained in equilibrium from using direct endorsements with the expected utility obtained from employing contributions. As is clear from the previous analysis, multiple equilibria for both routes exists, which makes a strict preference ordering over the two routes cumbersome. [9, 10] Note, in this context, that in both the endorsements and the contributions model pooling on no-expenditures by the interest group types is always an equilibrium. In these cases the voter takes her decision solely based on prior information, yielding the interest group exactly the same expected utility for both routes. Hence either route can, in principle, yield the interest group exactly the same utility in equilibrium.

Since both models incorporate the same "no-expenditures" pooling possibility, we take this equilibrium as a benchmark for comparing endorsements with contributions. For each model we investigate whether the possibilities for information transfer lead to an improvement for the interest group over this no-expenditures equilibrium. (This is done by separately considering the expected payoffs of the bad type interest group and those of the good type interest group.) This benchmark is also useful when one wants to address the question whether the interest group can gain at all from making direct endorsements or contributions. First we have to establish whether the interest group gains from information revelation anyway. Of course, this depends on the case we consider (case I or case III) and on the action the voter takes based on her prior belief ($p < v$: elect the incumbent, or $p > v$: elect the challenger). If information revelation is profitable for the interest group, it is assumed that the group will prefer to use the route where information revelation is indeed possible. In case both contributions and endorsements are potentially informative, the route with the most profitable hybrid equilibria and the largest possibilities for separation will be preferred (if these two criteria conflict, no conclusion will be drawn). On the other hand, when the interest group wants to conceal its identity the route with fewer possibilities for information revelation is preferred. This will also be the case when the voter's prior belief induces her to take the action preferred by the interest group, for in that case information transfer is only costly to the group. When we follow this procedure for the cases I and III, respectively, Proposition 4.3 results (Appendix 4.A contains a more elaborate justification of the procedure described above, as well as a proof of the proposition stated below).

Proposition 4.3
(a)   *Case I.* In case of no conflict of interests between the interest group and the voter it holds that:
         if (i) p<v: the interest group prefers to use endorsements.
         if (ii) p>v: when IRCO holds the bad type interest group prefers to use contributions and the good type prefers to use endorsements, in case IRCO does not hold the bad type prefers endorsements and the good type contributions.
(b)   *Case III.* When the interest group prefers the challenger it holds that:
         if (i) p<v: when the IRCO condition holds the interest group prefers to use contributions, whereas in case IRCO does not hold, but IRDE does, endorsements are preferred.
         if (ii) p>v: when the IRCO condition holds the interest group prefers to use endorsements, whereas in case IRCO does not hold, but IRDE does, contributions are preferred.   □

Proposition 4.3(a) states that in case I with p<v, the interest group prefers the use of direct endorsements from an ex ante perspective. That is, not knowing the type of the challenger and which equilibrium will result by using either contributions or endorsements, the interest group wants to focus on endorsements. The challenger only stands in the way here. However, when the voter a priori prefers the challenger (p>v), no definite conclusion can be drawn from an ex ante viewpoint.[11] Part (b) of the proposition says that when there is a partial conflict of interest between the voter and the interest group, and the voter is a priori inclined to elect the incumbent (p<v), the interest group prefers to use contributions when these are potentially informative. This follows because the bad type interest group does not lose from information transfer, whereas the good type interest group gains from information transfer. Since under the IRCO condition the possibilities for separation are the largest when contributions are used, contributions are preferred from an ex ante perspective. Only when contributions cannot reveal any information, the interest group prefers endorsements when these are potentially informative. On the other hand, when the voter a priori prefers the challenger the interest group prefers *not* to use contributions when these may be informative, whereas in case contributions cannot reveal any information, but

endorsements can, the interest group prefers the use of contributions (in order to avoid informative equilibria with wasteful expenditures). [12]

Due to the assumption of rational (Bayesian) decision making, *ex ante* the voter can never be made worse off by the possibility of expenditures on campaigning or direct endorsements. [13] Hence according to this model the voter would prefer that both endorsements and political campaigns are possible, for this maximizes the opportunities for separation. From Proposition 4.2 it follows that, generically, the voter does not prefer the one above the other.

In summary, the answer to the second question posed — why would an interest group prefer to use campaign contributions as indirect endorsements to support the challenger when it can also reach out to the voter via direct endorsements? — can be decomposed into two steps. First, it has to be established whether the group gains from information transfer anyway. In the most likely situation of an incumbency advantage (p<v) this is indeed the case. Contributions are then preferred whenever these are potentially informative, because the challenger may serve a useful function as an intermediary filter of the group's opposing interests. In case the group's interests are completely aligned with the voter's interests, however, endorsements are preferred.

4.4   DIRECTLY OBSERVABLE CONTRIBUTIONS

In the contributions model of Section 4.2 it is assumed that the voter can not directly observe donations, only costly campaigns are observed. Contributions are only indirectly observable, because only after seeing a costly campaign the voter can infer that contributions are made. Strictly speaking, this assumption does not correspond with reality for many countries. Candidates or parties are typically required to disclose information about donations received. [14] Information about donations may thus, in principle, be observable to the general voting public. Indeed, in the U.S. voters are increasingly made aware of the financial powers behind the political candidates. [15] Still, it remains rather time consuming, and thus costly, to become fully informed on campaign contributions. The assumption that contributions are unobservable can then be seen as a limit case where these costs are prohibitively high. Costly campaigns, on the other hand, are typically thrust upon voters and thereby freely and easily observable.

In order to establish whether the conclusions drawn from the basic contributions model of Section 4.2, particularly those in relation to the two main questions addressed in this chapter, crucially depend on the assumption that donations are not directly observable, we briefly analyze the case where contributions are freely observable. In addition, we investigate whether directly observable contributions enlarge the scope for information transfer, and thus, whether the electorate would benefit from imposing full disclosure requirements.

In the observable contributions model the voter knows, when she does not observe a costly campaign by the challenger, whether the challenger kept the funds received or whether he did not receive any contributions. In fact, when contributions are freely observable the voter receives one of three possible signals: (i) no contributions (and, thus, necessarily no costly campaign), (ii) contributions, but no costly campaign, and (iii) contributions and a costly campaign. Recall that in the basic contributions model of Section 4.2 the situations (i) and (ii) cannot be disentangled by the voter, because in both situations she only observes "no campaign". Put differently, in the basic contributions model there is uncertainty on the side of the voter concerning whether the challenger can command enough resources to run a costly campaign or whether he cannot. In Section 3.3 we already discussed a different kind of "receiver uncertainty", namely one where the uncertainty is about whether the sender is informed or not. There it was observed that this latter type of uncertainty may in fact increase the scope for information transfer. Contrarily, as we will observe below, in the context of the contributions model uncertainty about the resources the challenger can command reduces the possibilities for information revelation.

In order to establish under what circumstances information transmission through observable contributions is possible in equilibrium, the following observation is useful. From a formal point of view the observable contributions model is just an extension of the direct endorsements model with an additional round of signaling; the possibility of information transfer by the challenger at the campaigning stage after an observable expenditure by the interest group is added. Intuitively, a directly observable contribution incorporates two aspects: (i) it serves as a direct endorsement to the voter, and (ii) it provides the challenger with campaign funds which enable him to transfer information through campaigning.[16] By assuming that the challenger types pool on not campaigning were they to receive contributions (which can always be sustained as equilibrium behavior), it immediately follows that each equilibrium of the direct endorsements model is sustainable as (part of a larger) equilibrium in the observable contributions model. Hence the observable contributions model has at least the same scope for information transfer as the direct endorsements model has.

It can be concluded from Corollary 4.1 that the additional campaigning stage in the observable contributions model cannot be informative in case the IRCA condition does not hold. In that case direct endorsements and observable contributions in fact have the same possibilities for information transfer. However, there are cases where information revelation can occur by means of observable contributions and subsequent campaigning, but not through direct endorsements. For instance, consider the case $u_1 < c < u_2 < g_2 < g_1$. Because the IRDE condition is not met, information transfer through direct endorsements is not possible. But, information transfer is possible through observable contributions and subsequent campaigning. To illustrate, consider the following separating equilibrium: the good type interest group always contributes and the good type challenger always campaigns, the bad type interest group does not contribute. After observing the combination of a contribution and a campaign the voter elects the challenger, after observing no contribution she elects the incumbent. Out-of-equilibrium, when a contribution without a campaign is observed, the voter also elects the incumbent.

The last two paragraphs indicate that observable contributions have a strictly larger scope for (complete) information revelation than direct endorsements have. Interestingly, when comparing the observable contributions model with the model in which donations are not directly observable (cf. the model of Section 4.2), it appears that a similar result holds; at least the same amount of information can be revealed through observable contributions. This result is reflected in Proposition 4.4 below, which also summarizes the two other observations made above that are of interest (a formal proof is given in Appendix 4.A).

Proposition 4.4

(a)   When the IRCA condition does not hold all information that is revealed in an informative equilibrium of the observable contributions model is completely revealed through contributions alone; the campaigning stage is not informative.

(b)   Compared with direct endorsements at least the same amount of information can be revealed (in equilibrium) through observable contributions. Moreover, the scope for complete information revelation is strictly larger.

(c)   Compared with directly unobservable contributions at least the same amount of information can be revealed (in equilibrium) through observable contributions. Moreover, the scope for complete information revelation is strictly larger.   □

Combining part (b) and part (c) of Proposition 4.4 it follows that information transfer through observable contributions is possible when either the IRDE or the IRCO condition holds. In other words, both IRDE and IRCO are sufficient conditions for information revelation through observable contributions to be possible in equilibrium.[17] This last observation is useful when we want to establish whether our answers to the two main questions addressed in this chapter have to be altered when contributions are directly observable. It is to this topic that we turn now.

At the end of Section 4.2 we answered the first question posed in this chapter by comparing the IRCA with the IRCO condition. Because there are cases in which IRCO holds but IRCA does not, we concluded that under these circumstances political campaigns need (directly unobservable) contributions in order to be informative and, thus, effective. Above we observed that IRCO is also a sufficient condition for observable contributions to be potentially informative. Therefore, we immediately obtain that there are circumstances in which information revelation is possible in the observable contributions model, but not in the basic campaigning model where the challenger has to rely on his own funds. This can be taken as support for the contention that also when contributions are observable, political campaigns may need these contributions in order to be effective. However, as Proposition 4.4(a) indicates the campaigning stage itself reveals no additional information, and from that perspective is not "effective". In short, our answer to the first question posed remains the same when contributions are observable, but the interpretation of an "effective" campaign is somewhat flawed.

In view of the result reported in Proposition 4.4(b), our second main question of why an interest group may prefer contributions over direct endorsements reduces to the question under what circumstances the group benefits from the added possibilities for information transfer. In the most interesting case of an incumbency advantage (p<v) and the challenger preferred by the interest group (case III), the interest group prefers to transfer information, and thus (weakly) prefers observable donations over direct endorsements. So, even when contributions are observable the challenger can serve as a useful filter of the group's opposing interests.

Finally, Proposition 4.4(c) reveals that observable contributions have a larger scope for information transfer than directly unobservable contributions. More importantly, observable contributions have a strict larger scope for separation. Since the voter is strictly better off when all information is revealed compared with the situation where only some, or no information at all is transmitted (recall that in our model the voter does not benefit in expected payoff terms from partial information transmission), she prefers contributions being observable rather than unobservable. In other words, in the context of our model it is obtained that the electorate gains from contributions being readily observable, as is typically argued in policy oriented studies and popular publications. [18]

4.5   CONCLUDING DISCUSSION

When faced with the problem whom to vote for in an election, voters rely on several sources of information. Among these are: relevant past experience with the candidates or parties, political campaigns, and (in)direct endorsements by interest groups. Empirical evidence suggests that information from each of these sources influences voters' decisions. [19] Theoretically, the impact of past experience can be understood, and has indeed been modeled, from an adaptive behavior or learning perspective (cf. Calvert, 1986). The influence of political campaigns and direct endorsements can be explained with the basic signaling game presented in the previous chapter. The total expenditures made by political candidates and/or interest groups with respect to these activities may provide the voter indirect information about the policy stance of the candidate. This transmission of information is only possible when there is some congruence between the interests of the voter and the interests of the agent actually making the observable costly outlays (the political candidates in case of campaigns, and interest groups in case of direct endorsements).

In this chapter we started from a simple model of an election which incorporates the above ingredients. We directed our attention towards two questions. Firstly, we accounted for the observation that in a number of countries political campaigns are largely financed by interest groups, and we investigated whether this fact may in some instances be crucial for the costs of campaigning to potentially bear some information. The analysis presented in Section 4.2 shows that this question can be answered in the affirmative; when there is not enough congruence of interests between the candidate and the voter, campaigns may still be influential provided that the money for the campaign is donated by an interest group who has preferences that are congruent with the voter's. If the voter can have more faith in the interest group than in the candidate, the fact that the campaign money is contributed by the interest group may reassure her about the candidate's position. At the same time, the contributions model on which the analysis is based provides a position-induced model for the influence of campaign contributions, without having to rely on the *assumption* that voters respond positively to campaign spending.

Secondly, we questioned why an interest group would support a candidate through campaign contributions, instead of reaching out to the general voting public itself by means of direct endorsements. It appeared that when the preferences of the interest group and the voter are not sufficiently aligned, direct information transmission to the voter is not possible. In that case the interest group may exploit the stronger congruence between the preferences of the voter and those of the political candidate. However, the interest group has no unambiguous preference for either channel of political influence. Under a specified range of circumstances the interest group prefers to contribute to a candidate's campaign, at other instances the group is better off by reaching out to the voter directly. In general, interest groups may need, and indeed sometimes prefer to use, political candidates as an intermediary to filter their opposing interests.

The basic contributions model of Section 4.2 is based on the assumption that contributions are not directly observable. Though quite justifiable, in the previous section we investigated the consequences of dropping that assumption. It was argued that the answers to our two main questions remain valid even when donations are observable. Interestingly, it could also be derived that the voter benefits (in expected payoff terms) from campaign contributions being readily observable. Thereby the model provides a theoretical justification for the popular call for full disclosure laws (cf. Sloof, 1997, The Economist, 1994, 1997a, 1997b).

For an assessment of the relevance of our model from an empirical point of view, a number of empirical findings can be mentioned that are suggestive in this respect. First, the model reveals that contributions are more likely to be addressed to candidates which are on the same side as the interest group (i.e., when $g_1, g_2 > 0$). This result is in line with empirical evidence indicating that interest groups often give along ideological and partisan lines (cf. Chapter 2). Second, the model predicts that campaign expenditures and contributions are highest when the election is expected to be close. This result is also well documented in the empirical literature (e.g., Kau and Rubin, 1982, Poole and Romer, 1985, and the references in Chapter 2). Third, a finding concerning endorsements — albeit not very robust — is that business interests receive a less favorable hearing from voters than ideological and labor groups (Kau and Rubin, 1979, Lupia, 1994, Schneider and Naumann, 1982).[20] The latter type of groups are perhaps more encompassing and, for that reason, their preferences may be more aligned with the interests of a representative voter. In terms of the model of this chapter this might be reflected by the fact that the IRDE condition holds for these groups, whereas it does not hold for the business groups. The observation made by Schlozman and Tierney (1986) that endorsements are typically only used by labor unions may point in a similar direction.

Finally, some preliminary empirical support for the result obtained from the model that the source of campaign funding per se may have an impact on voter behavior can be found in Depken (1998), Palda (1996) and Palda and Palda (1998). Depken (1998), for instance, investigates how various types of campaign contributions affected outcomes in the U.S. Congressional elections of 1996. He finds that the level of campaign contributions received from PACs had a larger positive effect on the percentage of votes received than contributions donated by individuals. The impact of party contributions appeared to be equivocal. Unfortunately, personal donations were not identified as a separate source of funds in the empirical analysis. Candidates' own funds were considered by Palda (1996) and Palda and Palda (1998), though. They incorporated fractions of the candidates' total receipts rather than the level of contributions received from a particular source into the regressions, and found that the fraction of the candidate's own wealth either had a significant negative effect on the vote share, or was insignificant. On the other hand, voters did react positively to candidates who rely more heavily on contributions from individuals. In the Palda and Palda regressions the effect of contributions from special interests (corporations) turned out to be equivocal. In sum, although too few studies exist to draw firm conclusions yet, there is some empirical evidence that the source of campaign outlays may indeed act as a signal to voters about the quality of the candidate.

As this research project is financed by the Dutch taxpayer, a brief assessment of whether the election model presented in this chapter is of (any) relevance to the case of the Netherlands seems appropriate. In Dutch politics political parties dominate the elections. Except perhaps for some local municipality elections, individual candidates do not enter the race. In that respect, Dutch elections differ a lot from their U.S. counterparts. The election model discussed in this chapter seems more appropriate to describe the situation in the U.S. where basically a single incumbent runs against a single challenger, than the Dutch (European) situation with several political parties competing for office. Several other empirical facts also make the model less suitable for an application to the Netherlands. First, in comparison with other countries, especially the U.S., campaigns (conducted by the national headquarters of the parties) are very cheap.[21] Second, political parties, and thus their campaigns, are for the larger part financed out of the contributions of the parties' own members. Third, receiving money from business groups (and other interest groups), although fully legal and even promoted through tax exemptions on political gifts, is "not done" and only occurs on a very small scale (cf. Koole, 1990a, 1992).[22] An interesting hypothesis in this respect, put forward in Koole (1990a, p. 44), is that "..the politics of neo-corporatism in the Netherlands probably provides them [businessmen] with enough channels to influence politics: they do not need party organizations in this respect."[23] This hypothesis nicely links up with one of the main topics of this monograph, namely the endogenous choice between several means of influence, and suggests that the absence of campaign contributions does not imply little influence of interest groups.

Dutch interest groups only sporadically try to influence elections by reaching out to the voters directly. Van Praag (1987) provides a rare example and describes a case in which the FNV, a peak organization of labor unions, launched a campaign to influence the elections of 1986. Although the FNV intended to supply the voter "objective" information about the platforms of the several parties, its campaign was generally understood as an endorsement for the Labor party (PvdA). The costly campaign, about 3 million guilders were spent, was judged to be that unsuccessful that the FNV decided not to repeat such a campaign in the future.

Notwithstanding the rare occurrence of political gifts by interest groups, and the very small amounts of money involved, the Dutch government has recently considered designing regulation with respect to the disclosure of donator information (cf. Positie en subsidiëring van politieke partijen, 1996, and Regeling van de subsidiëring van politieke partijen, 1997/1998). For the time being, the government relies on the self-regulating capacity of the political parties; in the spring of 1995 the Minister of Domestic Affairs reached an agreement with the leadership of the political parties to fully disclose relevant information — viz. the amount of the gift, the identity of the donor and the date of the gift — about any gift over Dfl. 10,000.- received from interest groups. The agreement has been evaluated at the end of 1997. It was found that, by and large, the political parties did stick to this agreement. Nevertheless, the disclosure requirements appearing in the agreement are incorporated in a revised bill concerning the partial public funding of political parties after all (cf. Regeling van de subsidiëring van politieke partijen, 1997/1998). That bill also dictates, among other things, that public funds cannot be used for political campaigning.

We end this chapter with some suggestions for extending the election model considered here. Two obvious, but nevertheless very interesting extensions were already mentioned in Section 3.3. Firstly, it would be nice to allow the incumbent to engage in campaigning as well. Based on the analysis of other studies, it is reasonable to expect that the incumbent will mainly spend money to counteract the campaign of the challenger (cf. Chapter 3). Secondly, the model would become more realistic if the challenger and/or the interest group could decide on the level of expenditures themselves. If no upperbound is put on the level of expenditures, the scope for information transfer increases. A third, not already mentioned, extension relates to the assumption that the state of the world is exogenously fixed. Although it is quite reasonable to assume that a challenger is either better or worse for the voter than the incumbent, it would be more realistic to assume that the challenger's type is partly determined by the candidate himself (and the incumbent). Hence, an interesting extension would be to make the role of the challenger more substantive by making his policy stance endogenous (cf. Lupia, 1992). In case also the policy position of the incumbent would be endogenized, a model could be obtained in which political candidates actively compete for campaign contributions through their choice of platform. Such a model would be in line with the models of Austen-Smith (1987), Edelman (1992) and Prat (1997), and also with the model of Grossman and Helpman (1997) in which candidates take account of the influence of their policy choice on the endorsements received.

The result that the content of a political campaign (or similarly, an endorsement) has no informational value at all appears somewhat extreme. In a more general model where the candidate can decide on policy himself the content of the campaign may also have some informational value, next to the amount of campaign expenditures. For instance, Lupia (1993) extends his agenda setter model — where the amount of money paid by the setter gives the voter a cue about the minimum distance between the setter's privately known alternative and the commonly known status quo — with the possibility that the agenda setter sends a campaign message. The setter can indicate whether the alternative is either "left" or "right" of the status quo. In fact, Lupia (1993) investigates three different cases in which either (i) the campaign message is necessarily truthful, (ii) the message is just cheap talk, or (iii) lying bears an exogenous fixed cost. As noted in Chapter 3, the latter case can be interpreted as a reduced form analysis of reputational considerations. In a dynamic formulation the costs of lying may come about endogenously through the future loss of reputation (cf. Harrington, 1993).

Another extension concerns the voter in the model. For simplicity, it was just assumed that the voter could be considered as being "representative" or "decisive". It would be interesting to consider, for example, a spectrum of voters who differ according to the prior probability (p) with which they believe the challenger to be the best choice. Another, but related possibility, would be to consider a spectrum of voters who differ according to the stake (v_1, v_2) they have in selecting the best candidate available. An extension along these lines would allow the decisive (median) voter to be determined endogenously.

Finally, it would be interesting to account for the fact that candidates typically receive funds from a large number of interest groups. When campaign contributions serve the purpose of getting the favored candidate elected by informing voters, a free-rider problem may exist among the supporters of a particular candidate. A severe free-rider problem among the candidate's potential donors, then, may affect the candidate's campaign funds and hence his chances of electoral success. On the other hand, rational voters should take into account the different incentives for supporters to contribute to the campaign of their candidate, and thus the electoral consequences of the free-rider problem may be attenuated (cf. Lohmann, 1995). When a spectrum of interest groups is considered who differ in respect to their stake (g_1, g_2), it might be possible to assess whether the effect of the free-rider problem on the level of information transmission dominates the greater potential for information transmission due to the existence of multiple senders.

APPENDIX 4.A THE GENERAL CONTRIBUTIONS AND DIRECT ENDORSEMENTS MODEL

In this appendix we formally analyze the model in which the interest group's choice between (directly unobservable) contributions and direct endorsements is truly endogenous. Moreover, it is shown that the results reported in the main text follow from this more general model. First the general contributions and direct endorsements model is described. Then the equilibrium analysis of this model is presented, and all the equilibria of the model are stated in a proposition. To keep the discussion in this appendix well-running, the lengthy proof of this proposition is relegated to yet another appendix, viz. Appendix 4.B. After having listed all the equilibria the welfare implications of the general model are derived, and the consequences of imposing two equilibrium refinements (universal divinity and CFIEP, cf. Chapter 3) are elaborated upon. Finally, the propositions appearing in the main text are proved using the results obtained for the general model studied in this appendix.

Description of the game. The general contributions and direct endorsements model has the following setup. First, nature draws the type of the interest group-challenger combination $t \in \{t_1 \equiv \text{worse}, t_2 \equiv \text{better}\}$, with $P(t=t_2)=p$. Next, the interest group chooses from two or three options (the number of options available depends on the group's preferences). The interest group decides either to contribute a fixed amount c to the challenger (unobservable to the voter), or to spend c on direct endorsements (observable to the voter), or to spend no money at all. In case the interest group prefers the incumbent ($g_i<0$), only the first and third option are available to the group. Then the challenger decides whether to spend any contributions received on his campaign. In case the challenger does not receive contributions from the interest group he cannot run a (costly) campaign. The voter either observes a campaign by the challenger, a direct endorsement in support of the challenger, or none of these, and subsequently chooses between the incumbent and the challenger. Finally, payoffs are obtained. Concerning these payoffs, the value of c, and the prior belief p we make the assumptions as stated in Section 4.1; payoffs are given in Table 4.1, $c>0$, and no 'knife-edge' cases.[24]

In the sequel we refer to the voter as player V, to the challenger as player Ch and to the interest group as player G. The strategy of voter V is denoted as $\phi(s)$. Here $\phi(s)$ gives the probability that the voter will elect the challenger ($x=x_2\equiv$challenger), given that she has received signal s, for $s \in \{0, c_{cam}, c_{end}\}$. In our notation $s=c_{cam}$ indicates that the challenger runs a costly campaign, $s=c_{end}$ indicates that the interest group makes a direct and costly endorsement in support of the challenger, and $s=0$ indicates the absence of both a campaign and a direct endorsement. With $\gamma(t_i)$ we denote the (conditional) probability that the type t_i challenger engages in campaigning, given that he received contributions from the interest group. The challenger's strategy when he does not receive any contributions is trivially determined by the fact that in that case he has only one option to choose, viz. not to campaign, and thus need not be considered. Lastly, the strategy of the interest group is given by the two-tuple $(\delta(t_i), \varepsilon(t_i))$ for i=1,2, with $\delta(t_i)$ the probability that the type t_i interest group contributes a fixed amount c to the challenger, and $\varepsilon(t_i)$ the probability that the amount c is spent on a direct endorsement. Note that direct endorsements and donations are assumed to be equally costly. Since the interest group can either donate or endorse, but not both, we have $0 \leq \delta(t_i) + \varepsilon(t_i) \leq 1$. Moreover, because by assumption the interest group cannot endorse the incumbent we have $\varepsilon(t_i)=0$ when $g_i < 0$.

Equilibrium analysis. A perfect Bayesian equilibrium of the game described above is defined as a set of strategies $(\delta(\cdot), \varepsilon(\cdot))$, $\gamma(\cdot)$, $\phi(\cdot)$, and a set of beliefs $q(\cdot)$ that satisfy the following conditions:

(e1)   for i=1,2 the two tuple $(\delta(t_i), \varepsilon(t_i))$ must satisfy the following four conditions:

    (e1.1) $0 \leq \delta(t_i) + \varepsilon(t_i) \leq 1$.

    (e1.2)   if $\delta(t_i) > 0$, then both (e1.2.a) and (e1.2.b) must be satisfied:

        (e1.2.a) $\gamma(t_i)[\phi(c_{cam})-\phi(0)]g_i - c \geq 0$.

        (e1.2.b)   if $g_i > 0$, then $\gamma(t_i)[\phi(c_{cam})-\phi(0)] \geq [\phi(c_{end})-\phi(0)]$.

    (e1.3) $\varepsilon(t_i) > 0$ only when $g_i > 0$ and

        $\phi(c_{end})g_i - c \geq \phi(0)g_i + \max\{\gamma(t_i)[\phi(c_{cam})-\phi(0)]g_i - c, 0\}$.

    (e1.4)   if $\delta(t_i) + \varepsilon(t_i) < 1$, then

        $\max\{\gamma(t_i)[\phi(c_{cam})-\phi(0)]g_i, [\phi(c_{end})-\phi(0)] \cdot \max\{g_i, 0\}\} \leq c$.

(e2)   for i=1,2: if $\gamma(t_i) > 0$ (<1), then $[\phi(c_{cam})-\phi(0)]u_i \geq (\leq)$ c.

(e3)   for $s \in \{0, c_{cam}, c_{end}\}$: if $\phi(s) > 0$ (<1), then $q(s) \geq (\leq)$ v.

(e4) $q(s)$ is consistent with Bayes' rule whenever defined.

The first condition requires that the interest group chooses that feasible option that maximizes its payoff, taking the strategies of the challenger and the voter as given. Specifically, condition (e1.2) requires that when the interest group contributes in equilibrium, contributing must yield at least as much as not contributing (e1.2.a), and at least as much as a direct endorsement when the latter is feasible (e1.2.b). Condition (e1.3) reflects a similar requirement for direct endorsements to be used in equilibrium, and condition (e1.4) the requirement for the interest group spending no money at all in equilibrium. The first part (e1.1) reflects the assumption that the interest group cannot opt for both a contribution and a direct endorsement, but has to choose at most one of them. Condition (e2) states that for the challenger to spend the money received in equilibrium, the additional gain from doing so $[\phi(c_{cam})-\phi(0)]u_i$ must exceed its costs c. The third condition (e3) requires that the voter chooses the candidate that maximizes her expected payoff, given her posterior beliefs $q(s)$. According to the last condition (e4) these posterior beliefs have to be consistent whenever possible. (Explicit formulae for $q(s)$ based on Bayes' rule are given in Appendix 4.B.)

When Bayes' rule does not apply for signal s in equilibrium there is freedom in choosing the voter's posterior belief $q(s)$, thus her strategy $\phi(s)$, and possibly the challenger's strategy $\gamma(t_i)$. Due to this freedom, a specific equilibrium path may be compatible with various out-of-equilibrium strategies and beliefs. Below we only describe the strategies and beliefs along the equilibrium path, and do not specify all the possible out-of-equilibrium strategies and beliefs that sustain the specific path. Because every equilibrium path is sustainable by choosing $q(s)=0$ for each out-of-equilibrium signal s and subsequently letting $\phi(s)$ and $\gamma(t_i)$ for i=1,2 be determined by equilibrium conditions (e3) and (e2) above, we can always assume that out-of-equilibrium strategies and beliefs take this specific form. A brief consideration of two equilibrium refinements that restrict the choices of out-of-equilibrium beliefs is given further down this appendix.

We now list the equilibria of the general contributions and direct endorsements game. There are a large number of equilibria, and there seems no single obvious way to present them conveniently. We have chosen to list the equilibria in the following way. Besides the pooling on s=0 ("no-expenditures") equilibria, we distinguish equilibria in which (i) only contributions occur with positive probability ($\varepsilon(t_1)=\varepsilon(t_2)=0$ and $\delta(t_i)>0$ for some i=1,2), (ii) only direct endorsements occur with positive probability ($\delta(t_1)=\delta(t_2)=0$ and $\varepsilon(t_i)>0$ for some i=1,2), and (iii) equilibria in which both contributions and direct endorsements occur with positive probability ($\delta(t_i)\varepsilon(t_j)>0$ for some i and j in {1,2}). For each of the three possibilities we list the pooling (PE), hybrid (HE) and separating (SE) equilibria, together with the conditions under which they exist. This yields the following proposition (a proof is given in Appendix 4.B):

Proposition 4.5
The equilibrium paths of the perfect Bayesian equilibria of the general contributions and direct endorsements model are given by (with $v\equiv v_1/(v_1+v_2)$ and $w\equiv[p(1-v)]/[(1-p)v]$):

No expenditures equilibria:

p<v   PE1: $\delta(t_i)=\varepsilon(t_i)=0$ for i=1,2, $\phi(0)=0$;
                                             $q(0)=p$.

p>v   PE2: $\delta(t_i)=\varepsilon(t_i)=0$ for i=1,2, $\phi(0)=1$;
                                             $q(0)=p$.

Contributions equilibria:

$c<\min\{u_1,u_2,g_1,g_2\}$ PE3: $\delta(t_i)=1$, $\gamma(t_i)=1$ for i=1,2,
and p>v $\phi(c_{cam})=1$; $q(c_{cam})=1$.

$c<u_1<\min\{u_2,wg_1,g_2\}$ HE1: $\delta(t_1)=\delta(t_2)=1$, $\gamma(t_1)=w$, $\gamma(t_2)=1$,
and p<v $\phi(0)=0$, $\phi(c_{cam})=c/u_1$; $q(0)=0$,
                                             $q(c_{cam})=v$.

$\max\{c, wg_1\} < u_1 < \min\{u_2, g_1, g_2\}$
and $p < v$

HE2: $\delta(t_1) = wg_1/u_1$, $\delta(t_2) = 1$, $\gamma(t_1) = u_1/g_1$, $\gamma(t_2) = 1$, $\phi(0) = 0$, $\phi(c_{cam}) = c/u_1$; $q(0) = 0$, $q(c_{cam}) = v$.

$c < g_1 < \min\{u_1, u_2, g_2\}$
and $p < v$

HE3: $\delta(t_1) = w$, $\delta(t_2) = 1$, $\gamma(t_1) = \gamma(t_2) = 1$, $\phi(0) = 0$, $\phi(c_{cam}) = c/g_1$; $q(0) = 0$, $q(c_{cam}) = v$.

$\max\{c, \min\{u_1, g_1\}\} < u_2 < (1 - 1/w)g_2$
and $p > v$

HE4: $\delta(t_1) = 0$, $\delta(t_2) = 1$, $\gamma(t_2) = (1 - 1/w)$, $\phi(0) = 1 - c/u_2$, $\phi(c_{cam}) = 1$; $q(0) = v$, $q(c_{cam}) = 1$.

$\max\{c, \min\{u_1, g_1\}, (1 - 1/w)g_2\} < u_2 < g_2$
and $p > v$

HE5: $\delta(t_1) = 0$, $\delta(t_2) = (1 - 1/w)g_2/u_2$, $\gamma(t_2) = u_2/g_2$, $\phi(0) = 1 - c/u_2$, $\phi(c_{cam}) = 1$; $q(0) = v$, $q(c_{cam}) = 1$.

$\max\{c, \min\{u_1, g_1\}\} < g_2 < u_2$
and $p > v$

HE6: $\delta(t_1) = 0$, $\delta(t_2) = (1 - 1/w)$, $\gamma(t_2) = 1$, $\phi(0) = 1 - c/g_2$, $\phi(c_{cam}) = 1$; $q(0) = v$, $q(c_{cam}) = 1$.

$\min\{u_1, g_1\} < c < \min\{u_2, g_2\}$

SE1: $\delta(t_1) = 0$, $\delta(t_2) = 1$, $\gamma(t_2) = 1$, $\phi(0) = 0$, $\phi(c_{cam}) = 1$; $q(0) = 0$, $q(c_{cam}) = 1$.

Direct endorsements equilibria:

$c < \min\{g_1, g_2\}$ and $p > v$

PE4: $\varepsilon(t_1) = \varepsilon(t_2) = 1$, $\phi(c_{end}) = 1$; $q(c_{end}) = p$.

$c < g_1 < g_2$
and $p < v$

HE7: $\varepsilon(t_1) = w$, $\varepsilon(t_2) = 1$, $\phi(0) = 0$, $\phi(c_{end}) = c/g_1$; $q(0) = 0$, $q(c_{end}) = v$.

$\max\{c, g_1\} < g_2$
and $p > v$

HE8: $\varepsilon(t_1) = 0$, $\varepsilon(t_2) = (1 - 1/w)$, $\phi(0) = 1 - c/g_2$, $\phi(c_{end}) = 1$; $q(0) = v$, $q(c_{end}) = 1$.

$g_1 < c < g_2$

SE2: $\varepsilon(t_1) = 0$, $\varepsilon(t_2) = 1$, $\phi(0) = 0$, $\phi(c_{end}) = 1$; $q(0) = 0$, $q(c_{end}) = 1$.

Both contributions and direct endorsements equilibria:[25]

$c < \min\{u_2, g_1, g_2\}$
and $p > v$

PE5: $\gamma(t_i)\delta(t_i) + \varepsilon(t_i) = 1$ for $i = 1, 2$, with $w\delta(t_2) \geq \delta(t_1)$, $\delta(t_2) > 0$, $w\varepsilon(t_2) \geq \varepsilon(t_1)$, $\varepsilon(t_2) > 0$, $\gamma(t_1) = 1 (= 0)$ if $c < (>)u_1$, $\gamma(t_2) = 1$, $\phi(c_{cam}) = \phi(c_{end}) = 1$; $q(c_{cam}) \geq v$, $q(c_{end}) \geq v$.

$c<g_1<\min\{u_1,u_2,g_2\}$
and $p<v$

HE9: $\delta(t_1)=w\delta(t_2)>0$, $\varepsilon(t_1)=w\varepsilon(t_2)>0$, $\delta(t_2)+\varepsilon(t_2)=1$, $\gamma(t_1)=\gamma(t_2)=1$, $\phi(0)=0$, $\phi(c_{cam})=\phi(c_{end})=c/g_1$; $q(0)=0$, $q(c_{cam})=q(c_{end})=v$.

$\max\{c,g_1\}<g_2<u_2$
and $p>v$

HE10: $\delta(t_1)=\varepsilon(t_1)=0$, $\delta(t_2)+\varepsilon(t_2)=(1-1/w)$, $\delta(t_2)>0$, $\varepsilon(t_2)>0$, $\gamma(t_2)=1$, $\phi(0)=1-c/g_2$, $\phi(c_{cam})=\phi(c_{end})=1$; $q(0)=v$, $q(c_{cam})=q(c_{end})=1$.

$\max\{c,u_1,(1-1/w)g_1\}<u_2<g_1<g_2$
and $p>v$

HE11: $\delta(t_1)=0$, $\varepsilon(t_1)=w[1-(1-1/w)g_1/u_2]$, $\delta(t_2)=(1-1/w)g_1/u_2$, $\varepsilon(t_2)=1-(1-1/w)g_1/u_2$, $\gamma(t_2)=u_2/g_1$, $\phi(0)=1-c/u_2$, $\phi(c_{end})=1-(c/u_2)+(c/g_1)$, $\phi(c_{cam})=1$; $q(0)=q(c_{end})=v$, $q(c_{cam})=1$.

$g_1<c<\min\{u_2,g_2\}$

SE3: $\delta(t_1)=\varepsilon(t_1)=0$, $\delta(t_2)+\varepsilon(t_2)=1$ with $\delta(t_2)>0$ and $\varepsilon(t_2)>0$, $\gamma(t_2)=1$, $\phi(0)=0$, $\phi(c_{cam})=\phi(c_{end})=1$; $q(0)=0$, $q(c_{cam})=q(c_{end})=1$. □

In all but one equilibrium specification above in which the interest group both contributes and endorses with positive probability in fact a continuum of equilibria is described (viz. PE5, HE9, HE10 and SE3). These equilibria, however, do not differ in an essential way. When a specific type of interest group is indifferent between donating and endorsing, it may have some leeway in choosing one of the two options, leading to a continuum of equilibria. More generally, and more importantly, note that the restrictions on the parameters for the existence of specific types of equilibria overlap each other; for a large range of parameter values there are multiple equilibria.

Welfare implications. Next we consider the expected equilibrium payoffs for the three players in the game. The expected payoffs the voter, the challenger, and the interest group, respectively, obtain in equilibrium are given by:

$$U_V^* \equiv (1-p)[\gamma(t_1)\delta(t_1)(1-\phi(c_{cam}))+\varepsilon(t_1)(1-\phi(c_{end}))+(1-\gamma(t_1)\delta(t_1)-\varepsilon(t_1))(1-\phi(0))]v_1 +$$
$$p[\gamma(t_2)\delta(t_2)\phi(c_{cam})+\varepsilon(t_2)\phi(c_{end})+(1-\gamma(t_2)\delta(t_2)-\varepsilon(t_2))\phi(0)]v_2 \qquad (4.1)$$

$$U_{Ch}^*(t_i) \equiv [\gamma(t_i)\delta(t_i)\phi(c_{cam})+\varepsilon(t_i)\phi(c_{end})+(1-\gamma(t_i)\delta(t_i)-\varepsilon(t_i))\phi(0)]u_i +$$
$$\delta(t_i)(1-\gamma(t_i))c \qquad (4.2)$$

$$U_G^*(t_i) \equiv [\gamma(t_i)\delta(t_i)\phi(c_{cam})+\varepsilon(t_i)\phi(c_{end})+(1-\gamma(t_i)\delta(t_i)-\varepsilon(t_i))\phi(0)]g_i - [\delta(t_i)+\varepsilon(t_i)]c \qquad (4.3)$$

It can easily be verified that in all separating equilibria the voter's expected payoff U_V^* equals $U_V^* = (1-p)v_1 + pv_2$, whereas in all other (pooling and hybrid) equilibria the expected equilibrium payoff of the voter equals $U_V^* = \max\{(1-p)v_1, pv_2\}$. The expected equilibrium payoffs of the challenger and the interest group are given in Table 4.2 below.

Table 4.2 Expected equilibrium payoffs

Equilibrium	$U_{Ch}^*(t_1)$	$U_{Ch}^*(t_2)$	$U_G^*(t_1)$	$U_G^*(t_2)$
SE1/SE2/SE3	0	u_2	0	g_2-c
HE1	c	cu_2/u_1	$c(wg_1-u_1)/u_1$	$c(g_2-u_1)/u_1$
HE2	$c(wg_1)/u_1$	cu_2/u_1	0	$c(g_2-u_1)/u_1$
HE3/HE7/HE9	$c(wu_1)/g_1$	cu_2/g_1	0	$c(g_2-g_1)/g_1$
HE4	$u_1(1-c/u_2)$	u_2	$g_1(1-c/u_2)$	$g_2(1-c/wu_2)-c$
HE5	$u_1(1-c/u_2)$	$u_2+c[(1-1/w)(g_2/u_2)-1]$	$g_1(1-c/u_2)$	$g_2(1-c/u_2)$
HE6/HE8/HE10	$u_1(1-c/g_2)$	$u_2(1-c/wg_2)$	$g_1(1-c/g_2)$	g_2-c
HE11	$u_1[1-wc(g_1-u_2)/g_1u_2]$	$u_2-c[2+(g_1-u_2)/(wu_2)-(u_2/g_1)-(g_1/u_2)]$	$g_1(1-c/u_2)$	$g_2[1-(c/u_2)+(c/g_1)]-c$
PE1	0	0	0	0
PE2	u_1	u_2	g_1	g_2
PE3/PE4/PE5	u_1	u_2	g_1-c	g_2-c

As was the case for the basic signaling game of Section 3.1 the voter does not lose (in expected payoff terms) from the interest group having the possibilities of contributing and endorsing available. However, the voter only strictly benefits when all information is revealed in equilibrium. Both the challenger and the interest group (weakly) value the possibilities for information revelation when $p<v$. In case $p>v$ the challenger (weakly) dislikes the possibilities for information transfer. The same holds for the good type interest group. The bad type interest group, on the other hand, likes the opportunities for information revelation when $g_1<0$, and dislikes them in case $g_1>0$. Lastly, in the "full expenditures" pooling equilibria PE3, PE4, and PE5 political expenditures constitute a pure social waste.

Equilibrium refinements. We end our discussion of the general contributions and direct endorsements model with a brief elaboration on the use of equilibrium refinements (imposed to obtain the more plausible equilibria among all the PBE). Specifically, we apply similar equilibrium refinements as discussed in the context of the basic signaling game of Chapter 3, viz. universal divinity and CFIEP. Although both concepts cannot directly be applied to the general game considered here, straightforward extensions of the two concepts can be used. Instead of specifying the consequences of applying the two refinements in full detail, the following proposition establishes that their application leads to the same type of conclusions as those obtained for the basic signaling game of Chapter 3. (The proof of the proposition, relegated to Appendix 4.B, does list all the consequences.)

Proposition 4.6
(a)   When $p<v$ applying universal divinity or CFIEP deletes PE1 whenever information transfer is possible. In addition, HE7 is deleted when SE1 exists.
(b)   When $p>v$ applying CFIEP deletes the equilibria PE3, PE4 and PE5. Universal divinity deletes all these three equilibria only when both IRCO and IRDE do *not* hold.   □

In case $p<v$ the only hybrid equilibrium which may exist when separation is possible is HE7. Proposition 4.6(a) thus establishes that both refinements select the most informative equilibrium available when $p<v$. As was the case for the basic signaling game of Chapter 3, the likeliness of information revelation is emphasized when refinements are used. Part (b) of Proposition 4.6 questions the plausibility of pooling on costly messages, at least under certain circumstances. Hence the two equilibrium refinements weaken the prediction that expenditures on political campaigns and direct endorsements may constitute a pure social waste. Finally, from Proposition 4.6(a) it also follows that in case $p<v$ the type of equilibrium (pooling, hybrid or separating) is unique when applying either of the two refinements. But, there may exist several equilibria of the same type side by side. Although the two refinements have some bite when $p>v$, they do not even lead to a unique type of equilibrium in this case (cf. the proof of Proposition 4.6 in Appendix 4.B).

Proofs of the propositions appearing in the main text. The results presented in Section 4.1 through 4.5 can easily be obtained from the analysis of the general contributions and direct endorsements model given above. Recall that Corollary 4.1 (the IRCA condition) and Corollary 4.2 (the IRDE condition) resulted from the analysis of the basic signaling game presented in Chapter 3. However, Corollary 4.2 also directly follows from the direct endorsements equilibria reflected in Proposition 4.5. Proposition 4.1 can as easily be derived from this very same proposition, as the proof below indicates. With Proposition 4.5 at hand, the proof of Proposition 4.2 is trivial as well.

Proof of Proposition 4.1.
The equilibria of the basic contributions model of Section 4.2 where only contributions are possible follow from the contributions equilibria in Proposition

4.5, together with PE1 and PE2. Going through the conditions for existence of the various types of equilibria then yields the result.   *QED.*

Proof of Proposition 4.2.
Consider the equilibria listed in Proposition 4.5. Note that both the IRDE and the IRCO condition have to be satisfied for an informative equilibrium to exist where both contributions and direct endorsements are used with positive probability (cf. HE9, HE10, HE11 and SE3). For the informative direct endorsements equilibria of Proposition 4.5 (cf. HE7, HE8, SE2) to exist the IRDE condition necessarily has to hold, for the informative contributions equilibria (HE1 through HE6, SE1) to exist the IRCO condition must necessarily hold. Hence a simple comparison between the IRDE and IRCO condition is justified to conclude on differences in information revelation possibilities. The proposition now follows directly from comparing IRDE with IRCO, and the observation that when SE3 exists, both SE1 and SE2 exist as well.   *QED.*

Before we present the proof of Proposition 4.3 we first want to elaborate somewhat more on why the procedure described in Section 4.3 is used in constructing a preference ordering over the two routes. Recall that this procedure is based on a expected payoff comparison between (equilibria of) the contributions model and (equilibria of) the direct endorsements model. At first glance it seems that, in a specific equilibrium path of the general model with a really endogenous choice, the interest group reveals an unambiguous preference for the use of a specific route, or is indifferent between the two options. For instance, in a contributions equilibrium of this model (cf. Proposition 4.5) the interest group seems to reveal a preference for using contributions. However, contributions and direct endorsements equilibria, revealing conflicting preferences from this perspective, may exist side by side.

Alternatively, one might want to argue that when a contributions equilibrium exists, but a direct endorsements equilibrium does not, the group prefers to use contributions. Such a reasoning is also problematic, though. The interest group may namely prefer the no expenditures equilibrium PE2 over all the other equilibria. When for instance $c<u_1<g_2<u_2<g_1$ and $p>v$, an informative direct endorsements equilibrium does not exist, whereas an informative contributions equilibrium does (HE6). Because PE2 yields the interest group the highest expected payoff (cf. Table 4.2), it would prefer the contributions equilibria (and, thus, contributions) not to exist. In the contributions equilibrium HE6 the expectations of the voter (beliefs) forces the interest group to contribute with positive probability. In such a case, we can hardly argue that the interest group has a preference for contributions. In summary, when in a specific equilibrium path the interest group uses only one specific option, this does not necessarily indicate a preference for this specific route. A welfare comparison as suggested in Section 4.3 seems to take account of the above problems.

This welfare comparison to obtain the preference ordering of Proposition 4.3 can be summarized as follows. The no expenditures equilibrium PE1 (PE2) serves as a benchmark when p<v (p>v), and we determine whether the interest group gains from information revelation (we reason from an interim perspective, that is, after the type of the interest group is drawn and revealed to the group itself). If information revelation is indeed profitable for a specific type of interest group, it is assumed that this type prefers the route where information revelation is possible. When information revelation is possible by using either route, the group prefers the route with the largest possibilities for separation and the most profitable hybrid equilibria (if these two criteria conflict, no conclusion is drawn). A similar procedure is used when information revelation is unprofitable for a specific type of interest group.

Note that the procedure ignores the "full expenditures" pooling equilibria PE3 and PE4. This is completely justifiable when case I applies or when p<v. Only in case III with p>v these equilibria may exist. A problem with considering the full expenditures equilibria as well is that their existence cannot be directly related to the possibility of information transfer. That is, informative equilibria and the full expenditures pooling equilibria may, but need not exist at the same time. In order not to complicate the analysis we do not consider PE3 and PE4 in our procedure. However, note that when the former equilibrium exists the latter equilibrium exists as well, but that the reverse does not hold. From that perspective, contributions are somewhat more attractive in case III and p>v than our procedure indicates.

We now turn to the proof of Proposition 4.3. Throughout the proof use is made of Table 4.2, as well as of the results reported in Proposition 4.2.

Proof of Proposition 4.3.

(ai) The only possible equilibria besides PE1 are SE≡{SE1,SE2,SE3}. With $U_G^*(t_i|E)$ the expected payoff the interest group of type t_i (i=1,2) obtains in equilibrium E we get $U_G^*(t_1|PE1)=U_G^*(t_1|SE)$ and $U_G^*(t_2|PE1)<U_G^*(t_2|SE)$ (cf. Table 4.2). Both types thus (weakly) prefer the route with the largest possibilities for separation. By Proposition 4.2 it follows that the possibilities for separation are larger when direct endorsements are used.

(aii) The equilibria to consider are HE4, HE5, HE6, HE8, and SE. Now $U_G^*(t_1|PE2) <U_G^*(t_1|HEi)<U_G^*(t_1|SE)$ and $U_G^*(t_2|HEi)\leq U_G^*(t_2|SE)<U_G^*(t_2|PE2)$ for i=4,5,6,8. The bad type interest group likes the opportunities for information transfer, whereas the good type would rather see these opportunities not to exist. In addition, $U_G^*(t_1|HEj)>U_G^*(t_1|HE8)$ for j=4,5 when the equilibria exist side by side, whereas $U_G^*(t_2|HEj)<U_G^*(t_2|HE8)$ j=4,5 when the equilibria exist at the same time. The bad type prefers the partially informative contributions equilibria over the partially informative direct endorsements equilibrium, the good type the other way around. Note that when IRCO holds both routes are equally informative in this case. The result follows.

(bi) The equilibria to consider are HE1, HE2, HE3, HE7, and SE. From $U_G^*(t_1|PE1)=U_G^*(t_1|SE)=U_G^*(t_1|HEi)<U_G^*(t_1|HE1)$ for i=2,3,7, and

$U_G^*(t_2|PE1)<U_G^*(t_2|HEj)<U_G^*(t_2|SE)$ for $j=1,2,3,7$, together with $U_G^*(t_2|HE7)<U_G^*(t_2|HEi)$ for $i=1,2$ when these equilibria exist side by side, it follows that both types like the opportunities for information transfer through contributions. Proposition 4.2(b) then yields the result.

(bii) The equilibria to consider are HE4, HE5, HE6, HE8, and SE. The analysis for the good type interest group is the same as under (aii), so consider the bad type. In this case $U_G^*(t_1|SE)<U_G^*(t_1|HEi)<U_G^*(t_1|PE2)$ for $i=4,5,6,8$. Hence both types dislike the opportunities for information transfer. In addition, now $U_G^*(t_1|HEj)<U_G^*(t_1|HE8)$ for $j=4,5$ when the equilibria exist at the same time. The preferences of the bad and the good type interest group now coincide.   *QED.*

Observable contributions. The last proposition that requires a formal proof is Proposition 4.4 in which the model with observable donations is analyzed, and compared with directly unobservable contributions and direct endorsements. The following proof of this proposition ends the discussion of this appendix.

Proof of Proposition 4.4.
For the model in which contributions are directly observable the following notation is used. Let the two-tuple $s=(s_{con},s_{cam})\in\{(0,0),(c,0),(c,c)\}$ refer to the signal the voter observes, with s_{con} referring to observable contributions, and s_{cam} referring to costly campaigns. Let $q^{obs}(s)$ refer to the posterior belief of the voter after having received signal $s=(s_{con},s_{cam})$, and let $\gamma^{obs}(t_i)$, $\delta^{obs}(t_i)$ for $i=1,2$ and $\phi^{obs}(s)$ refer to the players' strategies in this model (in the observable contributions model the interest group only has the possibility of contributing to the challenger's campaign, direct endorsements are not possible). The definition of an equilibrium for the model in which contributions are observable follows from a straightforward adaption of equilibrium conditions (e1) through (e4), and is therefore omitted.

(a) For information being revealed in the campaigning stage, both $(s_{con},s_{cam})=(c,0)$ and $(s_{con},s_{cam})=(c,c)$ must occur in equilibrium with positive probability. Thus, the challenger must be prepared to spend contributions received; $\gamma^{obs}(t_i)>0$ for some i, requiring $\phi^{obs}(c,c)>\phi^{obs}(c,0)$. This requires $q^{obs}(c,c)\geq q^{obs}(c,0)$ in turn, and thus $\gamma^{obs}(t_1)\leq\gamma^{obs}(t_2)$ by Bayes' rule. Together with the previous observation $\gamma^{obs}(t_2)>0$. Now $\gamma^{obs}(t_2)>0$ requires $u_2>c$. Hence when $u_2<c$ the campaigning stage is not informative. On the other hand, when $u_1>u_2$ we have by the implication $\gamma^{obs}(t_2)>0$ $\Rightarrow \gamma^{obs}(t_1)=1$ (this implication follows from equilibrium condition (e2) adapted to observable contributions) that $\gamma^{obs}(t_1)=\gamma^{obs}(t_2)=1$, and again the campaigning stage is not informative. This establishes our claim.

(b) Choose $q^{obs}(c,c)$ and $\phi^{obs}(c,c)$ such that $q^{obs}(c,c)=q^{obs}(c,0)$ and $\phi^{obs}(c,c)=\phi^{obs}(c,0)$. Then necessarily $\gamma^{obs}(t_i)=0$ for $i=1,2$ in equilibrium, and (c,c) is indeed an out-of-equilibrium message, allowing the choice of $q^{obs}(c,c)$ and $\phi^{obs}(c,c)$ made above. Take a direct endorsements equilibrium specified in Proposition 4.5 and let $\delta^{obs}(t_i)=\varepsilon(t_i)$ for $i=1,2$, $\phi^{obs}(0,0)=\phi(0)$, $\phi^{obs}(c,0)=\phi(c_{end})$, $q^{obs}(0,0)=q(0)$, and

$q^{obs}(c,0)=q(c_{end})$. This yields an equilibrium of the model with observable contributions in which the same amount of information is being revealed as through direct endorsements. To establish the second claim, consider the following separating equilibrium described verbally in Section 4.3: $\delta^{obs}(t_1)=0$, $\delta^{obs}(t_2)=1$, $\gamma^{obs}(t_1)=0$, $\gamma^{obs}(t_2)=1$, $\phi^{obs}(0,0)=\phi^{obs}(c,0)=0$, $\phi^{obs}(c,c)=1$; $q^{obs}(0,0)=q^{obs}(c,0)=0$, $q^{obs}(c,c)=1$. This is an equilibrium when for instance $u_1<c<u_2<g_2<g_1$, a case for which no informative (and thus no separating) direct endorsements equilibrium exists (cf. Corollary 4.2).

(c) Consider the (directly unobservable) contributions equilibria of Proposition 4.5, and ignore the possibilities for direct endorsements. We first show that the separating equilibrium SE1, as well as the hybrid equilibria HE1, HE2 and HE3, are directly sustainable as an equilibrium when contributions are observed. It is easily seen that with $\delta^{obs}(t_i)=\delta(t_i)$, $\gamma^{obs}(t_i)=\gamma(t_i)$ for i=1,2, $\phi^{obs}(0,0)=\phi^{obs}(c,0)=\phi(0)$, $\phi^{obs}(c,c)=\phi(c)$, $q^{obs}(c,0)=q^{obs}(0,0)=q(0)$ and $q^{obs}(c,c)=q(c_{cam})$, SE1, HE1, HE2, and HE3 are sustainable as equilibrium when contributions are directly observed. The hybrid equilibria HE4, HE5 and HE6 are not immediately sustainable for the observable contributions case, but an equilibrium revealing the same amount of information also exist in these cases. For consider the following equilibrium: $\delta^{obs}(t_1)=0$, $\delta^{obs}(t_2)=(1-1/w)$, $\gamma^{obs}(t_1)=1$ (=0) if $u_1>u_2$ ($u_1<u_2$), $\gamma^{obs}(t_2)=1$, $\phi^{obs}(0,0)=1-c/g_2$, $\phi^{obs}(c,0)=\max\{0,1-c/\min\{u_1,u_2\}\}$ and $\phi^{obs}(c,c)=1$. Note that $q^{obs}(0,0)=v$ and $q^{obs}(c,c)=1$ by Bayes' rule, and we have chosen out-of-equilibrium belief $q^{obs}(c,0)=v$ to support $0<\phi^{obs}(c,0)<1$. This equilibrium exists when either HE4, HE5 or HE6 exists, and yields the voter the same expected payoff pv_2.

To support the second claim, note that from Proposition 4.1(c) follows that when contributions are unobservable a separating equilibrium only exists when $\min\{u_1,g_1\}<c<\min\{u_2,g_2\}$. Under these conditions the separating equilibrium of the observable contributions model described in the proof of part (b) also exists. Besides, when e.g. $u_1,u_2,g_1<c<g_2$ separation is still possible in the observable donations model, but not in the unobservable contributions model: $\delta^{obs}(t_1)=0$, $\delta^{obs}(t_2)=1$, $\gamma^{obs}(t_2)=0$, $\phi^{obs}(0,0)=0$, $\phi^{obs}(c,0)=1$; $q^{obs}(0,0)=0$, and $q^{obs}(c,0)=1$. Hence there is indeed an enlarged scope for separation when contributions are observed.

$QED.$

APPENDIX 4.B PROOFS OF PROPOSITIONS 4.5 AND 4.6

In order to complete the exact definition of a perfect Bayesian equilibrium of the general contributions and direct endorsements model, the formulae below give the posterior belief q(s) in case Bayes' rule applies (cf. equilibrium condition (e4)):

$$q(0) = P(t=t_2|s=0) = \frac{p[1-\gamma(t_2)\delta(t_2)-\varepsilon(t_2)]}{p[1-\gamma(t_2)\delta(t_2)-\varepsilon(t_2)]+(1-p)[1-\gamma(t_1)\delta(t_1)-\varepsilon(t_1)]}$$

$$(4.4)$$

$$q(c_{cam}) = P(t=t_2 \mid s=c_{cam}) = \frac{p\gamma(t_2)\delta(t_2)}{p\gamma(t_2)\delta(t_2)+(1-p)\gamma(t_1)\delta(t_1)} \tag{4.5}$$

$$q(c_{end}) = P(t=t_2 \mid s=c_{end}) = \frac{p\varepsilon(t_2)}{p\varepsilon(t_2)+(1-p)\varepsilon(t_1)} \tag{4.6}$$

In case the denominators in q(s) vanish, Bayes' rule does not apply and q(s) is an arbitrary probability distribution over $\{t_1,t_2\}$. Note that the probability that the type t_i (i=1,2) interest group-challenger combination sends signal s=0 is given by $1-\delta(t_i)-\varepsilon(t_i)+\delta(t_i)(1-\gamma(t_i))=1-\gamma(t_i)\delta(t_i)-\varepsilon(t_i)$.

In Proposition 4.5 three types of equilibria are distinguished besides the no expenditures equilibria: (i) contributions equilibria, (ii) direct endorsements equilibria, and (iii) both contributions and direct endorsements equilibria. The first type of equilibria satisfy $\varepsilon(t_1)=\varepsilon(t_2)=0$ and $\delta(t_i)>0$ for some i, the second $\delta(t_1)=\delta(t_2)=0$ and $\varepsilon(t_i)>0$ for some i, and the third type of equilibria satisfy $\delta(t_i)>0$ and $\varepsilon(t_j)>0$ for some $i,j\in\{1,2\}$. In a contributions equilibrium $q(c_{end})$ cannot be determined by Bayes' rule and we may choose $q(c_{end})\in[0,1]$. By choosing $q(c_{end})=0$, and thus $\phi(c_{end})=0$, we do not affect the equilibrium paths with $\varepsilon(t_i)=0$ for i=1,2. The intuition is that by choosing $q(c_{end})$ in such a way we do not affect the interest group's choice between contributing and not-contributing, since a direct endorsement is always worse than no political expenditures when it is feasible (i.e. when $g_i>0$). Generally, without loss of interesting generality we always choose q(s)=0 for out-of-equilibrium signal s, such that choices between equilibrium signals are not affected.

The proof of Proposition 4.5 takes several steps. We first present a lemma (Lemma 4.1) and its corollary (Corollary 4.4) that will appear useful in the process of proving the proposition. Then we derive the IRCO condition and the contributions equilibria (Lemmas 4.2 and 4.3), the IRDE condition and the endorsements equilibria (Lemma 4.4 and 4.5), and the combined equilibria (Lemma 4.6). With these lemmas Proposition 4.5 easily follows. Finally, a proof of Proposition 4.6 is provided.

Lemma 4.1
In any equilibrium it holds that:
(a)   if $g_i<c$, then $\delta(t_i)=\varepsilon(t_i)=0$ (i=1,2).
(b) $w\gamma(t_2)\delta(t_2)\geq\gamma(t_1)\delta(t_1)$.
(c) $w\varepsilon(t_2)\geq\varepsilon(t_1)$.
(d) $w+\gamma(t_1)\delta(t_1)+\varepsilon(t_1)\leq1+w[\gamma(t_2)\delta(t_2)+\varepsilon(t_2)]$. $\square$

Proof of Lemma 4.1.
(a)   Let $g_i<0$. $\varepsilon(t_i)=0$ trivially follows by the assumption that a direct endorsement is not possible (equilibrium condition (e1.3)). Suppose $\delta(t_i)>0$. This requires $[\phi(c_{cam})-\phi(0)]<0$. Now $[\phi(c_{cam})-\phi(0)]<0 \Rightarrow \gamma(t_i)=0 \Rightarrow \delta(t_i)=0$, contradicting $\delta(t_i)>0$. Next, let $0<g_i<c$. The maximal potential gain

from contributing or endorsing is g_i, strictly lower than the costs c. Trivially then $\delta(t_i)=\varepsilon(t_i)=0$.

(b)   When $\delta(t_1)=\delta(t_2)=0$ the weak inequality trivially holds, so suppose $\delta(t_i)>0$ for some $i \in \{1,2\}$. From equilibrium condition (e1.2) $\delta(t_i)>0$ only when $\gamma(t_i)>0$, and the latter only when $\phi(c_{cam})>0$. Hence $\phi(c_{cam})>0$. This requires $q(c_{cam})\geq v$ in turn. But, $q(c_{cam})\geq v \Leftrightarrow w\gamma(t_2)\delta(t_2)\geq\gamma(t_1)\delta(t_1)$ whenever (4.5) applies for $q(c_{cam})$, i.e. whenever $\delta(t_i)>0$ for some i.

(c)   When $\varepsilon(t_1)=\varepsilon(t_2)=0$ the weak inequality trivially holds, so suppose $\varepsilon(t_i)>0$ for some $i \in \{1,2\}$. $\varepsilon(t_i)>0 \Rightarrow \phi(c_{end})>0 \Rightarrow q(c_{end})\geq v$. From $q(c_{end})\geq v$, and the fact that (4.6) applies when $\varepsilon(t_i)>0$, we obtain $w\varepsilon(t_2)\geq\varepsilon(t_1)$.

(d)   When $\delta(t_1)+\varepsilon(t_1)=\delta(t_1)+\varepsilon(t_2)=0$ the weak inequality trivially holds, so suppose $\delta(t_i)+\varepsilon(t_i)>0$ for some $i \in \{1,2\}$. By part (a) $\delta(t_i)+\varepsilon(t_i)>0$ only when $g_i>0$, and thus $\delta(t_i)+\varepsilon(t_i)>0$ implies $\phi(0)<1$. This requires $q(0)\leq v$ in turn. When (4.4) does not apply for $q(0)$, i.e. when $\gamma(ti)\delta(t_i)+\varepsilon(t_i)=1$ for i=1,2, (d) holds (with equality). When (4.4) does apply for $q(0)$, (d) follows by $q(0)\leq v \Leftrightarrow w+\gamma(t_1)\delta(t_1)+\varepsilon(t_1)\leq 1+w[\gamma(t_2)\delta(t_2)+\varepsilon(t_2)]$.   *QED.*

Lemma 4.1(a) shows that when the interest group prefers the incumbent ($g_i<0$), it will never contribute to the challenger's campaign ($\delta(t_i)=0$). Combining the three remaining parts we obtain the following corollary of Lemma 4.1.

Corollary 4.4
In all equilibria necessarily holds $\gamma(t_2)\delta(t_2)+\varepsilon(t_2)\geq\gamma(t_1)\delta(t_1)+\varepsilon(t_1)$.   □

The probability that the good type combination sends a costly signal to the voter is weakly larger than the probability that the bad type combination sends such a signal.

Lemma 4.2
Consider only equilibria in which $\varepsilon(t_1)=\varepsilon(t_2)=0$. A necessary condition for information transfer to occur is $\max\{c, \min\{u_1,g_1\}\}<\min\{u_2,g_2\}$ (IRCO).   □

Proof of Lemma 4.2.
Suppose $g_2<c$. Then $\delta(t_2)=0$ (cf. Lemma 4.1(a)). This implies $\delta(t_1)=0$ by Lemma 4.1(b). Next, suppose $0<u_2<c$. Then $\gamma(t_2)=0$ by equilibrium condition (e2), implying $\delta(t_2)=0$ by equilibrium condition (e1), implying $\delta(t_1)=0$ in turn. So, when $c>\min\{u_2,g_2\}$ we have $\delta(t_2)=\delta(t_1)=0$, and information revelation cannot occur.
        From Corollary 4.4 follows that in any informative contributions equilibrium $\gamma(t_2)\delta(t_2)>\gamma(t_1)\delta(t_1)$. Suppose $\min\{u_1,g_1\}>u_2$. Then $\gamma(t_2)>0 \Rightarrow \gamma(t_1)=1 \Rightarrow \delta(t_1)=1$, implying $\gamma(t_2)=\delta(t_2)=1$ by Corollary 4.4. Information transfer cannot occur. Next, suppose $\min\{u_1,g_1\}>g_2$. Then $\delta(t_2)>0$ requires $\phi(c_{cam})-\phi(0)\geq c/g_2$, implying $\phi(c_{cam})-\phi(0)>c/u_1$. This yields $\gamma(t_1)=1$, and $\delta(t_1)=1$ in turn. By Corollary

4.4 information transfer is not possible. A necessary condition is thus $\min\{u_1,g_1\}\leq\min\{u_2,g_2\}$.

Taking both pieces together we get $\max\{c, \min\{u_1,g_1\}\}\leq\min\{u_2,g_2\}$, and the strict inequality by the assumption of no knife-edge cases (NKE). *QED.*

Lemma 4.3
Equilibria PE3, HE1 through HE6, and SE1 are the only contributions equilibria. □

Proof of Lemma 4.3.
We first consider informative equilibria, i.e. $\gamma(t_1)\delta(t_1)\neq\gamma(t_2)\delta(t_2)$. From Lemma 4.2 follows that necessarily IRCO must hold for information revelation to be possible, so assume this condition to hold. Suppose $\gamma(t_1)\delta(t_1)>0$. Then $\phi(c_{cam})-\phi(0)\geq c/\min\{u_1,g_1\}$, implying $\gamma(t_2)\delta(t_2)=1$ in turn. From the implication $\gamma(t_1)\delta(t_1)>0 \Rightarrow \gamma(t_2)\delta(t_2)=1$ follows that three cases have to be considered.

First, suppose $0<\gamma(t_1)\delta(t_1)<1$ and $\gamma(t_2)\delta(t_2)=1$. This requires $c<\min\{u_1,g_1\}$ besides IRCO. By (4.4) $q(0)=0$, implying $\phi(0)=0$. This requires $\phi(c_{cam})<1$, for otherwise $\gamma(t_1)=1$ and $\delta(t_1)=1$ (due to $c<\min\{u_1,g_1\}$), and $\phi(c_{cam})>0$. Hence $q(c_{cam})=v$, and thus $\gamma(t_1)\delta(t_1)=w<1$. By $q(c_{cam})>p$ this is only possible when $p<v$. From the implication $\gamma(t_1)=1 \Rightarrow \delta(t_1)=1$ when $u_1<g_1$ follows that necessarily $\gamma(t_1)<1$ when $u_1<g_1$. In that case $\phi(c_{cam})=1-c/u_1$. Either $\delta(t_1)=1$ or $\delta(t_1)<1$. When $\delta(t_1)=1$ necessarily $\gamma(t_1)=w$. To sustain $\delta(t_1)=1$ then, $wg_1>u_1$ is required (HE1). When $\delta(t_1)<1$ necessarily $\gamma(t_1)=u_1/g_1$ by the incentives of the interest group, yielding $\delta(t_1)=wg_1/u_1$ (HE2). In case $g_1<u_1$ we have by the implication $\delta(t_1)>0 \Rightarrow \gamma(t_1)=1$ that necessarily $\gamma(t_1)=1$ and $\delta(t_1)=w$ (HE3).

Second, suppose $\gamma(t_1)\delta(t_1)=0$ and $0<\gamma(t_2)\delta(t_2)<1$. By (4.5) $q(c_{cam})=1$, implying $\phi(c_{cam})=1$. Hence $0<\phi(0)<1$ by $\gamma(t_2)\delta(t_2)<1$, implying $q(0)=v$. By $q(0)<p$ this is only possible when $p>v$. From (4.4) $q(0)=v$ implies $\gamma(t_1)\delta(t_1)=(1-w)+w\gamma(t_2)\delta(t_2)$. From $\gamma(t_1)\delta(t_1)=0$ follows $\gamma(t_2)\delta(t_2)=1-1/w$. When $u_2<g_2$ we necessarily must have, from $\gamma(t_2)=1 \Rightarrow \delta(t_2)=1$ when $u_2<g_2$, that $\gamma(t_2)<1$ and thus $\phi(0)=1-c/u_2$. Note that $\gamma(t_1)=1$ (=0) when $u_1>u_2$ ($u_1<u_2$). Suppose $\delta(t_2)=1$. This is only possible when $(1-1/w)g_2\geq u_2$ (HE4). On the other hand $\delta(t_2)<1$ requires $\gamma(t_2)=u_2/g_2$ and $\delta(t_2)=(1-1/w)g_2/u_2$. By $\delta(t_2)\leq1$ this is only possible when $(1-1/w)g_2\leq u_2$ (HE5). When $u_2>g_2$ we must have, from the implication $\delta(t_2)>0 \Rightarrow \gamma(t_2)=1$ when $g_2<u_2$, that $\phi(0)=1-c/g_2$, $\gamma(t_2)=1$, and $\delta(t_2)=(1-1/w)$. Now $\delta(t_1)=0$ and $\gamma(t_1)=1$ (=0) when $g_2<u_1$ ($g_2>u_1$), yielding HE6.

Third, $\gamma(t_1)\delta(t_1)=0$ and $\gamma(t_2)\delta(t_2)=1$. Then $q(0)=0$ and $q(c_{cam})=1$ by (4.4) and (4.5), implying $\phi(0)=0$ and $\phi(c_{cam})=1$. This implies $\gamma(t_1)=1$ (=0) when $c<u_1$ ($c>u_1$). When $c<\min\{u_1,g_1\}$ we get $\gamma(t_1)\delta(t_1)=1$. Hence we must have $\min\{u_1,g_1\}<c$ (SE1).

Finally, consider non-informative (pooling) contributions equilibria, i.e. $\gamma(t_1)\delta(t_1)=\gamma(t_2)\delta(t_2)>0$. Suppose $\gamma(t_1)\delta(t_1)=\gamma(t_2)\delta(t_2)<1$. By (4.4) and (4.5) $q(0)=q(c_{cam})=p$, and by $p\neq v$ $\phi(c_{cam})=\phi(0)$. This implies $\gamma(t_i)=0$ for $i=1,2$, contradicting $\gamma(t_i)\delta(t_i)>0$ for $i=1,2$. Hence only $\gamma(t_1)\delta(t_1)=\gamma(t_2)\delta(t_2)=1$ is possible.

This requires $c < u_1, u_2, g_1, g_2$. By choosing out-of-equilibrium belief $q(0)=0$ such a pooling equilibrium is sustainable (PE3).   *QED.*

Lemma 4.4
Consider only equilibria in which $\delta(t_1)=\delta(t_2)=0$. A necessary condition for information transfer to occur is $\max\{c,g_1\}<g_2$ (IRDE).   □

Proof of Lemma 4.4.
When $g_2 < c$ we have $\varepsilon(t_2)=0$ (cf. Lemma 4.1(a)). By Lemma 4.1(c) follows $\varepsilon(t_1)=0$, and no information transfer occurs. So, $c > g_2$ is necessary. Next suppose $g_1 > g_2$. From Corollary 4.4 follows $\varepsilon(t_2) > \varepsilon(t_1)$ in any informative direct endorsements equilibrium. Now $\varepsilon(t_2)>0$ requires $\phi(c_{end})-\phi(0)\geq c/g_2$. When $g_1>g_2$ we thus have $\phi(c_{end})-\phi(0)>c/g_1$, implying $\varepsilon(t_1)=1$. An informative equilibrium is not possible. Together with NKE condition IRDE follows.   *QED.*

Lemma 4.5
Equilibria PE4, HE7, HE8 and SE2 are the only direct endorsements equilibria.
                                                                                      □

Proof of Lemma 4.5.
We first consider informative equilibria, so assume IRDE to hold. Suppose $\varepsilon(t_1)>0$. Then necessarily $\phi(c_{end})-\phi(0)\geq c/g_1$. By $g_1<g_2$ we have $\phi(c_{end})-\phi(0)>c/g_2$, implying $\varepsilon(t_2)=1$. Hence under IRDE the implication $\varepsilon(t_1)>0 \Rightarrow \varepsilon(t_2)=1$ holds. Three cases have to be considered.

First, $0<\varepsilon(t_1)<1$ and $\varepsilon(t_2)=1$. This requires $c<g_1$ besides IRDE. By (4.4) $q(0)=0$, implying $\phi(0)=0$. This requires $\phi(c_{end})<1$, for otherwise $\varepsilon(t_1)=1$ (due to $c<g_1$), and $\phi(c_{end})>0$. Hence $q(c_{end})=v$, and thus $\varepsilon(t_1)=w<1$ by (4.6). By $q(c_{end})>p$ this is only possible when $p<v$. Now $0<\varepsilon(t_1)<1$ requires $\phi(c_{end})=c/g_1$ (HE7).

Second, $\varepsilon(t_1)=0$ and $0<\varepsilon(t_2)<1$. By (4.6) $q(c_{end})=1$, implying $\phi(c_{end})=1$. Hence $0<\phi(0)<1$, implying $q(0)=v$. By $q(0)<p$ this is only possible when $p>v$. From (4.4) $q(0)=v$ follows $\varepsilon(t_1)=(1-w)+w\varepsilon(t_2)$, thus $\varepsilon(t_2)=1-1/w$. Now $0<\varepsilon(t_2)<1$ requires $\phi(0)=1-c/g_2$ (HE8).

Third, $\varepsilon(t_1)=0$ and $\varepsilon(t_2)=1$. Then $q(0)=0$ and $q(c_{end})=1$ by (4.4) and (4.6), implying $\phi(0)=0$ and $\phi(c_{end})=1$. When $c<g_1$ we get $\varepsilon(t_1)=1$. Hence we must have $g_1<c$ (SE2).

Finally, consider non-informative (pooling) direct endorsements equilibria, i.e. $\varepsilon(t_1)=\varepsilon(t_2)>0$. Suppose $\varepsilon(t_1)=\varepsilon(t_2)<1$. By (4.4) and (4.6) $q(0)=q(c_{end})=p$, and by $p\neq v$ $\phi(c_{end})=\phi(0)$. This implies $\varepsilon(t_i)=0$ for i=1,2, contradicting $\varepsilon(t_i)>0$ for i=1,2. Hence only $\varepsilon(t_1)=\varepsilon(t_2)=1$ is possible. This requires $c<g_1,g_2$. By choosing out-of-equilibrium belief $q(0)=0$ such a pooling equilibrium is sustainable (PE4). *QED.*

Lemma 4.6
Equilibria PE5, HE9, HE10, HE11 and SE3 are the only both contributions and direct endorsements equilibria.   □

Proof of Lemma 4.6.

From Lemma 4.1(b) and (c) follows that necessarily $\delta(t_2)>0$ and $\varepsilon(t_2)>0$. Using equilibrium condition (e1) the following equality is obtained:

$$\gamma(t_2)\phi(c_{cam})+(1-\gamma(t_2))\phi(0)=\phi(c_{end}) \tag{4.7}$$

Now $\delta(t_2)>0$ requires $\gamma(t_2)>0$, and hence (4.5) and (4.6) both apply. $\gamma(t_2)>0$ requires $\phi(c_{cam})>\phi(0)$ in turn. By (4.7) we then have $0<\phi(c_{end})\leq\phi(c_{cam})$. From Lemma 4.1(b) and (c) (and (4.4)) follows $q(0)<p$ when $w<1$, i.e. when $p<v$. So, $\phi(0)=0$ when $p<v$. Similarly, from Corollary 4.4 follows either $\phi(c_{cam})=1$ or $\phi(c_{end})=1$ (or both) when $p>v$. With $\phi(c_{cam})\geq\phi(c_{end})$ follows $\phi(c_{cam})=1$ when $p>v$. We consider the two cases $0<\phi(c_{end})<1$ and $\phi(c_{end})=1$ separately, for the cases $p<v$ and $p>v$, respectively.

First, suppose $0<\phi(c_{end})<1$. Then $q(c_{end})=v$ and thus $\varepsilon(t_1)=w\varepsilon(t_2)>0$. Now consider the case $p<v$ first. We have $\phi(0)=0$, implying $\gamma(t_2)\phi(c_{cam})=\phi(c_{end})$ by (4.7). Suppose $\delta(t_1)=0$. We get $\phi(c_{end})=c/g_1<1$ by $0<\varepsilon(t_1)<1$, and $\phi(c_{cam})=1$ by $q(c_{cam})=1$. But then $\gamma(t_2)=c/g_1$. This requires $\phi(c_{cam})=c/u_2$ by equilibrium condition (e2), and $c/u_2\neq1$ by NKE. Hence necessarily $\delta(t_1)>0$. Now $\delta(t_1)>0$ and $\varepsilon(t_1)>0$ requires $\gamma(t_1)\phi(c_{cam})=\phi(c_{end})$ (equality similar to (4.7), but now for the bad type). By $u_1\neq u_2$ (NKE) we get $\gamma(t_1)=\gamma(t_2)=1$. Thus $c<u_1$ is required. Now $0<\phi(c_{cam})=\phi(c_{end})<1$ implies $w\delta(t_2)=\delta(t_1)$ by (4.5). This yields $\delta(t_1)+\varepsilon(t_1)<1$ and $\phi(c_{end})=c/g_1$, requiring $c<g_1$. But $\delta(t_2)>0$ only when $\phi(c_{cam})>c/g_2$, hence necessarily $g_1<g_2$. Similarly, $\gamma(t_i)=1$ requires $g_i<u_i$ for $i=1,2$ (HE9). Next, let $p>v$. Then $\phi(c_{cam})=1$ and thus necessarily $0<\gamma(t_2)<1$ by (4.7). Now $\phi(c_{cam})=1$ together with $0<\gamma(t_2)<1$ requires by equilibrium condition (e2) that $\phi(0)=1-c/u_2$. By $u_1\neq u_2$ either $\gamma(t_1)=0$ or $\gamma(t_1)=1$. Suppose $\gamma(t_1)=1$. By $\phi(c_{end})<\phi(c_{cam})=1$ we get $\varepsilon(t_1)=0$, a contradiction. Hence necessarily $\gamma(t_1)=0$, requiring $u_1<u_2$. By $\delta(t_2)>0$ and $\gamma(t_2)<1$ (4.4) applies, and we must have $\varepsilon(t_1)<1$, for otherwise $q(0)=1$ and $\phi(0)=1$. Now $\delta(t_1)=0$, $0<\varepsilon(t_1)<1$, and $\phi(0)=1-c/u_2$ together require $\phi(c_{end})=1-c/u_2+c/g_1$ by equilibrium condition (e1), and thus $u_2<g_1$. In order to make the type t_2 indifferent between donating and endorsing we need $\gamma(t_2)=u_2/g_1$ (cf. (4.7)). $\delta(t_2)$ follows from $q(0)=v$, (4.4), $\gamma(t_1)=0$ and $w\varepsilon(t_2)=\varepsilon(t_1)$. We get $\delta(t_2)=(1-1/w)u_2/g_1$, requiring $(1-1/w)g_1<u_2$. Moreover, type t_2 must prefer $s=c_{end}$ above $s=0$ $\Rightarrow$ $g_1<g_2$ is needed. We get $\varepsilon(t_2)=1-\delta(t_2)$ and $\varepsilon(t_1)=w\varepsilon(t_2)$. This yields HE11.

Second, suppose $\phi(c_{end})=1$. From (4.7) necessarily $\phi(c_{cam})=1$ and $\gamma(t_2)=1$. Consider again first the case $p<v$. When $p<v$ we have $\phi(0)=0$ and thus $\delta(t_2)+\varepsilon(t_2)=1$. When $u_1<c$ we get $\gamma(t_1)=0$, in case $u_1>c$ we get $\gamma(t_1)=1$. Now $\gamma(t_1)=0$ $\Rightarrow$ $\delta(t_1)=0$ and either $\varepsilon(t_1)=0$ ($g_1<c$) or $\varepsilon(t_1)=1$ ($g_1>c$). The latter case is inconsistent with $\varepsilon(t_2)>\varepsilon(t_1)$ from Lemma 4.1(c). When $\gamma(t_1)=1$ we either get $\delta(t_1)+\varepsilon(t_1)=0$ ($g_1<c$) or $\delta(t_1)+\varepsilon(t_1)=1$ ($g_1>c$). Again, the second case is inconsistent with the combination of Lemma 4.1(b) and (c). When $g_1<c$ the first case applies and we get SE3. Next, let $p>v$. When $0<\delta(t_2)+\varepsilon(t_2)<1$ we have $\phi(0)=1-c/g_2$ by equilibrium condition (e1). By NKE either $\delta(t_1)+\varepsilon(t_1)=1$ or $\delta(t_1)+\varepsilon(t_1)=0$. The first case is inconsistent with Corollary 4.4. Hence $\delta(t_1)=\varepsilon(t_1)=0$, and $\delta(t_2)+\varepsilon(t_2)=(1-1/w)$ from (4.4) and $q(0)=v$. To secure $\gamma(t_2)=1$ for $\phi(c_{cam})-\phi(0)=c/g_2$ we must require $u_2>g_2$ (HE10). Now let

$\delta(t_2)+\epsilon(t_2)=1$. In case $0<\delta(t_1)+\epsilon(t_1)<1$ we get $q(0)=0$ from (4.4), implying $\phi(0)=0$. By NKE and equilibrium condition (e1) we either get $\delta(t_1)+\epsilon(t_1)=0$ or $\delta(t_1)+\epsilon(t_1)=1$, contradicting $0<\delta(t_1)+\epsilon(t_1)<1$. Hence we must either have $\delta(t_1)+\epsilon(t_1)=0$ or $\delta(t_1)+\epsilon(t_1)=1$. The first case requires $g_1<c$ and yields SE3. The second case requires $g_1>c$ and yields PE5.   *QED.*

Proof Proposition 4.5.
It is easily checked that pooling on s=0 (PE1 and PE2) always constitutes an equilibrium; just choose out-of-equilibrium beliefs equal to the prior such that $q(c_{cam})=q(c_{end})=p$, and thus $\phi(c_{cam})=\phi(c_{end})=\phi(0)$ (note that $p\neq v$ by assumption). (Alternatively, choose $q(c_{cam})=q(c_{end})=0$.) The proposition now follows from Lemmas 4.3, 4.5, and 4.6.   *QED.*

Proof of Proposition 4.6.
(a) From Proposition 4.5 follows that when $p<v$ signal s=0 is never an out-of-equilibrium signal, so we only have to consider out-of-equilibrium signals $s=c_{cam}$ and $s=c_{end}$. In this case the CFIEP concept for the game considered here can be defined as follows. We say that an equilibrium path $E^{\#}$ *deletes* another equilibrium path E iff for some signal $s\in\{c_{cam},c_{end}\}$ not occurring in E, but occurring in $E^{\#}$ with positive probability, either (A) or (B) holds:

(A) let $s=c_{cam}$. For the set $DEV(c_{cam}) = \{t\in\{t_1,t_2\}\mid c_{cam}$ is sent in $E^{\#}$ by interest group-challenger combination of type t, and *both* agents in this combination get a strictly higher utility in $E^{\#}$ than in E} (i), (ii) and (iii) hold:
    (i) $DEV(c_{cam})$ is non-empty.
    (ii) For the at most single type of combination not in $DEV(c_{cam})$ holds that at least one agent in the combination weakly prefers the equilibrium payoff in E over sending c_{cam}, assuming that c_{cam} is interpreted in the same way as in $E^{\#}$.
    (iii) Given that only combination types in $DEV(c_{cam})$ have sent signal c_{cam} and have sent c_{cam} according to the equilibrium strategies in $E^{\#}$, there exists a best response of the voter to c_{cam} such that all agents in $DEV(c_{cam})$ weakly prefer to send c_{cam} over following path E.

(B) let $s=c_{end}$. For the set $DEV(c_{end}) = \{t\in\{t_1,t_2\}\mid c_{end}$ is sent in $E^{\#}$ by interest group of type t, and this type of interest group gets a strictly higher utility in $E^{\#}$ than in E} (i), (ii) and (iii) hold:
    (i) $DEV(c_{end})$ is non-empty.
    (ii) For the at most single type of interest group not in $DEV(c_{end})$ holds that it weakly prefers the equilibrium payoff in E over sending c_{end}, assuming that c_{end} is interpreted in the same way as in $E^{\#}$ (assuming that sending c_{end} is not excluded for this type of interest group).
    (iii) Given that only types in $DEV(c_{end})$ have sent signal c_{end} and have sent c_{end} according to the equilibrium strategies in $E^{\#}$, there is a best response

to c_{end} such that all interest group types in $DEV(c_{end})$ weakly prefer to send c_{end}, rather than following the equilibrium path E.

Now, a PBE path is a CFIEP if it is *not* deleted by any other PBE path. Though notationally rather involved, the intuition behind the refinement and its actual application is quite simple. Loosely speaking, out-of-equilibrium signals are interpreted as proposals to follow another PBE path. When such a proposal is internally consistent, the original PBE path is deleted.

In regard to the actual application of the CFIEP concept in the context of the game considered here, the following observations are extremely useful. If $s=s_{cam}$ is not sent in E and there exists, for the very same parameter values, a contributions equilibrium $E^{\#}$ in which the type t_2 challenger and the type t_2 interest group both get a higher expected payoff, then E is not CFIEP. This follows from (A) above because in that case $t_2 \in DEV(c_{cam})$. Now, when $t_1 \in DEV(c_{cam})$ the response of the voter $\phi(c_{cam})$ as in equilibrium path $E^{\#}$ satisfies (Aiii), whereas in case $t_1 \notin DEV(c_{cam})$ response $\phi(c_{cam})=1$ yields the type t_2 combination at least the same as in $E^{\#}$. A similar reasoning with respect to $s=c_{end}$ yields that, If $s=s_{end}$ is not sent in E and there exists, for the very same parameter values, a direct endorsements equilibrium $E^{\#}$ in which the type t_2 interest group gets a higher expected payoff, then E is not CFIEP. The observation that $t_2 \in DEV(c_{cam})$ or $t_2 \in DEV(c_{end})$ implies that E is not CFIEP will be used below when applying the concept.

The universal divinity (UD) refinement for the game considered here comes down to the following. Let E be an equilibrium in which either $s=c_{cam}$ or $s=c_{end}$ is not sent in equilibrium, and let $U_G^*(t_i|E)$ be the expected payoff an interest group of type t_i gets in E. Suppose $s=c_{end}$ is an out-of-equilibrium signal in E. Define $RG_i(c_{end})=\{\phi(c_{end})\in[0,1] \,|\, g_i\phi(c_{end})-c \geq U_G^*(t_i|E)\}$, i.e. $RG_i(c_{end})$ denotes the set of (best) responses to c_{end} for which the type t_i interest group has a (weak) incentive to deviate from its equilibrium strategy to c_{end}. If $RG_1(c_{end}) \subset RG_2(c_{end})$ the good type interest group has an incentive to deviate whenever the bad type has, but not vice versa; UD requires $q(c_{end})=1$ in this case. Similarly, when $RG_2(c_{end}) \subset RG_1(c_{end})$ UD requires $q(c_{end})=0$. If neither case applies, the refinement does not put restrictions on $q(c_{end})$.

When $s=c_{cam}$ is out of equilibrium in E we make use of the fact that in case $p<v$ we have $\phi(0)=0$ in every PBE, thus also in E, when defining UD. Let $RG_i(c_{cam})=\{\phi(c_{cam})\in[0,1] \,|\, g_i\phi(c_{cam})-c \geq U_G^*(t_i|E)\}$ and $RCh_i(c_{cam})=\{\phi(c_{cam})\in[0,1] \,|\, u_i\phi(c_{cam})-c \geq U_{Ch}^*(t_i|E)\}$. Now if $(RCh_1(c_{cam}) \cap RG_1(c_{cam})) \subset (RCh_2(c_{cam}) \cap RG_2(c_{cam}))$ the good type combination has an incentive to deviate whenever the bad type combination has, but not vice versa; UD requires $q(c_{cam})=1$. Similarly, when $(RCh_2(c_{cam}) \cap RG_2(c_{cam})) \subset (RCh_1(c_{cam}) \cap RG_1(c_{cam}))$ UD requires $q(c_{cam})=0$. The idea behind this latter restriction is that for the type t_i combination to deviate to $s=c_{cam}$ the interest group of type t_i must deviate to contributing. In order to make such a deviation profitable, the interest group must think that $\phi(c_{cam})$ is such that the challenger will spend the funds unexpectedly received with sufficiently high probability (i.e. it has to believe that $\phi(c_{cam}) \in RCh_i(c_{cam})$), and it must find a

deviation profitable itself when the challenger spends all the money received ($\phi(c_{cam}) \in RG_i(c_{cam})$). Hence only when $\phi(c_{cam}) \in (RCh_i(c_{cam}) \cap RG_i(c_{cam}))$ a deviation is (weakly) profitable for the type t_i combination, and the standard requirement of UD can be used.

With the extended definitions at hand, we are able to proof Proposition 4.6(a). First we consider PE1. Suppose SE1 exists. Then signal $s=c_{cam}$ is dominated by signal $s=0$ for the bad type combination, and thus a clear signal of being of the good type (which indeed has an incentive to deviate for some best response of the voter). To apply CFIEP note that SE1 deletes PE1; c_{cam} is sent in SE1, $DEV(c_{cam})=\{t_2\}$ and thus non-empty, by $\min\{u_1, g_1\}<c$ there is at least one agent in the combination of type t_1 who does not want to spend c — and thus to send c_{cam} —, and $q(c_{cam})=\phi(c_{cam})=1$ indeed induces the good type combination to send $s=c_{cam}$. With respect to UD it easily follows that $RCh_1(c_{cam}) \cap RG_1(c_{cam})=\varnothing$ and $RCh_2(c_{cam}) \cap RG_2(c_{cam})=[c/\min\{u_2, g_2\}, 1]$, implying $q(c_{cam})=1$ by $c<\min\{u_2, g_2\}$, and upsetting PE1. Similarly, when SE2 and/or SE3 exists $s=c_{end}$ is dominated for the bad type interest group, but not for the good type interest group. UD requires $q(c_{end})=\phi(c_{end})=1$, and the good type interest group wants to deviate. In the context of CFIEP both SE2 and SE3 directly delete PE1 by realizing that $DEV(c_{end})=\{t_2\}$.

For a hybrid equilibrium to exist when $p<v$ we need $c<g_1, g_2$. Suppose a hybrid equilibrium exists and $g_1<g_2$ as well. With respect to UD note that when $g_1<g_2$ we have $RG_1(c_{end})=[c/g_1, 1] \subset [c/g_2, 1]=RG_2(c_{end})$, implying $q(c_{end})=1$ and upsetting PE1. CFIEP deletes PE1 in this case because the equilibrium is deleted by HE7 (cf. Table 4.2, $DEV(c_{end})=\{t_2\}$), which exists when $g_1<g_2$. Next, consider the case $g_1>g_2$, and suppose either HE1 or HE2 exists. From Table 4.2 it is easily seen that HE1 and HE2 delete PE1 in the context of CFIEP, because $t_2 \in DEV(c_{cam})$. In respect to UD we get $RCh_1(c_{cam}) \cap RG_1(c_{cam})=[c/\min\{u_1, g_1\}, 1]$ and $RCh_2(c_{cam}) \cap RG_2(c_{cam})=[c/\min\{u_2, g_2\}, 1]$, implying $q(c_{cam})=1$ by the IRCO condition, and upsetting PE1. In summary, PE1 is deleted by both refinements whenever information revelation is possible in equilibrium.

Equilibrium HE7 is deleted by both CFIEP and UD when either SE1, HE1 or HE2 exists at the same time. Note that when HE7 exists $c<g_1<g_2$. With respect to CFIEP note that SE1 and HE2 delete HE7 by $DEV(c_{cam})=\{t_2\}$ (cf. Table 4.2). HE1 directly deletes HE7 in the context of CFIEP by realizing that $DEV(c_{cam})=\{t_1, t_2\}$ in that case, and that the best response of the voter as in HE1 induces both types of combination to deviate. Applying UD with respect to HE7 we get $RCh_1(c_{cam}) \cap RG_1(c_{cam})=[c/\min\{u_1, g_1\}, 1]$, $RCh_2(c_{cam})=[c/u_2, 1]$, and $RG_2(c_{cam})=[c/g_1, 1]$. Hence $RCh_2(c_{cam}) \cap RG_2(c_{cam})=[c/\min\{u_2, g_1\}, 1]$. By $\min\{u_1, g_1\}<\min\{u_2, g_1\}$ whenever SE1, HE1, or HE2 exists side by side with HE7, we must have $q(c_{cam})=\phi(c_{cam})=1$, upsetting the equilibrium. From the above reasoning it also follows that SE1, SE2, HE1 and HE2 are never deleted in the context of CFIEP. With respect to UD it holds for HE1 that $RG_1(c_{end})=[cw/u_1, 1]$ and $RG_2(c_{end})=[c/u_1, 1]$, for HE2 that $RG_1=[c/g_1, 1]$ and $RG_2(c_{end})=[c/u_1, 1]$. In both cases $RG_2(c_{end}) \subset RG_1(c_{end})$, thus $q(c_{end})=\phi(c_{end})=0$, and UD does not affect HE1 and HE2. In regard to SE1 (SE2) it suffices to note that $RG_2(c_{end})=\varnothing$ ($RG_2(c_{cam})=\varnothing$).

Finally, when HE3 exists there does not exist a PBE which yields the type t_2 combination strictly more, so HE3 is not deleted by CFIEP. For UD, we have $RG_1(c_{end})=[c/g_1,1]=RG_2(c_{end})$, and HE3 is not affected.

(b) When $p>v$ also equilibria exist, viz. PE3, PE4 and PE5, where $s=0$ constitutes an out-of-equilibrium message. The definition of CFIEP given in part (a) thus has to be extended in a straightforward way to cover $s=0$ as well. However, by realizing that the type t_2 interest group always gets g_2-c in these equilibria, and $g_2>g_2$-c in PE2, it follows directly that all the three equilibria PE3, PE4 and PE5 are deleted by PE2 under the CFIEP concept. A formal extension is thus omitted. To apply UD with respect to the pooling equilibria PE3, PE4, and PE5 we need the following two sets: $RCh_i(0)=\{\phi(0)\in[0,1]\,|\,u_i\phi(0)\geq U_{Ch}^*(t_i\,|\,E)\}$ and $RG_i(0)=\{\phi(0)\in[0,1]\,|\,g_i\phi(0)\geq U_G^*(t_i\,|\,E)\}$. First consider PE4. In this equilibrium a deviant signal must necessarily be initiated by the interest group, so we solely focus on the sets $RG_1(0)$ and $RG_2(0)$. Note that $RG_i(0)=[1-c/g_i,1]$. When $RG_1(0)\subset RG_2(0)$, i.e. IRDE does not hold, we get $q(0)=1$, upsetting PE4. When IRDE holds, UD does not affect PE4. Next consider PE3. Here a deviation to $s=0$ can be either initiated by the interest group, or by the challenger. We thus have to consider the sets $RCh_i(0)\cup RG_i(0)=[1-c/min\{u_i,g_i\},1]$ (in this case the union rather than the intersection has to be considered because now a deviation can be induced *separately* by either agent in the combination). Under IRCO we have $(RCh_2(0)\cup RG_2(0))\subset(RCh_1(0)\cup RG_1(0))$, implying $q(0)=0$ by UD, sustaining PE3. When IRCO does not hold, however, $(RCh_1(0)\cup RG_1(0))\subset(RCh_2(0)\cup RG_2(0))$, and PE3 is deleted. From the considerations with respect to PE3 and PE4 it follows that PE5 is also deleted when both IRDE and IRCO do *not* hold.

   Turning briefly to the equilibria in which $s=0$ is not an out-of-equilibrium signal, it follows directly from Table 4.2 that the payoffs for the good type interest group in PE1, HE6, HE8, SE1, and SE2 are such that it has never a strong incentive to send a costly (contributions or direct endorsements) out-of-equilibrium signal. Hence, these equilibria are certainly not deleted by CFIEP or UD. On the other hand, the remaining equilibria HE4 and HE5 are affected by applying these two refinements. With respect to CFIEP it follows from Table 4.2 that HE4 and HE5 are deleted by SE2 and HE8 when the latter equilibria exist ($t\in DEV(c_{end})$). In the context of UD we again make use of $RG_i(c_{end})$. First, consider HE4. We get $RG_1(c_{end})=[1-c(g_1-u_2)/g_1u_2,1]$ and $RG_2(c_{end})=[1-c/wu_2,1]$. Now $RG_1(c_{end})\subset RG_2(c_{end})$ when $g_1(1-1/w)<u_2$, upsetting the equilibrium for that case. In HE5 the incentives for the bad type are the same as in HE4, for the good type we have $RG_2(c_{end})=[1-c(g_2-u_2)/g_2u_2,1]$. Therefore, HE5 is deleted by UD when $g_1<g_2$.   QED.

NOTES

1. This chapter draws heavily from Potters, Sloof and Van Winden (1997).

2. Technically, this amounts to depriving the (type t_i) interest group for which $g_i < 0$ of the possibility of sending a costly endorsement message $m \in M$. When $g_i > -c$ for $i=1,2$ this restriction is immaterial because in that case the interest group will never endorse the incumbent; either the group prefers the challenger ($g_i > 0$), or the group prefers the incumbent, but an endorsement is too expensive ($-c < g_i < 0$).

3. Helsley and O'Sullivan (1994) and Lohmann (1993) also suggest that the size of campaign contributions can be informative to voters. Alternatively, the size of a candidate's campaign chest can also be informative to other potential candidates. Epstein and Zemskey (1995) focus on the latter role of campaign contributions. In their signaling model the incumbent employs strategic fundraising to deter strong challengers from entering the political fray. By raising a lot of funds the incumbent tries to convince potential challengers that he is of "high" quality, and thus very hard to beat in an election. As opposed to our setup, in the Epstein and Zemskey model contributors are not incorporated as explicit (strategic) actors. Interestingly, Box-Steffenmeier (1996) obtains from a duration analysis some empirical evidence that large war chests indeed scare off high quality challengers.

4. To the extent that campaign contributions are motivated by "quid pro quo" considerations, and recognized as such by voters, the informational value of contributions is likely to be attenuated (cf. Ferguson, 1976). Thus, our focus on the electoral motive of campaign contributions may somewhat overstate their informational role. Moreover, by taking also the influence motive into account a more balanced view can be obtained on the ultimate value (to the voters) of campaign contributions as a source of relevant information. In a setting like ours where campaigns are not directly informative but only have informational value through their costs this is illustrated in the recent models of Gerber (1996) and Prat (1997). They allow for the possibility of the explicit buying of favors through campaign contributions and find that the indirect costs to the voter of the politicians' supply of "quid pro quos" to interest groups may exceed the voters' informational benefits from political campaigns financed by these interest groups. In other words, although campaign contributions incorporate relevant information for the voters, in some instances they would be better off were this kind of information transmission not possible (see note 8 for an illustration in the context of the model presented in this section).

5. Campaign contributions may also be used directly for personal purposes. As reported in Lewis (1996, p. 3), since 1979 112 former members of Congress converted to personal use over $10 million in unspent campaign contributions. Relatedly, in their study of the 1979 Canadian federal election, Palda and Palda (1985) estimate an equation that relates candidates' campaign expenditures to the amount of contributions received (as well as other variables). The coefficient of the contributions variable was either insignificant (2SLS estimation) or significant (OLS) but very small (about .1), possibly indicating that not all contributions were used for campaigning.

6. With $e_i \equiv \min\{u_i, g_i\}$, where the l.h.s. of the identity refers to the notation employed in Chapter 3, Proposition 3.2 also applies for the contributions model when $\min\{u_1, g_1\} < 0$. For instance, the non-pooling equilibria E3 and E4 also exist when $\min\{u_1, g_1\} < 0$ (cf. Appendix 4.A). In case $\min\{u_2, g_2\} < c$ (and, thus, in cases II and IV) only pooling on no-campaigning can occur and either E2 or E5 applies, depending on the value of p relative to $d \equiv v$.

7. A pooling equilibrium no longer always exists when we employ an equilibrium refinement concept like universal divinity or CFIEP. In case $p < v$ both refinements delete the single pooling equilibrium whenever an informative (non-pooling) equilibrium exists. As a consequence, when using these refinements information revelation becomes more likely. As was the case for the basic signaling game of

Section 3.1, the pooling on contributions (and subsequent campaigning) equilibrium is at some instances deleted by the two refinements, weakening the prediction that donations may constitute a pure social waste. A complete discussion of the consequences of the two equilibrium refinements is given in Appendix 4.A.

8. This conclusion does not hold any longer when receiving a campaign message bears a cost to the voter (cf. Chapter 3). To illustrate this, consider the contributions model under the alternative assumption that, when the interest group does not contribute it does not get any payoff at all, irrespective of which candidate is elected, whereas when the group of type t_i contributes and the challenger is elected, the interest group's gross payoff g_i is paid for by the voter (such that the voter gets $-g_1$ when the interest group appears to be of the bad type and v_2-g_2 when the interest group appears to be of the good type). In such an alternative setting contributions effectively serve as "quid pro quos", where the favors supplied by the challenger to the donating interest group when elected are paid for by the voter (cf. Gerber, 1996, Prat, 1997). For this case the separating equilibrium (SE1 in Proposition 4.5 in Appendix 4.A) still exists whenever $g_2 < v_2$. The voter then gets an expected payoff of $(1-p)v_1 + p(v_2-g_2)$ in this separating equilibrium, instead of $(1-p)v_1 + pv_2$ as in the original setup of the model. Were campaigning not possible, the voter would base her decision on a priori information and get an expected utility of $\max\{(1-p)v_1, pv_2\}$. Now, when $\{(1-p)/p\}v_1 < g_2 < v_2$ the latter exceeds the former. So, in the alternative setup of the model a separating equilibrium exists in which the voter is worse off compared with the situation where campaigning is not possible. More generally, taking also the possibility of favors buying through campaign contributions into account, a less optimistic view is obtained about the (informational) value of campaign contributions to the general voting public.

9. Again, we want to emphasis that this problem is not caused by the fact that in the text we consider the contributions and endorsements models in isolation. For every specific choice of the parameters, every equilibrium path described in an equilibrium of one of the two separate models can be sustained as an equilibrium in the more general model in which the choice between endorsing and contributing is truly endogenous (see Appendix 4.A).

10. In this respect the assumption that endorsing and contributing are equally costly is also rather restrictive. When we drop this assumption it may appear to be easier to construct a strict preference ordering. However, such a preference ordering would not only be based on differences in information revelation possibilities, which is the focus of this chapter, but also on cost differences between endorsements and contributions.

11. The bad type interest group likes to transfer information and, in addition, weakly prefers the hybrid contributions equilibria over the hybrid endorsement equilibria. So, when IRCO holds the bad type prefers contributions. From the perspective of the good type interest group the voter already takes the right decision based on her prior belief. Expenditures on endorsements or contributions that the good type interest group makes in equilibrium are then, in a sense, only wasteful. When IRCO does not hold the IRDE condition may still hold, and the good type prefers contributions. In case IRCO holds, the IRDE condition holds as well and the good type prefers route with the most profitable hybrid equilibria, i.e. endorsements.

12. When the interest group can earmark its donations, i.e. the interest group can require that donations are used for a specific purpose (i.e. the campaign), Proposition 4.3 can interpreted alternatively. Because earmarked donations reduce to direct endorsements in our setup, the proposition then indicates under what circumstances the interest group wants to earmark its contributions.

13. As already indicated in note 8, this conclusion would no longer hold true when the voter — either directly or indirectly — pays for the political advertising by the candidate and/or the interest group.

14. The Economist (1994) provides a short overview of campaign finance regulation in several countries. Although it is quite likely that in the near future legislation will be forthcoming that will require disclosure of the size and source of donations (cf. The Economist, 1997b), in the spring of 1998 Britain was still among the rare countries where parties do not have to reveal where they get their money. See also Section 4.5 for a short discussion of the case of the Netherlands.

15. To give an example, Charles Lewis' book *The buying of the President* lists the interest groups that provided the largest financial support to the several potential candidates for the presidential election of 1996. On the cover page the reader is advised: "Before you vote, find out about the favors candidates do for the wealthy corporations and individuals who back them." According to the writer himself, the book "provides information [..] that the electorate can, and dare we say, should use in determining which candidate it will allow to 'obtain its suffrages' [..]." (Lewis, 1996, p. 20).

16. Similar to the exclusion of endorsing the incumbent in the direct endorsements model, we exclude the possibility of giving an observable donation to the challenger when in fact the incumbent is preferred. Hence the interest group with $g_i<0$ is excluded from supporting the challenger financially in the observable donations model.

17. In fact, it can be shown that either IRDE or IRCO has to hold for information transmission through observable contributions to be an equilibrium phenomenon. Hence a necessary and sufficient condition is the 'Information Revelation by OBServable COntributions (IROBSCO) condition: $\max\{c,g_1\}<g_2$ or $\max\{c, \min\{u_1,g_1\}\}<\min\{u_2,g_2\}$.

18. See, for instance, Lewis (1996), Sabato and Simpson (1996) and The Economist (1994, 1997a, 1997b), and for the Netherlands in particular, Koole (1990a, 1992) and Waarborg van Kwaliteit (1991). In Sloof (1997) a more elaborate justification is presented of how the model discussed in this section can be used to argue the desirability of full disclosure laws on information revelation grounds.

19. Concerning past experience, see the overview and the results provided in Schram (1989). The influence of political campaigns follows from the empirically observed positive impact of campaign expenditures on the probability of winning (cf. Morton and Cameron, 1992). Endorsements are found to be influential by Lupia (1994), Rapoport et al. (1991), Schneider and Naumann (1982), and to a lesser extent by Williams (1994) in a laboratory experiment.

20. Page, Shapiro and Dempsey (1987) find some weak evidence that the reported statements and actions of interest groups on the television news had a negative impact on public opinion when it concerned a group with a narrow interest (e.g. business groups), whereas some public interest groups, like environmental groups, influenced public opinion positively.

21. This is not to say that the "costs of democracy" in the Netherlands are low. In fact, when roughly measuring the costs of democracy as the total expenditures of all parties at the national level per person entitled to vote, the Dutch democracy appears to be somewhat more expensive than the American one, at least when the period 1977-1980 is considered. The costs of democracy are significantly higher in e.g. Belgium and Germany, though (Koole, 1990a).

22. Only the Liberal Conservative Party (VVD), and its affiliated scientific foundation, now and then received significant, yet small amounts from business circles (cf. Koole, 1990a, 1990b). Businessmen did, however, mainly provide the financial foundation for the VVD when it was established in 1948. A list of early business contributors (april 1946-march 1948) to the VVD can be found in Koole (1985). The figures mentioned in HP/De Tijd (October 11, 1996, Shadows around the Inner Circle) show that even in more recent years (1995/1996) the amounts involved in political gifts are extremely low.

23. Besides, in his (1992) publication Koole argues that, firstly, corporations typically also have influence through maintaining direct contacts with governmental departments, and secondly, that the effect of (financially) supporting a single party in a coalitional government is most likely extremely small. In that way, the buying of political influence through individual political parties is not needed, or not efficient (Koole, 1992, p. 197).

24. More specifically, the 'no-knife-edge cases' (NKE) assumption entails: $c \neq u_i$; $c \neq g_i$; $u_j \neq g_i$ for i,j=1,2 ; $u_1 \neq u_2$; $g_1 \neq g_2$ and $p \neq v$.

25. Although PE5 is strictly speaking not a pooling equilibrium, as by using the equilibrium strategies of the agents the voter may improve upon her prior belief concerning the challenger's type, in contrast to the hybrid equilibria the updated belief in PE5 never leads to a decision different from the decision based on the prior belief. Moreover, the combined campaigning and endorsing strategy $(\gamma(t_i)\delta(t_i)+\varepsilon(t_i))$ of the two types of interest group-challenger combination is the same in this equilibrium. Strictly speaking, it would be best to characterize PE5 as a *non-influential* or *essentially uninformative* equilibrium.

REFERENCES

Austen-Smith, D., 1987, Interest groups, campaign contributions, and probabilistic voting, *Public Choice* 54, 123-39.

Box-Steffensmeier, J.M., 1996, A dynamic analysis of the role of war chests in campaign strategy, *American Journal of Political Science* 40, 352-371.

Calvert, R.L., 1986, Models of imperfect information in politics (Harwood, Chur).

Depken, C.A., 1998, The effect of campaign contribution sources on the Congressional elections of 1996, *Economics Letters* 58, 211-215.

Edelman, S., 1992, Two politicians, a PAC, and how they interact: Two extensive form games, *Economics and Politics* 4, 289-305.

Epstein, D. and P. Zemskey, 1995, Money talks: Deterring quality challengers in congressional elections, *American Political Science Review* 89, 295-308.

Ferguson, J.M., 1976, Comment on political information, *Journal of law and economics* 19, 341-346.

Gerber, A., 1996, Rational voters, candidate spending, and incomplete information: A theoretical analysis with implication for campaign finance reform, Working Paper.

Grossman, G.M. and E. Helpman, 1994, Protection for sale, *American Economic Review* 84, 833-850.

Grossman, G.M. and E. Helpman, 1997, Competing for endorsements, Working Paper (Princeton University, Princeton).

Harrington, J.E., 1992, The revelation of information through the electoral process: An exploratory analysis, *Economics and Politics* 4, 255-275.

Harrington, J.E., 1993, The impact of reelection pressures on the fulfillment of campaign promises, *Games and Economic Behavior* 5, 71-97.

Helsley, R.W. and A. O'Sullivan, 1994, Altruistic voting and campaign contributions, *Journal of Public Economics* 55, 107-119.

Kau, J.B. and P.H. Rubin, 1979, Public interest lobbies: Membership and influence, *Public Choice* 34, 45-54.

Kau, J.B. and P.H. Rubin, 1982, Congressmen, constituents and contributors: Determinants of roll call voting in the House of Representatives (Martinus Nijhoff, Boston).

Koole, R.A., 1985, Partijfinanciën en bedrijfsleven: De giften vanuit ondernemerskringen aan de Partij van de Vrijheid 1946-1948, in: Jaarboek 1984 Documentatiecentrum Nederlandse Politieke Partijen (Rijksuniversiteit Groningen, Groningen) 108-128.

Koole, R.A., 1990a, Political parties going Dutch: Party finance in the Netherlands, *Acta Politica* 25, 37-65.

Koole, R.A., 1990b, De arme VVD: Partijfinanciën in historisch perspectief, in: Jaarboek 1989 Documentatiecentrum Nederlandse Politieke Partijen (Rijksuniversiteit Groningen, Groningen) 60-72.

Koole, R.A., 1992, De opkomst van de moderne kaderpartij. Veranderende partijorganisatie in Nederland 1960-1990 (Het Spectrum, Utrecht).

Lewis, C., 1996, The buying of the President (Avon Books, New York).

Lohmann, S., 1993, A signaling model of informative and manipulative political action, *American Political Science Review* 87, 319-333.

Lohmann, S., 1995, A signaling model of competitive political pressures, *Economics and Politics* 7, 181-206.

Lupia, A., 1992, Busy voters, agenda control and the power of information, *American Political Science Review* 86, 390-403.

Lupia, A., 1993, Credibility and the responsiveness of direct legislation, in: W.A. Barnett, M.J. Hinich and N.J. Schofield, eds., Political economy: Institutions, competition, and representation (Cambridge university press, Cambridge) 379-404.

Lupia, A., 1994, Shortcuts versus encyclopedias: information and voting behavior in California insurance reform elections, *American Political Science Review* 88, 63-76.

Morton, R. and C. Cameron, 1992, Elections and the theory of campaign contributions: A survey and critical analysis, *Economics and Politics* 4, 79-108.

Olson, M., 1982, The rise and decline of nations (Yale University Press, New Haven).
Page, B.I., Shapiro, R.Y. and G.R. Dempsey, 1987, What moves public opinion?, *American Political Science Review* 81, 23-43.
Palda, K.F., 1996, Are campaign contributions a form of speech? The case of the U.S. House elections, Working Paper (École Nationale d'Administration Publique, Montreal).
Palda, K.F. and K.S. Palda, 1985, Ceilings on campaign spending: Hypothesis and partial test with Canadian data, *Public Choice* 45, 313-331.
Palda, K.F. and K.S. Palda, 1998, The impact of campaign expenditures on political competition in the French legislative elections of 1993, *Public Choice* 94, 157-174.
Poole, K.T. and T. Romer, 1985, Patterns of PAC contributions to the 1980 campaigns for the U.S House of Representatives, *Public Choice* 47, 63-111.
Positie en subsidiëring van politieke partijen, 1996, Kamerstuk 24 688, nrs. 1 t/m 3.
Potters, J., Sloof R. and F. van Winden, 1997, Campaign expenditures, contributions, and direct endorsements: The strategic use of information and money to influence voter behavior, *European Journal of Political Economy* 13, 1-31.
Prat, A., 1997, Campaign advertising and voter welfare, Discussion Paper No. 97118 (Center for Economic Research, Tilburg).
Rapoport, R.B., Stone, W.J. and A.I. Abramowitz, 1991, Do endorsements matter? Group influence in the 1984 Democratic caucuses, *American Political Science Review* 85, 193-203.
Regeling van de subsidiëring van politieke partijen (Wet subsidiëring politieke partijen), 1997/1998, Kamerstuk 25 704, nrs. 1 t/m 6 en A.
Sabato, L.J. and G.R. Simpson, 1996, Dirty little secrets. The persistence of corruption in American politics (Times Books, New York).
Schlozman, K. and J. Tierney, 1986, Organized interests and American democracy (Harper and Row, New York).
Schneider, F. and J. Naumann, 1982, Interest groups in democracies - how influential are they? An empirical examination for Switzerland, *Public Choice* 38, 281-303.
Schram, A.J.H.C., 1989, Voter behavior in economic perspective (Springer Verlag, Berlin).
Sloof, R., 1997, Campaign contributions and the desirability of full disclosure laws, Working Paper TI 97-068/1 (Tinbergen Institute, Amsterdam).
Snyder, J.M., 1991, On buying legislatures, *Economics and Politics* 3, 93-109.
The Economist, 1994, Paying the piper, April 16.
The Economist, 1997a, Money and politics. Politicians for rent, February 8.
The Economist, 1997b, Fuelling the political machine, November 15.
Van Praag, P., 1987, Een mislukte campagne: De FNV en de verkiezingen van 1986, *Namens* 7, 368-370.
Waarborg van kwaliteit, 1991, Rapport van de Commissie subsidiëring politieke partijen.
Williams, K.C., 1994, Spatial elections with endorsements and uninformed voters: Some laboratory experiments, *Public Choice* 80, 1-8.

5 LOBBYING OR PRESSURE?[1]

Unlike the situation today, workers involved in strikes over wage increases in the 1880s were typically not members of a secure union with recognized bargaining rights. Instead, the outcome of the strike determined their collective bargaining status. In modern [..] terminology, we interpret strikes over wage increases in the 1880s as primarily strikes over 'union recognition'. The main question resolved by strikes was whether the employer would recognize employees' bargaining power. Card and Olson (1995, p. 33)

As observed in Chapter 2 interest groups typically use several means to influence government policy. For instance, interest groups may use words (threats) to convince a policymaker. However, when such threats are not successful, it becomes necessary for the interest group to carry out the threat in order to influence policy. Casual observations suggest, just like the quote above, that especially newly organized (starting) interest groups first have to show their teeth, like a strike in case of a union, before they arrive at an established position. Interest groups with an already recognized position, on the other hand, typically seem to rely on verbal persuasion and informal contacts. In the present chapter we try to provide a rigorous game-theoretical underpinning for these observations obtained from casual empiricism.

The game models presented in this chapter study within one framework lobbying and pressure as different means to influence a single policymaker. Both are modeled as means of transmitting information, albeit as different ones. Lobbying is associated with a kind of verbal persuasion ("words"), and pressure with the explicit enforcement of a threat ("actions" or "deeds"). In Section 5.1 the formal difference between lobbying and pressure employed in our models is made precise. Besides, in this section a general description of the different models analyzed in subsequent sections is given, together with an elaboration on the rationale for studying these different models. In Sections 5.2 through 5.4 the equilibria of the different models are analyzed in turn. In accordance with the opening quote the equilibrium analysis yields that an interest group (e.g. a union) which does not have an established position yet must first show its teeth (e.g. a successful strike) in order to gain recognition. That is, pressure is typically indispensable in getting an established position. Lobbying, on the other hand, is mainly used to maintain such an established position. A welfare comparison between the models, leading to a discussion of the issue of institutional design, is presented in Section 5.5. Finally, Section 5.6 concludes.

5.1   DESCRIPTION OF THE GENERAL MODEL

The question of interest that runs through this chapter is whether actions (explicit enforcement of threats) or rather words (lobbying) are used by interest groups to persuade a policymaker, and to build up and maintain a reputation. In this section we describe the general setup behind the different models that are used to study this question. First, it is illustrated in Subsection 5.1.1 how the basic signaling game of Section 3.1 is used as a building block of the models to be discussed. Moreover, the formal distinction between lobbying and pressure employed in these models is elaborated upon. Subsection 5.1.2 gives a detailed description of the dynamic models that are analyzed in subsequent sections, and provides a justification of why all these models are of interest. Finally, Subsection 5.1.3 introduces the notation used and informally discusses the equilibrium concept employed.

5.1.1   Lobbying, pressure, and the connection to the basic signaling game

In Section 3.2 it was argued that the basic signaling game can be interpreted, and indeed originally has been introduced, as a model of informational lobbying. In this application the lobbyist/interest group replaces the sender, and the policymaker substitutes for the decisionmaker. It was said that the kind of information transmitted may for instance be about the electoral salience of the group's cause, or about the to be expected consequences of a certain policy. Of course, the abstract setup of the basic signaling game allows a wider range of interpretations of the kind of information transferred. For instance, the information may also refer to the to be expected reaction of the interest group after the policymaker has taken her decision (cf. Potters, 1992). It is this latter interpretation that is used in this chapter. For this interpretation to make sense the basic signaling game has to be extended to allow for a reaction by the interest group after the decisionmaker has chosen her action.

A verbal description of the extension we have in mind runs as follows. The policymaker has to choose between Conceding to the group's wishes (x=C), or Not-conceding to these wishes (x=N). In case the policymaker does not concede, the interest group may either Enforce its threat (y=E), or nOt enforce its threat (y=O). When the policymaker concedes the interest group is committed not to enforce. Only a strong type of interest group ($t=t_2$) prefers to choose y=E after x=N, a weak type of group ($t=t_1$) prefers not to enforce after no-concession. Both types of interest group, however, prefer the policymaker to concede rather than not to concede. The policymaker, on the other hand, only prefers to concede if no-concession would induce the group to enforce, that is, in case the group is strong.[2] Unfortunately for her, a priori the policymaker does not know whether the interest group is weak or strong. It follows from the above assumptions that the interest group has an incentive to let the policymaker belief that it is strong, irrespective of whether this true or not. Now, by sending a fixed cost lobbying message to the policymaker the interest group may try to affect her belief that it is strong.

When we define $x_1 \equiv N$ and $x_2 \equiv C$ in Section 3.1, the game described above extends the setup of the basic signaling game in a very simple way. The equilibrium reaction of the interest group to the policymaker's choice of $x=N$ is trivially determined by its dominant action, and the equilibrium analysis presented in Section 3.1 immediately applies. Specifically, because the interest group and the policymaker have partially conflicting interests, Result 3.1 can be used. Again, in the setup described above the informational value of a lobbying message lies merely in its costs, and not in its content. This result somewhat weakens our association of lobbying with "words", but its close connection to direct communication and verbal persuasion remains present. In the sequel we will not consider the content of a lobbying message, and identify the lobbying message with its costs.[3]

The introduction of a reaction by the interest group suggests a different possibility of information transfer, namely by means of the actual enforcement of a threat ("actions"). Given that the interest group cannot commit itself beforehand to carry out the threat, such a kind of information transfer will only occur in a repeated setting. In a single shot play of the game described above the game ends after the reaction of the interest group, and hence its reaction cannot favorably affect the behavior of the policymaker. In a repeated setting, on the other hand, the present enforcement of a threat may increase the belief of the policymaker that the interest group is strong, i.e. may increase the interest group's reputation, and thus favorably affect her future behavior. In this chapter such repeated interactions are indeed considered, and the focus is on whether actions (threat enforcement) or words (lobbying) are used to build up and maintain a reputation for being strong.

In the models to be considered the formal difference between lobbying and pressure is reflected in two important assumptions: (i) contrary to lobbying, pressure is directly costly to the policymaker, and (ii) contrary to the opportunities for exerting pressure, the opportunities to lobby are independent of the policymaker's actions. Of course, in practice giving access to lobbyists and listening to their speeches bears some (opportunity) costs for the policymaker. However, these costs are typically low in comparison with the costs of the actual enforcement of a threat, and therefore they are neglected. Assumption (ii) entails that pressure can only occur after the policymaker has taken an unfavorable action from the group's perspective (i.e. after $x=N$). In other words, the policymaker must provide the interest group an opportunity for pressure by taking an unfavorable decision in the first place.[4] Lobbying, on the other hand, may already occur before the policymaker decides which action to take. How these two differences exactly work out in our models is illustrated in the next subsection where these models are described.

5.1.2   Exact description of the models to be analyzed

Each of the dynamic game models considered in this chapter consists of two periods. Periods are numbered in the normal order, such that the game starts with period 1 and ends after period 2. Total payoffs are the undiscounted sum of the per period payoffs. The preferences of the policymaker are common knowledge, the preferences of the interest group are only privately known to the group itself. As noted, the latter preferences can take two forms; the interest group is either of the weak ($t=t_1$), or of the strong type ($t=t_2$). This type is drawn by nature prior to period 1 with $P(t=t_2)=p$ ($0<p<1$), and remains fixed during the complete game.[5] Nature's draw is shown to the interest group, but not to the policymaker. However, the policymaker knows the prior probability p that the interest group is strong.

In each period, one of two possible period (stage) games is played. The first possible period game is given by the setup described verbally in the previous subsection. This Lobbying period game L can be summarized as follows:

period game L:
(o)   The interest group decides whether to send a costly lobbying message c or not (signal 0).
(i)   The policymaker observes the lobbying signal sent by the interest group (0 or c) and chooses between N (No-concession) and C (Concession).
(ii)   After the policymaker's choice of N the interest group chooses between E (Enforce threat) and O (nOt enforce threat). In case the policymaker chooses C the interest group is committed to choose O.[6]

With respect to the preferences of the agents in period game L it is assumed that they fit the general description given in Subsection 5.1.1. Thus, the interest group always prefers the policymaker to concede, whereas the policymaker only prefers to concede in case not conceding induces the interest group to enforce its threat. In case the interest group is strong, the enforcement of the threat is not costly to the group itself; given the policymaker's future actions, the strong type prefers to enforce rather than not to enforce. In the opposite case of a weak type, the interest group prefers — ceteris paribus — not to enforce the threat. *Pressure* is defined to occur when the weak type decides to enforce its threat. In this case the interest group sends a costly signal to the policymaker, by means of an in itself (i.e. in the context of a single shot play of L) costly action, in order to induce the policymaker to give in in future periods. Since enforcing the threat is in itself a profitable action for the strong type of interest group, we prefer not to speak of pressure in this case.[7]

A specific parameterization of the payoffs in game L that fits the general description given above, and which will be used throughout this chapter, is given in Table 5.1 below.

Table 5.1 Gross payoffs in period game L (and in period game A)

actions		payoffs		
policymaker	interest group	policymaker	weak type $(t=t_1)$	strong type $(t=t_2)$
N	E	a-1	g	g
N	O	a	1	0
C	O	0	2	2

Additional assumptions: $0 < a < 1$ and $0 < g < 1$

By assumption, sending a lobbying message bears a fixed cost of c, with $0<c<1$, and refraining from sending a lobbying message bears no direct costs. To get the net payoffs in period game L we have to subtract the costs c from the gross payoffs given in Table 5.1 in case the interest group indeed sent a lobbying message.

Though the levels of the payoffs used in Table 5.1 are somewhat arbitrary, relative payoffs are chosen with the conditions for information transfer in mind. Note that by the assumptions $0<g<1$ and $0<c<1$ the Information Revelation condition for informational lobbying is satisfied (cf. Corollary 3.1);[8] max$\{c,1\}<$ 2-g. Hence the sorting condition (2-g>1), requiring that the gain from persuading the policymaker to concede is larger for the type of interest group with the 'right' information (the strong type), is assumed to hold from the outset. This is done to preserve the informational role of lobbying even in a single shot play of period game L (cf. Chapter 3). The assumption $c<1$ is made in order not to make lobbying prohibitively costly for both types of interest group in period game L. The assumption excludes complete separation through lobbying in a single play of L (cf. Proposition 3.2). Under these assumptions concerning the relative payoffs, changing the parameterization does not qualitatively affect our results.

The second possible period game simplifies period game L by just excluding the possibility of sending a lobbying message. This game is referred to as game A because only Actions (threat enforcements) can be used to influence the future actions of the policymaker. Payoffs in period game A are the same as in period game L, and thus reflected in Table 5.1 as well. In order to keep our exposition conveniently arranged, the setup in period game A is summarized below:

period game A:
(i)   The policymaker chooses between N and C.
(ii)   After the policymaker's choice of N the interest group chooses between
         E and O. In case the policymaker chooses C the interest group is
         committed to choose O.

The parameters a and g appearing in the payoffs of period games A and L (cf.
Table 5.1) can be interpreted in the following ways. An increase in the parameter
a raises the potential gain (a) for the policymaker from not conceding. Put
differently, parameter a denotes the potential costs of concession. At the same
time, an increase in a lowers the potential loss, that is the net cost of being
punished 1-a, of not conceding to the group's wishes. Hence, the value of a
represents the relative attractiveness of not-conceding (the choice of N) for the
policymaker. Because the interest group always prefers the policymaker to
concede, parameter a can be interpreted as a measure of conflict of interests; the
larger a, the larger the disagreement over preferable outcomes between the interest
group and the policymaker. Parameter g affects both the costs of exerting pressure
(1-g) for the weak type of interest group, as well as the potential gain (2-g) to the
strong type from persuading the policymaker to concede rather than not to
concede.[9] Hence, the value of g, given the costs of lobbying c, influences the
willingness of the weak type to exert pressure, and the willingness of both types
to lobby. A higher value of g makes it easier for the weak type to pretend being
strong by mimicking the behavior of the strong type.

With two different period games and two periods, four different models
can be constructed by making different combinations of the two period games A
and L. This results in the four different models uv∈{AA,AL,LA,LL}, with u (v)
in uv referring to the type of period game played in period one (two). The model
in which period game A is repeated a finite number of times is already discussed
at length in Potters and Van Winden (1990). Therefore, the results obtained for
model AA are only briefly discussed in this chapter, at the beginning of Section
5.2. Subsequently, we analyze the three remaining models AL (Section 5.2), LA
(Section 5.3), and LL (Section 5.4) in more detail.

The rationale for investigating the three models AL, LA, and LL is
fourfold. Firstly, the three models can be seen as capturing different situations.
Model AL tries to describe the situation in which an interest group, which has not
yet gained an established position, has little or no access to lobby the policymaker.
Intuitively, model AL seems to fit the case of a starting interest group which has
yet to gain recognition, and for which not all channels for political influence are
open yet. In order to be able to lobby, the policymaker must be prepared to receive
the interest group and to listen to its message. Lobbying thus requires having
access to the "right" people, and a starting interest group typically lacks such
access. Once the interest group and the policymaker have interacted through
"actions", the position of the interest group may be recognized and the group gets
access to lobbying possibilities (cf. the opening quote at the beginning of this
chapter). Loosely put, in model AL access has to be earned.[10]

On the other hand, an interest group with an established position already has acquired access to the policymaker, and may directly communicate with the policymaker, either informally through lobbying contacts or formally through institutionalized consultations. For this case models LA and LL provide more appropriate descriptions. The LA model seems to apply when the policymaker organizes a hearing (public enquiry procedure) at which interest groups may present testimony before she takes a decision. Model LL nicely takes into account the fact that in practice lobbying generally has an intertemporal character, and that interest groups may lobby or advice the policymaker on a regular basis. For instance, a situation captured by model LL is when hearings are held at several different stages in a step-wise decision procedure. The difference between model AL and model LA can also be related to the policymaker's term. Above we already suggested that in the beginning of a policymaker's (initial) term, lobbying contacts between interest groups and policymakers may be scarce. Only at the end of her term channels for contacts may open up. On the other hand, it could be argued that at the beginning of a policymaker's term interest groups can still have a voice in determining public policy. Once the policymaker has got her policy going, there may be no opportunities left for (public) comments.[11]

Secondly, by analyzing the three different models AL, LA and LL, interesting questions in the context of institutional design can be addressed. For instance, a comparison between the different models may indicate the gain to the policymaker of granting lobbying access to the interest group at a particular stage. This issue of institutional design will be taken up in Section 5.5. Thirdly, the study of the three different models allows us to explore the sensitivity of equilibrium phenomena to certain modeling aspects. In view of our main question whether actions or words are typically used to build up and/or maintain a reputation, it is of particular interest to find out to what extent our conclusions are robust to specific modeling assumptions. We can have more faith in conclusions that are supported by all three models, because they are not completely dependent then on the specific structure of interaction.

A final argument for discussing the three models is an expositional one. Although model LL may be considered to be the most interesting, the equilibrium analysis of this model is quite complex. By first studying the more simple models AL and LA in Section 5.2 and 5.3, respectively, considerable intuition can be obtained for the results of model LL. This greatly simplifies the discussion of this model. Some conclusions with respect to the building up and the maintenance of reputation will be presented already when discussing the AL and LA model. This is done for expositional reasons, and to keep the focus on the main questions posed. To avoid confusion, it is immediately added here that these "intermediate" conclusions (Results 5.1 and 5.2) hold for all three models.

5.1.3   Notation and equilibrium concept

In this subsection first the (additional) notation employed throughout this chapter is introduced. The action taken by the policymaker in period i (i=1,2) is represented by $x_i \in \{N,C\}$, the response of the interest group in period i is represented by $y_i \in \{E,O\}$. The lobbying signal sent in period i is represented by $s_i \in \{0,c,\varnothing\}$, with $s_i=0$ indicating that no lobbying message is sent in period i, $s_i=c$ indicating that a costly lobbying message is received by the policymaker in period i, and $s_i=\varnothing$ denoting that lobbying is not possible at all in period i (because game A is played in this period). The three-tuple $h_1=(s_1,x_1,y_1)$ is used to describe the history of play in period 1.

Turning to the strategies of the players, $\lambda_i(t_j;\cdot)$ denotes the probability that the type t_j (j=1,2) interest group sends lobbying message c in period i, $\mu_i(\cdot)$ the probability that the policymaker takes action $x_i=C$ in period i, and $\pi_i(t_j;\cdot)$ gives the probability that the type t_j interest group responds to $x_i=N$ with $y_i=E$ in period i.[12] These (behavioral) strategies are, in general, functions of the initial belief p and the whole history of play up to the information set for which the strategy is specified. In most cases (Bayes' rule) updated beliefs act as a sufficient statistic for the combination of p and the history of play, and can be used as a state variable to replace them as an argument in the relevant strategies (cf. Kreps and Wilson, 1982, Van Damme, 1991). However, it appears that in some cases updated beliefs are not sufficient, and that such a substitution is not possible. Therefore, in the propositions we always specify strategies as functions of p and the history of play. Finally, $q_1(p,s_1)$ denotes the updated belief of the policymaker that the interest group is strong at the time she has to decide on her first period action x_1, given that she has received lobbying signal s_1 in period 1. Note that by definition $q_1(p,\varnothing)=p$. Similarly, $q_2(p,h_1,s_2)$ denotes the updated belief of the policymaker when she has to decide on her second period action x_2, given history of play (h_1,s_2) and prior belief p. $q_2(p,h_1)$ denotes the policymaker's belief at the beginning of period 2.

The different models analyzed in this chapter all belong to the class of repeated signaling games, potentially allowing a large number of perfect Bayesian equilibria. Some of these can only be sustained with rather implausible beliefs off-the-equilibrium path. For example, that the observation of $y_1=E$ after $x_1=N$ makes it more likely that the interest group is weak. We do not want to focus on such equilibria. Therefore, we use a version of the STABAC procedure described by Cho (1993) to select the more plausible equilibria. Equilibrium paths of perfect Bayesian equilibria selected by this procedure are denoted *plausible* equilibrium paths. The STABAC procedure entails that STAbility (forward induction) like arguments are used in a BACkward inductive fashion. The procedure is used in many other economic models dealing with repeated signaling games.[13]

In the LA model our application of STABAC comes down to the following. We reason backwards from the end of the game and start by solving the period 2 game (A). At the beginning of period 2 the belief of the policymaker equals $q_2(p,h_1)$, and for each value of $q_2(p,h_1)\in[0,1]$ the perfect Bayesian equilibria for period game A are determined. By substituting for each value of $q_2(p,h_1)$ the equilibrium payoff obtained in a specific perfect Bayesian equilibrium of the continuation game, we get a reduced game. This reduced game is solved by using the *universal divinity* restriction (cf. Chapter 3); for each value of $q_1(p,s_1)$, the universally divine equilibria of the continuation game that starts just after the first lobbying stage is determined.[14] Subsequently the total two period game is reduced to a one period signaling game by substituting for each $q_1(p,s_1)$ the equilibrium payoff of the second component game. In a last step, the universally divine equilibria for this reduced signaling game are determined for every value of p.

In case period L is played in the second period (cf. AL and LL), the interest group takes two subsequent decisions after the policymaker has chosen x_1. It first chooses reaction y_1, then it decides on its lobbying signal s_2. Only after these two actions the policymaker has to decide on x_2. Therefore, the continuation game that starts after action x_1 of the policymaker can be interpreted as a multi-dimensional signaling game with the interest group choosing multi-dimensional signal (y_1,s_2). When applying STABAC we make use of this interpretation. We start by determining for every value of $q_2(p,h_1,s_2)$ all perfect Bayesian equilibria of the continuation game after the second lobbying stage. Substitution of the equilibrium payoffs leads to a reduced game. We again reason backwards in this reduced game; for each value of $q_1(p,s_1)$ the universally divine equilibria of the multi-dimensional signaling component game are determined. This yields the plausible equilibria of model AL. In case we consider model LL we have to add a final step. By substituting for each $q_1(p,s_1)$ the equilibrium payoff obtained in the equilibrium continuation path we get a reduced game and the universally divine equilibria for this (standard) signaling game are determined.

In all models the equilibrium reaction (threat enforcement) strategy of the interest group in the second and last period is trivially determined and independent of the history of play. The interest group reacts to the policymaker's choice $x_2=N$ with its dominant response, for the weak (strong) type given by reaction $y_2=O$ ($y_2=E$). Hence in all equilibria $\pi_2(t_1)=0$ and $\pi_2(t_2)=1$, and in subsequent specifications of equilibrium paths we refrain from listing the second period reaction strategy of the interest group. Besides, instead of completely specifying the plausible equilibria, our propositions will only present their equilibrium paths. A specification of out-of-equilibrium strategies and beliefs that satisfy the STABAC refinement and sustain the plausible equilibrium paths can be found in the proofs of the propositions (cf. Appendix 5.A).

5.2   THE AL GAME: PRESSURE AS A MEANS TO BUILD UP REPUTATION

Does a starting interest group without a recognized position typically use "actions" or "words" in building up a reputation for being strong? In this section we tackle this question by examining the AL model, and find that typically "actions" are used to obtain an established position. However, first we show, by means of briefly reviewing the results for the AA model, how pressure (threat enforcements) can be used to build up such a reputation. Simultaneously, the exposition of the equilibria of model AA provides some feeling for the results to be expected for the extended model AL, and a benchmark for comparing our newly obtained results.

The plausible equilibrium paths of the AA model — that is, the paths satisfying our refinement concept STABAC — are denoted AAE and given in the following proposition:

Proposition 5.1[15]
The plausible equilibrium paths of the AA game are given by:

$p \leq a^2$:   AAE1: $\mu_1 = 0$, $\pi_1(t_1) = (1-a)p/(1-p)a$, $\pi_1(t_2) = 1$, $\mu_2(N,O) = 0$,
                $\mu_2(N,E) = 1-g$; $q_2(N,O) = 0$, $q_2(N,E) = a$.

$a^2 < p \leq a$:   AAE2: $\mu_1 = 1$, $\mu_2(C,O) = 0$; $q_2(C,O) = p$.

$a \leq p$:   AAE3: $\mu_1 = 1$, $\mu_2(C,O) = 1$; $q_2(C,O) = p$.   □

It should be noted that equilibrium path strategies and beliefs depend on p, but for ease of exposition this parameter is not incorporated as an argument. Similarly, in model AA we have $s_1 = s_2 = \emptyset$ by definition, and therefore both s_1 and s_2 are left out as an argument. In short, the arguments in $\mu_2(\cdot)$ and $q_2(\cdot)$ refer to the (relevant) history of play in the first period (x_1, y_1).

In case the initial reputation of the interest group is low ($p < a^2$), the policymaker does not concede in the first period (in AAE1 we have $\mu_1 = 0$). In order to induce a favorable decision in the second period, the interest group must enforce its threat. Specifically, the weak type must exert pressure to build up a reputation for being strong, and so to influence the second period decision of the policymaker. Because the weak type enforces its threat with a lower probability $(\pi_1(t_1) < \pi_1(t_2) = 1)$ than the strong type does, an enforcement raises the reputation of the group to $q_2(N,E) = a$. When $p > a^2$ the policymaker, due to the credible threat of enforcement with a large probability, chooses to concede in the first period.[16] For intermediate levels of the group's initial reputation ($a^2 < p < a$) the policymaker does not give in in the second period, for high values ($p > a$) she does. Although in the first case the weak interest group type is willing to exert pressure to alter the second period decision of the policymaker, she is not willing to provide the interest group an opportunity to exert pressure, because pressure is directly costly to her. In summary, the AA model predicts that pressure (the enforcement of a threat) only occurs in case the reputation of the interest group is rather low.

Of course, in the AA model the only means by which the interest group can build up a reputation for being strong is by enforcing its threat. But what happens if we add the possibility of (second period) lobbying in order to influence the second period decision of the policymaker? Note that information obtained through the lobbying process is in a certain sense cheaper for the policymaker because lobbying is not directly costly to her (whereas the enforcement of a threat is). Therefore it could be conjectured that in model AL the policymaker will not give the interest group the opportunity to enforce its threat any longer, i.e. will always choose $x_1=C$, and will just rely on the information transmitted through the lobbying process instead to determine her second period choice. On the other hand, by always choosing $x_1=C$ the policymaker forgoes the potential benefits (to the value of a) of not conceding in the first period. It appears that this latter effect outweighs the former, because when $p<a^2$ the policymaker chooses $x_1=N$ for sure in equilibrium. This can be seen from our next proposition, which specifies all equilibrium paths for the AL model (ALE) that satisfy STABAC.

Proposition 5.2
The plausible equilibrium paths of the AL game are given by:

$\underline{p<a^2}$:
c<g

ALE1: $\mu_1=0$, $\pi_1(t_1)=(1-a)p/(1-p)a$, $\pi_1(t_2)=1$, $\lambda_2(t_1;N,O)=0$, $\lambda_2(t_1;N,E)=1$, $\lambda_2(t_2;N,E)=1$, $\mu_2(N,O,0)=0$, $\mu_2(N,E,c)=1-g+c$; $q_2(N,O,0)=0$, $q_2(N,E,c)=a$.

g<c<(2-g)g

ALE2: $\mu_1=0$, $\pi_1(t_1)=0$, $\pi_1(t_2)=1$, $\lambda_2(t_1;N,O)=0$, $\lambda_2(t_2;N,E)=1$, $\mu_2(N,O,0)=0$, $\mu_2(N,E,c)=1$; $q_2(N,O,0)=0$, $q_2(N,E,c)=1$.

(2-g)g<c

ALE3: $\mu_1=0$, $\pi_1(t_1)=(1-a)p/(1-p)a$, $\pi_1(t_2)=1$, $\lambda_2(t_1;N,O)=$ $\lambda_2(t_1;N,E)=0$, $\lambda_2(t_2;N,E)=0$, $\mu_2(N,O,0)=0$, $\mu_2(N,E,0)=1-g$; $q_2(N,O,0)=0$, $q_2(N,E,0)=a$.

$\underline{a^2<p<a}$:
g<c<(2-g)g

ALE2.

c$\notin$[g,(2-g)g]

ALE4: $\mu_1=1$, $\lambda_2(t_1;C,O)=p(1-a)/a(1-p)$, $\lambda_2(t_2;C,O)=1$, $\mu_2(C,O,0)=0$, $\mu_2(C,O,c)=c$; $q_2(C,O,0)=0$, $q_2(C,O,c)=a$.

$\underline{a<p<2a/(a+1)}$:
g<c<(2-g)g

ALE2.

ALE5: $\mu_1=1$, $\lambda_2(t_1;C,O)=0$, $\lambda_2(t_2;C,O)=0$, $\mu_2(C,O,0)=1$; $q_2(C,O,0)=p$.

ALE6: $\mu_1=1$, $\lambda_2(t_1;C,O)=1$, $\lambda_2(t_2;C,O)=1$, $\mu_2(C,O,c)=1$; $q_2(C,O,c)=p$.

ALE7: $\mu_1=1$, $\lambda_2(t_1;C,O)=0$, $\lambda_2(t_2;C,O)=(p-a)/(1-a)p$, $\mu_2(C,O,0)=1-c/(2-g)$, $\mu_2(C,O,c)=1$; $q_2(C,O,0)=a$, $q_2(C,O,c)=1$.

$\underline{2a/(a+1)<p}$:

ALE5, ALE6, ALE7. □

Again, it should be noted that all equilibrium path strategies and beliefs depend on p, but that this parameter is suppressed as an argument. The arguments in $\lambda_2(t_j;\cdot)$ refer to (x_1,y_1), the arguments in $\mu_2(\cdot)$ and $q_2(\cdot)$ represent (x_1,y_1,s_2). (Because in model AL $s_1=\varnothing$ by definition, it is left out.)

In contrast with the AA model (cf. Proposition 5.1), a separating equilibrium exists for the AL model. In ALE2 the policymaker takes her second period decision x_2 under complete information. Note that $g<c$ implies $1-g+c>1$; in this case the combined act of pressure and lobbying is prohibitively costly for the weak type, because the sum of the costs of pressure $(1-g)$ and lobbying (c) exceeds the maximal possible gain (1) the combined influence attempt may lead to in the second period.[17] In that case the weak type is not prepared to choose $y_1=E$ and $s_2=c$ after $x_1=N$ to induce the policymaker to choose $x_2=C$. Due to the assumptions made, in particular $0<g<1$ and $0<c<1$, we have $2-g>c$ and the strong type is prepared to choose $(y_1,s_2)=(E,c)$ after $x_1=N$ to let the policymaker take action $x_2=C$. Hence when $g<c$ complete separation through the combined act of lobbying and the enforcement of a threat is possible in the AL model. In the separating equilibrium, lobbying ("words") and threat enforcements ("actions") act as complements in the production of reputation.

For two reasons the possibility of full separation does not imply its occurrence as an equilibrium phenomenon. Firstly, when the costs of lobbying are too high, the additional gain to the strong type interest group from complete separation is outweighed by the lobbying costs. In that case the strong type prefers to only partially separate itself out through enforcing the threat, and refrains from lobbying. To illustrate this, consider the case $p<a^2$, and in particular ALE3. From the perspective of the payoffs in this equilibrium, the additional gain to the strong type from complete separation through $(y_1,s_2)=(E,c)$ equals $(2-g)g$; the utility gain $(2-g)$ from having the policymaker choosing $x_2=C$ rather than $x_2=N$ times the increase (g) in the probability that $x_2=C$ is chosen. When $(2-g)g<c$ these gains fall short of the costs of lobbying, and the strong type prefers to partial pool with the weak type. In the opposite case, on the other hand, the strong type prefers to completely separate. These considerations provide some intuition for the conditions on g and c under which ALE2 and ALE3 exist. Secondly, the policymaker must be prepared to provide the strong type the opportunity to separate itself out. Since the strong type enforces its threat for sure, the policymaker is prepared to choose $x_1=N$ only when the initial reputation of the interest group is not too high $(p<2a/(a+1))$.

An interesting, and in respect to other models of reputation building uncommon,[18] feature of the separating equilibrium ALE2 is that the policymaker chooses $x_1=N$ even in cases where the expected payoffs in period 1 from doing so are negative. That is, when $a<p<2a/(a+1)$, choosing $x_1=N$ yields the policymaker an expected payoff in period 1 of a-p (<0), given the interest group types equilibrium strategies $\pi_1(t_1)=0$ and $\pi_1(t_2)=1$. Choosing $x_1=C$ instead would yield a first period payoff of 0. The policymaker bears this short-run loss just to learn more about the type of interest group, and so whether no-concession in the second period might be worthwhile. Would the policymaker in period 1 only take account of short-run considerations (period 1 payoffs), she would choose $x_1=C$. In short, when $p>a$ equilibrium ALE2 crucially requires that the policymaker takes long-run considerations into account. Were the policymaker in period 1 and the policymaker in period 2 in fact different players taking only their period payoffs into account, the strategies in ALE2 would not describe an equilibrium for the case $p>a$. Therefore, Proposition 5.2 does not immediately apply when the policymaker in period 1 is different from the policymaker in period 2. The proposition would only slightly change in such a case,[19] though, and our main results remain valid when we consider short-run rather than long-run policymakers. Proposition 5.1 does directly apply for the AA model with two short-run policymakers (cf. Potters and Van Winden, 1990).

   Contrary to ALE2, in all other, non-separating, equilibria also the weak type threatens to enforce its threat with positive probability. This induces the policymaker to put the interest group to the test ($x_1=N$) only when its initial reputation is low ($p<a^2$). After enforcing its threat the group's reputation for being strong increases to $q_2(N,E)=a$, and the lobbying stage does not yield additional information. The types either pool on lobbying (ALE1) or on no-lobbying (ALE3). In comparison with equilibrium AAE2 of model AA, when $a^2<p<a$ the interest group now does get an opportunity to increase its own reputation. In equilibrium (ALE4) a costly lobbying message raises the reputation of the group to $q_2(C,O,c)=a$, inducing the policymaker to concede with positive probability in period 2. Silence, on the other hand, leads to a complete loss of reputation ($q_2(C,O,0)=0$). Finally, when $p>a$ the policymaker is already predisposed towards conceding in both periods. Hence lobbying is not necessary to convince the policymaker (cf. ALE5). However, lobbying may still be needed when the policymaker expects it to occur (ALE6), or may be used to convince her even more fully of the toughness of the group (ALE7).

   From Proposition 5.2 and the discussion above it immediately follows that, like in model AA, in model AL pressure only occurs in equilibrium in case the reputation of the interest group is sufficiently low. Pressure only occurs with positive probability in ALE1 and ALE3, and does not occur when $p>a^2$. On the other hand, lobbying by both types occurs in ALE1, ALE4 and ALE6, and by only the strong type in ALE2 and ALE7. Note that at the beginning of the lobbying stage the reputation of the interest group equals $q_2(x_1,y_1)$, and from the equilibria with lobbying follows that lobbying may occur whenever $q_2(x_1,y_1) \geq a^2$. So, for high(er) values of the interest group's actual reputation lobbying may indeed be observed in equilibrium. The occurrence of lobbying with positive probability is not inevitable in equilibrium, though, as the existence of ALE3 and ALE5 demonstrates. We summarize these observations in the following result:

Result 5.1
Pressure (more general, the enforcement of a threat) only occurs in equilibrium in case the actual reputation of the interest group is sufficiently low (weakly below a^2). Lobbying, on the other hand, may occur in equilibrium for high values of the interest group's actual reputation (weakly above a^2).   □

It is important to note that when $p<a^2$ the only way for the weak type to get a higher reputation is to exert pressure. Similarly, the strong type interest group necessarily enforces its threat in that case. Together with the results reported in Result 5.1 we therefore tentatively conclude that "actions" rather than "words" are used to build up a reputation for being strong.

Result 5.1 is a consequence of our two assumptions that (i) pressure can only occur when the policymaker provides the interest group an opportunity to enforce its threat by not-conceding in first instance, and the assumption that (ii) pressure — in contrast to lobbying — is directly costly to the policymaker. These costs, relative to the costs of concession, are important in determining whether the policymaker provides the interest group such an opportunity. The interest group needs less reputation to induce the policymaker to concede in all periods (and thus getting no opportunity to exert pressure) when the costs of concession (a) are low relative to the costs of being punished (1). In other words, when there is only a low degree of conflict of interests between the interest group and the policymaker (a is small), already a low reputation suffices to convince the policymaker not to challenge the interest group.

Finally, one other comment is made concerning Proposition 5.2. In ALE6 lobbying is required in order to induce the policymaker to take the preferred action of conceding in the second period, although after the first period the reputation of the interest group has not altered and is still high. A similar remark applies when the interest group's initial reputation is rather low, and the group endogenously builds up a high reputation by enforcing its threat. In equilibrium ALE1 the interest group must, after enforcing its threat, lobby in the second period to keep a high reputation ($q_2(N,E,c)=a$). In a sense lobbying serves as a kind of maintenance costs the interest group has to bear to preserve the policymaker's faith. However, these maintenance costs are not always necessary after the enforcement of the threat, as the existence of ALE3 shows. Specifically, it follows from the previous discussion that when the combined act of lobbying and pressure is not prohibitively costly for the (weak type of) interest group, the group is forced to lobby in order to maintain its reputation built up in the past through the enforcement of a threat. Note that contrary to lobbying, exerting pressure (an enforcement) always improves the reputation of the group.

5.3   THE LA GAME: LOBBYING AND THE MAINTENANCE OF REPUTATION

The results of the previous section tentatively suggest that typically "actions" rather than "words" are used to build up a reputation for being strong. The AL model studied there, however, is at least very conducive to obtain such a conclusion, and seems almost tailored to yield the result; in the sequence of the AL model the possibilities for using "actions" precede the possibilities for using "words". Therefore we verify in this section whether the same conclusion holds true when we reverse that order, that is, when we consider model LA in which the interest group may already lobby from the outset. As mentioned, the LA model seems to apply when an interest group already has lobbying access to the policymaker. The LA model thus also allows us to check whether lobbying might indeed be needed to preserve an established position, as suggested at the end of the previous section. As will become clear the results obtained for the LA model corroborate our conclusions drawn there. "Actions" are typically used to obtain a reputation, "words" to maintain a reputation.

In model LA lobbying is only possible at the beginning of period 1. After this lobbying stage there are two rounds of the policymaker choosing her action x and the interest group subsequently choosing its reaction y. The interest group can, in principle, directly influence the decision of the policymaker in both the first (x_1) and the second period (x_2) through lobbying (s_1). Pressure (enforcing a threat), on the other hand, can only be used to directly influence the policymaker's decision in the second period. Hence lobbying and pressure have a different reach in this setup. The equilibrium paths for the LA model are denoted LAE and given in the following proposition.

Proposition 5.3
The plausible equilibrium paths of the LA game are given by:

$p<a^2$:

> LAE1: $\lambda_1(t_1)=\lambda_1(t_2)=0$, $\mu_1(0)=0$, $\pi_1(t_1;0)=(1-a)p/(1-p)a$, $\pi_1(t_2;0)=1$, $\mu_2(0,N,O)=0$, $\mu_2(0,N,E)=1-g$; $q_1(0)=p$, $q_2(0,N,O)=0$, $q_2(0,N,E)=a$.

> LAE2: $\lambda_1(t_1)=(1-a^2)p/(1-p)a^2$, $\lambda_1(t_2)=1$, $\mu_1(0)=0$, $\mu_1(c)=c$, $\pi_1(t_1;0)=0$, $\pi_1(t_1;c)=a/(1+a)$, $\pi_1(t_2;c)=1$, $\mu_2(0,N,O)=\mu_2(c,N,O)=\mu_2(c,C,O)=0$, $\mu_2(c,N,E)=1-g$; $q_1(0)=0$, $q_1(c)=a^2$, $q_2(0,N,O)=q_2(c,N,O)=0$, $q_2(c,C,O)=a^2$, $q_2(c,N,E)=a$.

$a^2<p<a$:

> LAE3: $\lambda_1(t_1)=(p-a^2)/a(1-p)$, $\lambda_1(t_2)=(p-a^2)/p(1-a)$, $\mu_1(0)=1-c/(2-g)$, $\mu_1(c)=1$, $\pi_1(t_1;0)=a/(1+a)$, $\pi_1(t_2;0)=1$, $\mu_2(0,N,O)=\mu_2(0,C,O)=0$, $\mu_2(0,N,E)=1-g$, $\mu_2(c,C,O)=c(1-g)/(2-g)$; $q_1(0)=a^2$, $q_1(c)=a$, $q_2(0,N,O)=0$, $q_2(0,C,O)=a^2$, $q_2(0,N,E)=q_2(c,C,O)=a$.

$a<p$:

> LAE4: $\lambda_1(t_1)=\lambda_1(t_2)=0$, $\mu_1(0)=1$, $\mu_2(0,C,O)=1$; $q_1(0)=p$, $q_2(0,C,O)=p$.

> LAE5: $\lambda_1(t_1)=\lambda_1(t_2)=1$, $\mu_1(c)=1$, $\mu_2(c,C,O)=1$; $q_1(c)=p$, $q_2(c,C,O)=p$.

> LAE6: $\lambda_1(t_1)=0$, $\lambda_1(t_2)=(p-a)/(1-a)p$, $\mu_1(0)=\mu_1(c)=1$, $\mu_2(0,C,O)=1-c/(2-g)$, $\mu_2(c,C,O)=1$; $q_1(0)=a$, $q_1(c)=1$, $q_2(0,C,O)=a$, $q_2(c,C,O)=1$. $\square$

Once more, to avoid an overload of notation, p is suppressed as an argument in the equilibrium strategies and updated beliefs. The argument in $\mu_1(\cdot)$, $\pi_1(t_j;\cdot)$ for j=1,2, and $q_1(\cdot)$ refers to the lobbying message $s_1 \in \{0,c\}$ the interest group sends in the first period. The arguments in $\mu_2(\cdot)$ and $q_2(\cdot)$ refer to the history of play in the first period, i.e. to $h_1=(s_1,x_1,y_1)$. By definition $s_2=\varnothing$ in model LA, and s_2 is therefore left out.

From Proposition 5.3 follows that, contrary to model AL, no completely separating plausible equilibrium for the LA game exists. With positive probability the policymaker has to base her decision in the final period (x_2) on incomplete information about the type of the interest group. The intuition behind this result is that in the LA model the policymaker never wants to give the strong type interest group the opportunity to separate itself out. Due to the assumptions c<1 and g<1 complete separation through lobbying only, or just once enforcing the threat ($y_1=E$), cannot be an equilibrium.[20] Although in case that 1-g+c>1 the weak type is never prepared to both lobby and enforce the threat to induce a change to $x_2=C$ (in contrast with the strong type), separation through lobbying *and* the enforcement of a threat cannot occur in equilibrium. In case of full separation the weak type would abstain from both lobbying and pressure, and the policymaker would already know the type of interest group after the lobbying stage in the first period. In case the policymaker receives a lobbying message this would be a clear signal that the interest group is strong ($q_1(c)=1$), and she would not want to choose $x_1=N$. Hence, separation cannot occur in equilibrium. The above considerations also explain why, contrary to the equilibria of the AL model, the existence of the various equilibria reflected in Proposition 5.3 does not depend on the relative values of g and c. As explained in the previous section, in the AL model the value of g relative to c determines whether separation is possible and, if so, profitable. In that way these parameters determine what type of equilibria exist in the AL model. In model LA separation is not possible, however, irrespective of the values of g and c. As an indirect consequence of separation being not possible in equilibrium, Proposition 5.3 immediately applies when the long-run policymaker is replaced by two short-run policymakers.[21]

When the initial reputation is rather low (p<a) the interest group types partially separate, either through enforcements alone (LAE1) or through both lobbying and enforcements (LAE2 and LAE3). In LAE1 both types of interest group do not send a lobbying message. Due to the fact that the reputation of the group remains low after receiving equilibrium signal $s_1=0$ ($q_1(0)=p<a^2$), the policymaker does not give in and plays $x_1=N$. In response, the strong type enforces its threat for sure, the weak type with probability strictly between 0 and 1. Not enforcing is thus a clear sign of being weak, and the policymaker chooses $x_2=N$ afterwards. In case the interest group chooses $y_1=E$ the policymaker updates her belief to $q_2(0,N,E)=a$, and randomizes between conceding and not-conceding in the second period. The game ends by the weak type choosing $y_2=O$ and the strong type choosing $y_2=E$ in case $x_2=N$.

In equilibrium LAE2 the interest group types already partly separate at the lobbying stage. The weak type mixes between sending and not sending a lobbying message, the strong type always sends a lobbying message. Silence reveals that the interest group is weak and leads to $x_1=x_2=N$. After being exposed to initial lobbying the policymaker mixes between $x_1=N$ and $x_1=C$. In case the interest group lobbied and is put to the test ($x_1=N$), the strong type always enforces its threat and the weak type mixes. Only in case the interest group passes this test, i.e. only after history $h_1=(c,N,E)$, the policymaker updates to $q_2(c,N,E)=a$ and mixes between N and C in the second period. Otherwise, she chooses not to concede. Finally, the interest group chooses its dominant reaction after $x_2=N$, with the weak type choosing $y_2=O$ and the strong type choosing $y_2=E$.

Compared with model AA also first period lobbying options may yield the interest group additional possibilities for increasing its reputation when the initial reputation is not too low ($a^2<p<a$). In equilibrium LAE3 both interest group types play a mixed lobbying strategy. The mixing probabilities are such that after receiving no lobbying message the interest group's reputation falls to $q_1(0)=a^2$. On the other hand, the receipt of a lobbying message increases the reputation of the interest group to $q_1(c)=a$. In the latter case the policymaker concedes in the first period with probability 1, and in the second period with positive probability. No lobbying leads the policymaker to mix between N and C in the first period, and enforcing the threat after $x_1=N$ leads to a renewed increase in reputation ($q_2(0,N,E)=a$). Interestingly, in LAE3 the histories $h_1=(0,N,E)$ and $h_1^{\#}=(c,C,O)$ both lead to $q_2=a$, but to different mixing probabilities for the policymaker. In other words, in this case the updated belief is not a sufficient statistic for the history of play in the first period.

One other feature of equilibrium LAE3 is also worth mentioning. In other game-theoretic models of reputation building, reputation is typically monotonically (weakly) increasing, up to a point where it is completely lost and cannot be regained.[22] In LAE3, on the other hand, reputation may increase over time in a non-monotonic way. When the interest group does not lobby its reputation falls from p to $q_1(0)=a^2$. Subsequently enforcing its threat after $x_1=N$ again increases its reputation to $q_2(0,N,E)=a>p>a^2$. Loosely put, this equilibrium suggests that established interest groups may still, once in a while, have to show their teeth to keep a reputation for being strong.

For high initial reputations such that the policymaker is already a priori predisposed to give in in both periods ($p>a$), next to a partially separating equilibrium (LAE6) two pooling equilibria exist (LAE4 and LAE5). Lobbying only occurs when the policymaker expects it to occur (LAE5), or when the strong type wants to convince the policymaker completely (LAE6). These equilibria resemble the equilibria of the AL model of the previous section for the same case of $p>a$.

It is interesting to note that also in the LA model pressure does not actually occur in equilibrium when $p>a$. Of course, this is caused by the fact that the interest group is never put to the test; due to the credible threat of enforcement the policymaker always chooses to give in the first period. On the other hand, when $p<a$ pressure does occur with positive probability in equilibrium. In fact, pressure only occurs when the actual reputation of the interest group $q_1(s_1)$ is weakly below a^2. Only when $q_1(s_1)\leq a^2$ the policymaker is prepared to choose $x_1=N$

with positive probability. Contrary to pressure, lobbying is an equilibrium phenomenon for (now) any value of the reputation of the interest group (cf. LAE2, LAE3, LAE5, LAE6). Lobbying does not necessarily occur in equilibrium, though (cf. LAE1 and LAE4). In short, Result 5.1 also applies when we consider model LA. This robustness over the two models strengthens our faith in the conclusions drawn in Result 5.1.

When discussing the AL model we already observed that the interest group may be forced to lobby in order to maintain its high reputation. The same conclusion can be obtained from model LA. In LAE5, for instance, the interest group is forced to sustain its high reputation through lobbying. When the policymaker expects the interest group to lobby, "silence" by the interest group induces a choice of $x_2 = N$ for sure. For the AL model we obtained a similar result when a high reputation was endogenously built up in the past through the enforcement of a threat. Besides, we also find for the LA model that, in contrast to lobbying, exerting pressure (or more general, the enforcement of a threat) always improves the reputation of the interest group. Based on these observations from both models (AL and LA) we get the following result:

Result 5.2
The interest group may be forced to lobby in order to maintain its reputation built up in the past. Contrary to lobbying, exerting pressure (the enforcement of a threat) always leads to an increase in reputation.   □

Combining Results 5.1 and 5.2 we conclude that, in contrast to pressure, lobbying may be an important means to sustain reputation. Whereas the enforcement of a threat is only used in equilibrium to improve upon an initially rather low reputation, lobbying may function as a way to preserve reputation. In this respect it is interesting to note that also in model LA, as in AL, when $p < a^2$ the only way for the weak type to get a sufficiently high reputation to induce the policymaker to give in in the second period, is to exert pressure. So, even in the unlikely case that a starting interest group with a low reputation already has lobbying access to the policymaker, the group must show its teeth first in order to gain an established position.

Next, we turn to the *expected occurrence* of pressure and lobbying in both models AL and LA, and relate it to the interest group's initial reputation. It appears that in case $p < a^2$ the expected occurrence of pressure is weakly increasing in initial reputation p, and when $a^2 < p < a$ weakly decreasing in initial reputation p.[23] In case $p > a$ pressure (the enforcement of the threat) does not occur. The expected occurrence of lobbying, on the other hand, weakly increases with initial reputation when $p < a$.[24] When $p > a$ lobbying either does not occur (ALE5, LAE4), occurs with certainty (ALE6, LAE5), or its expected occurrence is increasing in p (ALE2, ALE7, LAE6). Taking all the considerations from the last two paragraphs together, we conclude that pressure is mainly used to build up a reputation, whereas lobbying is merely used to maintain a reputation.[25] In other words, "actions" are typically used to obtain an established position, and when such a position is reached interest groups switch over to using "words" in order to preserve this position.

In both the AL and LA model it is observed that when the initial reputation of the interest group is fairly moderate, viz. $a^2 < p < a$, just lobbying may be sufficient to convince the policymaker. In this case lobbying rather than the enforcement of a threat is used to increase the reputation of the interest group. Roughly put, it appears that lobbying may be used to increase an — already notable — reputation when the policymaker only needs a small "push", and the enforcement of an explicit threat is too coarse a method to give such a little push. Also when $p < a^2$ initial lobbying may lead to an increase in reputation (cf. LAE2), but this increase is not large enough to completely convince the policymaker such that she concedes in both periods. Taken together these observations suggest that lobbying only leads to "small" improvements in reputation. This immediately leads to the question whether perhaps several small improvements finally add up to a significant increase in reputation. In the next section we investigate, by means of studying the LL game, whether repeated lobbying may lead to larger improvements in reputation, i.e. whether repeated lobbying may do the "trick" even in cases where the initial reputation is rather low ($p < a^2$).

5.4   THE LL GAME: REPUTATION BUILDING WHEN REPEATED LOBBYING IS POSSIBLE

In practice lobbyists interact frequently with policymakers, and lobbying is usually ascribed to have intertemporal characteristics. In this section we account for this observation by allowing lobbying in both the first and the second period in our model. That is, we study model LL in which in each period stage game L is played. Our focus is on the two questions whether (i) repeated lobbying may yield a starting (low initial reputation) interest group an established position, as was suggested at the end of the previous section, and (ii) whether the conclusions drawn on the basis of the AL and LA model carry over to the LL model. The results show that, indeed, under specific circumstances repeated lobbying can be used to build up a reputation. However, they still support the conclusions (i.c. Results 5.1 and 5.2) drawn in the previous two sections.

The LL model has a large number of perfect Bayesian equilibria satisfying STABAC. Therefore we refrain from listing all equilibria for every value of p and we mainly focus on the more interesting case $p < a^2$ instead. Even for this specific case the presentation of the plausible equilibria is already quite involved, and not very helpful. For that reason the formal description of all plausible equilibrium paths for the case $p < a^2$ is relegated to Appendix 5.A (cf. Proposition 5.4). That appendix also contains a formal description of the single two plausible equilibria for the case $p > a$ in which pressure occurs with positive probability (cf. Proposition 5.5). In the main text we verbally describe the equilibria presented in Propositions 5.4 and 5.5, and elaborate on them in regard to the two questions addressed in this section.

Based on the interest group's lobbying strategy in the first period two types of equilibria can be distinguished. In the *lobbying equilibria* the interest group lobbies with positive probability, in the *no-lobbying equilibria* the interest group does not lobby at all in the first period. In the latter type of equilibria the policymaker does not alter her belief after the first lobbying stage ($q_1(0)=p$), and the continuation game equals the AL game. Using Proposition 5.2 it is easily seen that we get three different equilibria with $\lambda_1(t_1)=\lambda_1(t_2)=0$, followed by the equilibrium continuation path ALE1, ALE2 or ALE3 (depending on the values of g and c).[26] The no-lobbying equilibria neither provide insight whether repeated lobbying may be used for building up a reputation, nor invalidate our Results 5.1 and 5.2. Therefore we focus in the sequel on the lobbying equilibria in which lobbying does occur with positive probability in the first period.

In all lobbying equilibria the weak type mixes between sending and not sending a lobbying message in the first period. The strong type always sends such a message. Silence reveals that the interest group is weak and leads to $x_1=x_2=N$. In that case the enforcement of a threat and/or lobbying in the second period cannot alter the period 2 decision of the policymaker anymore. After being exposed to initial lobbying the policymaker mixes between $x_1=N$ and $x_1=C$. The lobbying equilibria essentially differ in the continuation paths after the policymaker's choice of x_1. With respect to the path after $x_1=N$ three types of equilibria can be distinguished: (i) partial separation through both an enforcement and lobbying, (ii) full separation through both an enforcement and lobbying, and (iii) partial separation only through the enforcement of a threat.[27] Class (i) and (iii) contain only one equilibrium, whereas class (ii) contains two equilibria.[28] These latter two equilibria differ in their continuation path after $x_1=C$, based on ALE5 and ALE6, respectively (continuation path ALE7 is not possible).

In the two equilibria of class (ii) only the strong type enforces its threat after $x_1=N$, and pressure does not occur. In the lobbying equilibria belonging to the other two classes, on the other hand, the weak type now and then enforces its threat as well, and thus pressure does occur with positive probability. Another difference is that more information is revealed after $x_1=N$ in the complete separation path of class (ii). Consequently, $x_1=N$ is a "more attractive" option in the equilibria belonging to class (ii), and the interest group needs a higher reputation to make the policymaker indifferent between $x_1=N$ and $x_1=C$. Therefore, in the equilibria belonging to this class the weak type of interest group lobbies in the first period with a much lower probability than in the equilibria of the other two classes (cf. Proposition 5.4). In the first case the policymaker is more convinced that the interest group is of the strong type in case she receives a lobbying message.[29] In short, in the lobbying equilibria of class (ii) a first period lobbying message has a stronger impact on the reputation of the interest group than in the other lobbying equilibria.

As follows from our informal description of the lobbying equilibria, initial lobbying may persuade the policymaker not to put the interest group to the test in the first period, but to concede to its wishes. A second round of lobbying at the beginning of the second period may subsequently do the same in regard to the period 2 decision of the policymaker. Indeed, in the equilibria belonging to class (i) and (iii) repeated lobbying may finally yield the interest group a high reputation for being strong. Given the fact that in the equilibria of class (ii) initial lobbying has a larger impact, it is not surprising that in one equilibrium in this class (cf. LLE4 in Proposition 5.4), lobbying only once may already persuade the policymaker to give in in both periods. This shows that our conjecture at the end of the previous section is false; lobbying does not only lead to "small" improvements in reputation. However, we do find that in a repeated lobbying setting it is possible — for any feasible value of g and c — that the interest group builds up a high reputation by (repeated) lobbying.

In the lobbying equilibria first period lobbying may forestall that the interest group is put to the test. It must be noted, though, that initial lobbying cannot forestall such a test for sure, because after a lobbying message the policymaker mixes between conceding and not-conceding. When the interest group is put to this test ($x_1=N$) it has to pass ($y_1=E$) in order to increase its reputation. When it fails, reputation is completely lost and cannot be regained through lobbying. So, when the group's initial reputation is low ($p<a^2$) there always is a positive probability that the weak type interest group has to exert pressure in order to increase its reputation. Put differently, (repeated) lobbying cannot completely substitute for pressure in building up a reputation. Note that this conclusion holds even stronger for the AL and LA model, where it was found that exerting pressure was necessarily needed to get the weak type a sufficiently high reputation when its initial reputation is low. We summarize the observations of the previous two paragraphs in the following result:

Result 5.3
Only under specific circumstances (repeated) lobbying can be used for building up a reputation. However, when the initial reputation of the interest group is low (below a^2), there always is a positive probability that the group has to exert pressure (enforce its threat) to increase its reputation. □

Contrary to pressure, lobbying in the first period with positive probability is typically not required in order to build up a reputation for being strong (cf. the *no-lobbying equilibria*).[30] We conclude that in building up a reputation opportunities for enforcing a threat necessarily have to be utilized by the interest group, in contrast with opportunities for lobbying (in general).

Next we turn to our second question of whether Results 5.1 and 5.2 still obtain under model LL. This leads us to (partly) consider the case $p>a$. It appears that, as announced, Result 5.1 remains true; pressure (the enforcement of a threat) occurs with positive probability only when the actual reputation of the interest group is weakly below a^2 (cf. proof of Proposition 5.5). Lobbying, on the other hand, is an equilibrium phenomenon for (now) any value of the reputation of the group. Note that Result 5.1 relates the occurrence of pressure and lobbying to the actual reputation of the interest group, not just to its initial reputation.

144

Chapter 5

Interestingly, in the LL model pressure may also occur in case an interest group starts off with a high initial reputation (p>a), but loses this reputation over time. In that case the interest group may have to enforce its threat again to regain its established position. In both LLE8 and LLE9 of Proposition 5.5, for instance, lobbying in the first period fully convinces the policymaker ($q_1(c)=1$). When the interest group does not lobby in the first period, its reputation falls to $q_1(0)=a^2$. After $s_1=0$ the policymaker mixes between $x_1=C$ and $x_1=N$. When $x_1=N$ is chosen the weak type engages in pressure with positive probability. After history $h_1=(0,N,E)$ the group's reputation equals $q_2(0,N,E)=a$. Thus, in both equilibria pressure occurs with positive probability, and the reputation of the interest group may vary non-monotonically over time. Again, interest groups which have obtained an established position still may have to exert pressure once in a while (cf. Section 5.3).[31]

In both LLE8 and LLE9, although $s_1=c$ leads to $q_1(c)=1$, this conviction on the side of the policymaker has to be maintained through lobbying in the second period. Together with the results of Proposition 5.4 it follows that also Result 5.2 remains valid when considering the LL model; the interest group may be forced to lobby in order to maintain its reputation built up in the past and, contrary to lobbying, exerting pressure always leads to an increase in reputation. We conclude that also the results for the LL model corroborate our assertion that "actions" are typically used to obtain an established position, and "words" to maintain such a position.

Finally we briefly consider an alternative setup of the LL model that allows a focus on repeated lobbying without there being an opportunity for the weak type to exert pressure. To that purpose, suppose that reaction y_1 to $x_1=N$ is exogenously given for each type of interest group, with the weak type committed to choose $y_1=O$ after $x_1=N$ and the strong type committed to choose $y_1=E$ after $x_1=N$. In that case the two equilibria LLE4 and LLE5 are the only possible equilibria when $p<2a/(a+1)$ (the restriction $g<c<(2-g)g$ appearing in Proposition 5.4 has to be dropped). Hence, also in this alternative setup repeated lobbying may yield the (weak type of) interest group an established position, but there always is the possibility that the policymaker chooses $x_1=N$ (and then only reaction $y_1=E$ convinces the policymaker).

Somewhat surprisingly, when $p<a^2$ the weak type of interest group does not lose from a deprivation of the possibility to exert pressure in expected payoff terms. This follows because in all equilibria of the original LL model for the case $p<a^2$ (cf. Proposition 5.4) the weak type either prefers not to enforce ($y_1=O$), or is indifferent when it is put to the test ($x_1=N$). In other words, when $p<a^2$ $y_1=E$ after $x_1=N$ is never strictly preferred by the weak type, and thus the exclusion of this possibility does not hurt this type of interest group. (When $p>a^2$ the weak type does lose from a deprivation to exert pressure, though.) The strong type may either gain or lose when the weak type does not have the possibility to exert pressure (cf. Table 5.5 in Appendix 5.B). The strong type obtains gross gains because it becomes more difficult for the weak type to pretend being strong, and the strong type incurs gross losses because not conceding becomes a more attractive option for the policymaker. As noted above, the overall result may either be positive (a net gain) or negative (a net loss). Lastly, the policymaker invariably gains in expected payoff terms when the weak type cannot exert pressure (cf. Table 5.5 in Appendix 5.B). This follows because one option (viz. x=N) becomes more attractive in expected payoff terms, whereas the expected value of the other option (x=C) remains the same.

The next section considers the welfare implications of different opportunities for lobbying by comparing the four original models AA, AL, LA and LL.

5.5   WELFARE COMPARISON AND INSTITUTIONAL DESIGN

In this section it is briefly reported upon the results of a welfare comparison between the four models. We first consider the ex ante expected payoff the policymaker obtains in the equilibria of the different models (cf. Table 5.5 in Appendix 5.B). The welfare comparison reveals that the policymaker prefers that lobbying is certainly possible after her first period decision x_1. For any value of p the policymaker is at least as well off in the AL and LL models compared with the AA and LA models,[32] and due to the possibility of separation in the former models sometimes strictly better off. Surprisingly, the expected payoffs of the policymaker in model AA equal her expected payoffs in model LA. In a way, the added lobbying stage in the first period does not yield the policymaker additional information. The same holds true for the first lobbying stage when we compare the models AL and LL for the case $p < a^2$. For low values of the interest group's initial reputation these two models are payoff equivalent to the policymaker.[33]

As was already touched upon in Section 3.2 of Chapter 3, Ainsworth (1993, p. 49) informally discusses the idea that "..in anticipation of their interaction with lobbyists, legislators structure the institutional features under which they will play." He argues that policymakers control undue influence of better informed lobbyists by means of restricting access through the costliness of their signals. In an empirical study Brinig et al. (1993) also argue that lobbying regulations are introduced by policymakers to increase the costs of lobbying to the lobbyist, and thereby the value of the lobbyists to themselves. They find some empirical evidence that policymakers regulate lobbyists in such a way as to benefit those in the legislature (but see Lowery and Gray, 1997, for a critique on their empirical evidence).

Turning now to the question of institutional design in the context of the models presented in this chapter, it follows from the results of the welfare comparison discussed above that the policymaker should design its institutions in such a way as to allow lobbying in the second period. Establishing the possibility to lobby already in the first period, however, does not yield the policymaker additional gain, at least when starting interest groups with a low initial reputation ($p < a^2$) are considered. Hence, if providing the opportunity to lobby bears some costs for the policymaker, especially for groups that have not reached an established position yet and are not very well-known to the policymaker, she should not provide such an opportunity.[34] Roughly put, the welfare comparison suggests that the policymaker should design institutions in such a way as to force a starting interest group to show its teeth first before giving access for lobbying activities.[35] Moreover, after having reached an established position, the interest group should be given the opportunity to maintain this position through lobbying. In order to take full advantage of providing this second period lobbying possibility, the costs of lobbying c should be set by the policymaker in such a way as to enable separation (viz. $g<c<(2-g)g$).

The expected payoffs of the interest group are studied from an interim perspective, that is, after its type has become known to the group itself. (An analysis from an ex ante perspective would lead to the same type of conclusions, though.) The preferences of the interest group over the four models are less straightforward, and for some cases depend on the value of p, as well as on the possibility of separation in the AL and LL model. However, irrespective of the value of p model AA is (weakly) preferred over model LA by both types of the interest group (cf. Table 5.5 in Appendix 5.B). When $p<a^2$ the interest group (weakly) prefers model AL over AA. So, for low values of p the interest group's relative preferences concerning the two models AL and LA correspond with those of the policymaker ($AL\geq LA$). This is surprising, because only in the latter model the group can already directly influence the first period decision of the policymaker. The relative position of the LL model in the preference ordering of the interest group depends on the values of g and c (we only consider the case $p<a^2$). For both low values of the costs of lobbying ($c<g$), as well as fairly high values of c ($(2-g)g<c<1-g$), the preference ordering of the interest group is given by $AL\geq LL\geq AA\geq LA$. When $g<c<(2-g)g$ the relative preference concerning AL and LL reverses, and the ordering becomes $LL>AL>AA\geq LA$. For high values of c ($c>\max\{(2-g)g,1-g\}$), again LL is preferred over AL. The result that LA is the worst possible institutional setting for the interest group when its initial reputation is low supports our finding that the enforcement of a threat is especially suitable in building up a reputation, whereas lobbying is more appropriate for maintaining a high reputation.

5.6 CONCLUDING DISCUSSION

In this chapter a stylized game-theoretic model of lobbying and pressure is presented. Both lobbying and pressure are modeled as means of information transmission, with the former being interpreted as persuasion through the use of "words", and the latter as persuasion through the use of "actions". Formally, the difference between lobbying and pressure is reflected by two important assumptions. A first difference is that pressure, in contrast to lobbying, is assumed to be directly costly to the policymaker. A second difference is that the possibilities for lobbying are assumed to be independent of the policymaker's actions, whereas the opportunities for exerting pressure are not. Three different models are analyzed which differ in the opportunities an interest group has for lobbying. Our conclusions with respect to the building up and the maintenance of reputation are an overall evaluation of all three models. They are robust to alterations in the specific assumptions made with respect to the exact opportunities for lobbying.

The equilibrium analysis of the three different models yields that, in contrast to lobbying, pressure only occurs in case the reputation of the interest group is rather low, and always leads to an increase in reputation. The enforcement of a threat (pressure) is never used to maintain a reputation. It is also shown that only under rather specific circumstances (repeated) lobbying can be used for building up a reputation, and that it cannot completely substitute for pressure. Based on these results it is concluded that in order to build up a reputation, interest groups must first show their teeth and signal their importance through "actions". They must use heavy weaponry to get their favored policies implemented. Once a high reputation is established and the policymaker has acknowledged the importance of the group, just lobbying ("words") may be enough to convince the policymaker. Also, in order to maintain a good relationship with the policymaker, the interest group may be forced to lobby.

The theoretical predictions obtained from the model are hard to test empirically, because notions like stakes and, especially, reputation (agents' beliefs) are difficult to measure. Rigorous empirical support for the model may thus be difficult to obtain, as holds for any stylized game-theoretic model. (This remark applies when field data is used. Game-theoretic models of the type considered in this monograph are possibly more easy to test by using laboratory experiments.) However, the model yields some intuitions which, roughly, seem to correspond with reality. For instance, the prediction that lobbying in certain cases serves as a kind of maintenance costs is supported by the observation that lobbyists sometimes lobby legislators who are already inclined to take the decision preferred by the group ("friends"; cf. Chapter 2).

Although cast in terms of the interactions between an interest group and a governmental policymaker, our model seems to be relevant for the analysis of other forms of social interaction as well. For instance, the interactions between an interest group and a private company may also directly fit the general description of our model. In fact, the opening quote of this chapter already suggests this broader applicability.[36] In their successful attempt to prevent the sinking of the *Brent Spar* (summer 1995), Greenpeace pressured Shell by getting consumers to boycott Shell gas stations. This established a reputation of public interest groups being able to mobilize public opinion. Arguably, later attempts to persuade Shell not to invest in Nigeria were taken far more seriously by the oil company.[37]

The welfare comparison between the several models studied indicates that the policymaker should design its institutions in such a manner as to force an interest group which has not yet reached an established position to show its teeth first before giving access for lobbying activities (given that the policymaker cannot prohibit pressure). In view of this and our other results, a codification of the behavior of interest groups, as proposed by a Dutch director of Shell in his 1996 new year's speech, would not be effective until an interest group has reached an established position.[38] Therefore, it might not be as useful as Shell thinks it is.

Though it is easy to think of several interesting ways to extend the present model,[39] we only want to elaborate briefly on just three of them. First, it might be interesting to model a situation in which also the government (policymaker) can show its teeth and build up a reputation for being strong when interacting with interest groups. This would entail considering the case of two-sided asymmetric information (cf. Carlsen, 1994, Kreps and Wilson, 1982). Given that policymakers typically have a short time horizon, however, such a model might not be very well in accordance with existing practice. Second, in order to take account of changing economic environments (e.g. the business cycle), it might be interesting to extend the model to a two dimensional state space. In that case the policymaker is not only uncertain about the fixed underlying preferences of the interest group (weak or strong), but also about the specific economic circumstances under which they interact and that vary from period to period. The interest group, of course, knows both these aspects. Such a model would be in line with the reputation building model of Sobel (1985).

Third, a nice feature of the present model is that, without having to assume that preferences are stochastic and vary over the two periods (cf. Calvert, 1987), in equilibrium reputation may vary non-monotonically over time. The interest group is sometimes able to regain its reputation lost in the past. In our model this loss of reputation is a consequence of the interest group's failure to maintain its relationship with government through lobbying. It would be nice to account for the fact that, in reality, reputation is also typically subject to decay due to turnover among policymakers (think of elections). When a new policymaker comes into office, established interest groups may have to regain their recognized position because a political newcomer is likely to be incompletely informed on the record of the interest group (cf. Konrad and Torsvik, 1997).[40] Therefore, a model in which the policymaker has no complete knowledge about the past history of play seems more realistic. For a first attempt to model these kind of aspects in the context of a game with reputation building see Massó (1996).[41]

In this chapter we focused on the choice of an interest group between two different ways to influence a single policymaker. An important determinant of this choice was the behavior of the policymaker herself. The next chapter focuses on the choice an interest group has between trying to influence the legislative branch and/or the executive branch of government. It will be seen that, again, an important determinant of this choice is the actual behavior of politicians and bureaucrats.

APPENDIX 5.A FORMAL DESCRIPTION OF THE EQUILIBRIUM PATHS OF THE LL GAME

Proposition 5.4

In case $p<a^2$ the plausible equilibrium paths of the LL game are given by:

$c<g$ LLE1: $\lambda_1(t_1)=0$, $\lambda_1(t_2)=0$, $\mu_1(0)=0$, $\pi_1(t_1;0)=(1-a)p/(1-p)a$, $\pi_1(t_2;0)=1$, $\lambda_2(t_1;0,N,O)=0$, $\lambda_2(t_1;0,N,E)=1$, $\lambda_2(t_2;0,N,E)=1$, $\mu_2(0,N,O,0)=0$, $\mu_2(0,N,E,c)=1-g+c$; $q_1(0)=p$, $q_2(0,N,O,0)=0$, $q_2(0,N,E,c)=a$.

$c<g$ LLE2: $\lambda_1(t_1)=(1-a^2)p/(1-p)a^2$, $\lambda_1(t_2)=1$, $\mu_1(0)=0$, $\mu_1(c)=c$, $\pi_1(t_1;0)=0$, $\pi_1(t_1;c)=a/(1+a)$, $\pi_1(t_2;c)=1$, $\lambda_2(t_1;0,N,O)=\lambda_2(t_1;c,N,O)=0$, $\lambda_2(t_1;c,C,O)=a/(1+a)$, $\lambda_2(t_1;c,N,E)=1$, $\lambda_2(t_2;c,C,O)=\lambda_2(t_2;c,N,E)=1$, $\mu_2(0,N,O,0)=\mu_2(c,N,O,0)=\mu_2(c,C,O,0)=0$, $\mu_2(c,C,O,c)=c$, $\mu_2(c,N,E,c)=1-g+c$; $q_1(0)=0$, $q_1(c)=a^2$, $q_2(0,N,O,0)=q_2(c,N,O,0)=q_2(c,C,O,0)=0$, $q_2(c,C,O,c)=q_2(c,N,E,c)=a$.

$g<c<(2-g)g$ LLE3: $\lambda_1(t_1)=0$, $\lambda_1(t_2)=0$, $\mu_1(0)=0$, $\pi_1(t_1;0)=0$, $\pi_2(t_2;0)=1$, $\lambda_2(t_1;0,N,O)=0$, $\lambda_2(t_2;0,N,E)=1$, $\mu_2(0,N,O,0)=0$, $\mu_2(0,N,E,c)=1$; $q_1(0)=p$, $q_2(0,N,O,0)=0$, $q_2(0,N,E,c)=1$.

$g<c<(2-g)g$ LLE4: $\lambda_1(t_1)=p(1-a)/2a(1-p)$, $\lambda_1(t_2)=1$, $\mu_1(0)=0$, $\mu_1(c)=\frac{1}{2}c$, $\pi_1(t_1;0)=\pi_1(t_1;c)=0$, $\pi_1(t_2;c)=1$, $\lambda_2(t_1;0,N,O)=\lambda_2(t_1;c,N,O)=\lambda_2(t_1;c,C,O)=0$, $\lambda_2(t_2;c,C,O)=0$, $\lambda_2(t_2;c,N,E)=1$, $\mu_2(0,N,O,0)=\mu_2(c,N,O,0)=0$, $\mu_2(c,C,O,0)=\mu_2(c,N,E,c)=1$; $q_1(0)=0$, $q_1(c)=2a/(a+1)$, $q_2(0,N,O,0)=q_2(c,N,O,0)=0$, $q_2(c,C,O,0)=2a/(a+1)$, $q_2(c,N,E,c)=1$.

$g<c<(2-g)g$ LLE5: $\lambda_1(t_1)=p(1-a)/2a(1-p)$, $\lambda_1(t_2)=1$, $\mu_1(0)=0$, $\mu_1(c)=c/(2-c)$, $\pi_1(t_1;0)=\pi_1(t_1;c)=0$, $\pi_1(t_2;c)=1$, $\lambda_2(t_1;0,N,O)=\lambda_2(t_1;c,N,O)=0$, $\lambda_2(t_1;c,C,O)=1$, $\lambda_2(t_2;c,C,O)=\lambda_2(t_2;c,N,E)=1$, $\mu_2(0,N,O,0)=\mu_2(c,N,O,0)=0$, $\mu_2(c,C,O,c)=\mu_2(c,N,E,c)=1$; $q_1(0)=0$, $q_1(c)=2a/(a+1)$, $q_2(0,N,O,0)=q_2(c,N,O,0)=0$, $q_2(c,C,O,c)=2a/(a+1)$, $q_2(c,N,E,c)=1$.

$(2-g)g<c<1-g$ LLE6: $\lambda_1(t_1)=0$, $\lambda_1(t_2)=0$, $\mu_1(0)=0$, $\pi_1(t_1;0)=(1-a)p/(1-p)a$, $\pi_1(t_2;0)=1$, $\lambda_2(t_1;0,N,O)=\lambda_2(t_1;0,N,E)=0$, $\lambda_2(t_2;0,N,E)=0$, $\mu_2(0,N,O,0)=0$, $\mu_2(0,N,E,0)=1-g$; $q_1(0)=p$, $q_2(0,N,O,0)=0$, $q_2(0,N,E,0)=a$.

(2-g)g<c　　　LLE7: $\lambda_1(t_1)=(1-a^2)p/(1-p)a^2$, $\lambda_1(t_2)=1$, $\mu_1(0)=0$, $\mu_1(c)=c$,
$\pi_1(t_1;0)=0$, $\pi_1(t_1;c)=a/(1+a)$, $\pi_1(t_2;c)=1$, $\lambda_2(t_1;0,N,O)=$
$\lambda_2(t_1;c,N,O)=0$, $\lambda_2(t_1;c,C,O)=a/(1+a)$, $\lambda_2(t_1;c,N,E)=0$,
$\lambda_2(t_2;c,C,O)=1$, $\lambda_2(t_2;c,N,E)=0$, $\mu_2(0,N,O,0)=\mu_2(c,N,O,0)=$
$\mu_2(c,C,O,0)=0$, $\mu_2(c,C,O,c)=c$, $\mu_2(c,N,E,0)=1-g$; $q_1(0)=0$, $q_1(c)=a^2$,
$q_2(0,N,O,0)=0$, $q_2(c,N,O,0)=0$, $q_2(c,C,O,0)=0$,
$q_2(c,C,O,c)=q_2(c,N,E,0)=a$. $\qquad\qquad\qquad\qquad\qquad\qquad$ □

The argument in $\mu_1(\cdot)$, $\pi_1(t_j;\cdot)$ for j=1,2, and $q_1(\cdot)$ refers to s_1, $\lambda_2(t_j;\cdot)$ for j=1,2
depends on $h_1=(s_1,x_1,y_1)$, and $\mu_2(\cdot)$ and $q_2(\cdot)$ depend on (h_1,s_2). For ease of
exposition p is left out as an argument in the equilibrium strategies and beliefs.
The additional condition c<1-g on LLE6 is a consequence of the refinement
concept employed. When c>1-g the strong type has a stronger incentive to deviate
to out-of-equilibrium message s_1=c. Universal divinity then implies $q_1(c)$=1, and
both types want to deviate to s_1=c. This upsets LLE6 when c>1-g.

Proposition 5.5
When p>a the plausible equilibrium paths of the LL game in which pressure
occurs with positive probability are given by:

max{c,(2-3c)/(1-c)}<g
LLE8: $\lambda_1(t_1)=0$, $\lambda_1(t_2)=(p-a^2)/(1-a^2)p$, $\mu_1(0)=[2-2c+(g-c)(1-g)]/(2-g)g$, $\mu_1(c)=1$,
$\pi_1(t_1;0)=a/(1+a)$, $\pi_1(t_2;0)=1$, $\lambda_2(t_1;0,N,O)=0$, $\lambda_2(t_1;0,C,O)=a/(1+a)$, $\lambda_2(t_1;0,N,E)=1$,
$\lambda_2(t_2;0,C,O)=\lambda_2(t_2;0,N,E)=\lambda_2(t_2;c,C,O)=1$, $\mu_2(0,N,O,0)=\mu_2(0,C,O,0)=0$,
$\mu_2(0,C,O,c)=c$, $\mu_2(0,N,E,c)=1-g+c$, $\mu_2(c,C,O,c)=1$; $q_1(0)=a^2$, $q_1(c)=1$,
$q_2(0,N,O,0)=q_2(0,C,O,0)=0$, $q_2(0,C,O,c)=q_2(0,N,E,c)=a$, $q_2(c,C,O,c)=1$.

(2-3c)/(1-c)<g and (2-g)g<c
LLE9: $\lambda_1(t_1)=0$, $\lambda_1(t_2)=(p-a^2)/(1-a^2)p$, $\mu_1(0)=[2-2c+g(1-g)]/[(2-g)g+c(1-g)]$, $\mu_1(c)=1$,
$\pi_1(t_1;0)=a/(1+a)$, $\pi_1(t_2;0)=1$, $\lambda_2(t_1;0,N,O)=0$, $\lambda_2(t_1;0,C,O)=a/(1+a)$, $\lambda_2(t_1;0,N,E)=0$,
$\lambda_2(t_2;0,C,O)=1$, $\lambda_2(t_2;0,N,E)=0$, $\lambda_2(t_2;c,C,O)=1$, $\mu_2(0,N,O,0)=\mu_2(0,C,O,0)=0$,
$\mu_2(0,C,O,c)=c$, $\mu_2(0,N,E,0)=1-g$, $\mu_2(c,C,O,c)=1$; $q_1(0)=a^2$, $q_1(c)=1$,
$q_2(0,N,O,0)=q_2(0,C,O,0)=0$, $q_2(0,C,O,c)=q_2(0,N,E,0)=a$, $q_2(c,C,O,c)=1$. $\qquad$ □

Both LLE8 and LLE9 already exist when p>a^2, but when a^2<p<a other equilibria
exist in which pressure occurs with positive probability. The condition g>(2-3c)/(1-
c) is equivalent to c>(2-g)/(3-g); the costs of lobbying must be sufficiently high
for both equilibria to exist. The maximal gain the strong type gets from not
separating itself out and choosing s_1=0 is 2+g+c(1-g). The payoff the strong type
obtains when separating itself out by lobbying twice is 4-2c. In case c<(2-g)/(3-g)
the costs of separation are so low for the strong type that it always prefers to
separate itself out.

APPENDIX 5.B PROOFS OF THE PROPOSITIONS

In general, a perfect Bayesian equilibrium of the different models considered in the main text is a set of strategies $(\mu_1,\mu_2,\lambda_1,\lambda_2,\pi_1,\pi_2)$ such that in every period each player maximizes expected payoffs given the strategy of the other player, and a set of beliefs $(q_1(p,s_1),q_2(p,s_1,x_1,y_1,s_2))$ that is consistent with Bayes' rule along the equilibrium path. A precise definition of a PBE for each of the models is straightforward, and omitted here (it is implicitly given in the proofs, though). STABAC puts additional restrictions on out-of-equilibrium beliefs for which Bayes' rule does not apply. These restrictions, based on the (universal) divinity refinement concept of Banks and Sobel (1987), are reviewed and discussed below. Appendix 5.C contains a supplementary discussion of additional equilibrium refinements.

In proving the propositions we make use of three lemmas. Lemma 5.1 supplements Proposition 5.1 and gives the plausible equilibria of the AA model also for the knife-edge beliefs $q_1(p,s_1) \in \{0,a^2,a,1\}$. This lemma is used in the proof of Proposition 5.3 when deriving the plausible equilibria of the LA model. In the models AL and LL the continuation game after $x_1=C$ equals period game L, which is a trivial extension of the basic signaling game already analyzed at length in Chapter 3. We state the universally divine equilibria for this one-dimensional signaling game in Lemma 5.2, for any value of $q_1(p,s_1)$ ($0 \le q_1(p,s_1) \le 1$). In the same two models the continuation game starting after $x_1=N$ can be envisaged as a multi-dimensional signaling game. In Lemma 5.3. we derive the universally divine equilibria for this continuation game.

The (universal) divinity equilibrium refinement. In all models the equilibrium reaction strategy of the interest group in the second period after $x_2=N$ is independent of the history of play and equals $\pi_2(t_1)=0$ and $\pi_2(t_2)=1$. Given this reaction, the equilibrium strategy of the policymaker in period 2 is given by:

$$\mu_2(p,h_1,s_2) = \begin{cases} 1 & > \\ \in [0,1] & q_2(p,h_1,s_2) = a \\ 0 & < \end{cases} \tag{5.1}$$

Expression (5.1), together with the reaction strategy of the interest group, describes all perfect Bayesian equilibrium paths for period game A.

The signaling continuation games analyzed in Lemmas 5.1, 5.2 and 5.3 are so-called monotonic signaling games (Cho and Sobel, 1990); given a specific signal, all types prefer the policymaker to give in (in period 2) with a probability as high as possible. Therefore, the application of universal divinity in the three lemmas is equivalent to the application of *D1* (Cho and Kreps, 1987) and *never a weak best response*, and generically equivalent to *strategic stability* (cf. Cho and Sobel, 1990).

The universal divinity refinement requires the following restriction on out-of-equilibrium beliefs. Let $U_G^{sub}(t_j|x_1)$ be the expected payoff of the type t_j interest group (j=1,2) in a supposed equilibrium path of the continuation game that starts after the policymaker's choice of x_1.[42] Note that these payoffs by definition do not incorporate any expenditures on first period lobbying, but do include the (expected) gross payoffs (cf. Table 5.1) obtained in both periods, as well as any expenditures on second period lobbying. Let $s=(y_1,s_2)$ be a signal not sent by any type in equilibrium, with $s \in \{(E,\varnothing),(O,\varnothing)\}$ in Lemma 5.1, $s \in \{(O,0),(O,c)\}$ in Lemma 5.2, and $s \in \{(O,0),(O,c),(E,0),(E,c)\}$ in Lemma 5.3. Let BR(s) the set of second period strategies of the policymaker $\mu_2(p,s_1,x_1,s)$ which are potentially a best response to s. With expression (5.1) above, and any inference $q_2(p,s_1,x_1,s) \in [0,1]$ possible after receiving signal s, we have BR(s)=[0,1]. Let $R_j(s) \subseteq BR(s)$ be the set of (best) responses for which type t_j (j=1,2) has a weak incentive to deviate (as in t_j, the subscript j in $R_j(s)$ refers to the type of interest group and not to the period):

$$R_1(s) = \{\mu_2(p,s_1,x_1,s) \equiv \mu_2 \in BR(s) \mid 3 + \mu_2 - c(s|x_1;t_1) \geq U_G^{sub}(t_1|x_1)\} \quad (5.2)$$

$$R_2(s) = \{\mu_2(p,s_1,x_1,s) \equiv \mu_2 \in BR(s) \mid 2+g+(2-g)\mu_2 - c(s|x_1;t_2) \geq U_G^{sub}(t_2|x_1)\}$$

$$(5.3)$$

in which $c(s|x_1;t_j)$ denotes the cost of sending signal s for type t_j (j=1,2) given the policymaker's choice of x_1. These costs are made specific in Lemmas 5.1, 5.2 and 5.3. In case $0 < q_1(p,s_1) < 1$, universal divinity requires the following restriction on out-of equilibrium beliefs $q_2(p,s_1,x_1,s)$:

$$0 < q_1(p,s_1) < 1 \qquad \begin{array}{l} R_2(s) \subset R_1(s) \ \Rightarrow \ q_2(p,s_1,x_1,s) = 0 \\ \\ R_1(s) \subset R_2(s) \ \Rightarrow \ q_2(p,s_1,x_1,s) = 1 \end{array} \qquad (5.4)$$

The universal divinity refinement procedure is not defined for degenerate prior beliefs $q_1(p,s_1) \in \{0,1\}$. Simply extending the refinement to cover also degenerate priors is not possible, for an equilibrium satisfying the restriction does not always exist.[43] Therefore, when $q_1(p,s_1) \in \{0,1\}$ we apply the following related, but somewhat weaker divinity restriction:

$$q_1(p,s_1) \in \{0,1\} \qquad \begin{array}{l} R_2(s) \subset R_1(s) \ \Rightarrow \ q_2(p,s_1,x_1,s) \leq q_1(p,s_1) \\ \\ R_1(s) \subset R_2(s) \ \Rightarrow \ q_2(p,s_1,x_1,s) \geq q_1(p,s_1) \end{array} \qquad (5.5)$$

In Propositions 5.3, 5.4 and 5.5 the universal divinity restriction given in expression (5.4) is extended to the multi-period signaling game in the following way (cf. Cho, 1993). For any out-of-equilibrium message s_1 it is verified whether type t_j (j=1,2) gains — in comparison with the expected payoff in the supposed equilibrium path — when message s_1 leads the policymaker to belief that the group is strong with probability $q_1(p,s_1)$, and after s_1 an equilibrium path for the continuation game based on this belief $q_1(p,s_1)$ follows. Note that there may be multiple equilibrium continuation paths after $q_1(p,s_1)$. In case the weak type always has a weak incentive to deviate whenever the strong type has (that is, when for any inference $q_1(p,s_1)$ and subsequent equilibrium continuation path for which the strong type wants to deviate, the weak type also wants to deviate), but not vice versa, we conclude that $q_1(p,s_1)=0$. In the opposite case we conclude that $q_1(p,s_1)=1$. If both cases do not apply, universal divinity does not put any restrictions on $q_1(p,s_1)$. Because by assumption a degenerate prior belief does not occur, i.e. $0<p<1$ is assumed, we need not consider an additional condition like (5.5) in this case.

     In the specification of the following three lemmas, we drop the arguments p and s_1 in all equilibrium strategies and beliefs (except in $q_1(p,s_1)$ where we only drop p to get $q_1(s_1)$). Because $s_2=\varnothing$ in the AA and LA model, s_2 is dropped as an argument in Lemma 5.1. In addition, since in Lemma 5.2 necessarily $x_1=C$ and $y_1=O$, we drop the argument (x_1,y_1) in $\mu_2(p,h_1,s_2)$ and $q_2(p,h_1,s_2)$ as well, and we get $\mu_2(s_2)$ and $q_2(s_2)$. In Lemma 5.3 necessarily $x_1=N$ and we do not incorporate $x_1=N$ as an argument in the equilibrium strategies and beliefs. Specifically, we get $\pi_1(t_j)$, $\lambda_2(t_j;y_1)$, $\mu_2(y_1,s_2)$ and $q_2(y_1,s_2)$ (j=1,2). As noted, the interest group's reaction strategy in period 2 is always given by $\pi_2(t_1)=0$ and $\pi_2(t_2)=1$, and is left implicit in the statement of the lemmas and propositions.

Lemma 5.1
In model LA the plausible equilibrium paths after the policymaker has received message s_1 are given by:

$\underline{q_1(s_1)=0:}$ AAE4: $\mu_1=0$, $\pi_1(t_1)=0$, $\pi_1(t_2)=1$, $\mu_2(N,O)=0$, $\mu_2(N,E)\leq 1-g$; $q_2(N,O)=0$, $q_2(N,E)\leq a$.

$\underline{q_1(s_1)\leq a^2:}$ AAE1: $\mu_1=0$, $\pi_1(t_1)=(1-a)q_1(s_1)/(1-q_1(s_1))a$, $\pi_1(t_2)=1$, $\mu_2(N,O)=0$, $\mu_2(N,E)=1-g$; $q_2(N,O)=0$, $q_2(N,E)=a$.

$\underline{q_1(s_1)=a^2:}$ AAE5: $\mu_1\in[0,1]$, $\pi_1(t_1)=a/(1+a)$, $\pi_1(t_2)=1$, $\mu_2(N,O)=\mu_2(C,O)=0$, $\mu_2(N,E)=1-g$; $q_2(N,O)=0$, $q_2(C,O)=a^2$, $q_2(N,E)=a$.

$\underline{a^2<q_1(s_1)<a:}$ AAE2: $\mu_1=1$, $\mu_2(C,O)=0$; $q_2(C,O)=q_1(s_1)$.

$\underline{q_1(s_1)=a:}$ AAE6: $\mu_1=1$, $\mu_2(C,O)\in[0,1]$; $q_2(C,O)=a$.

$\underline{a<q_1(s_1):}$ AAE3: $\mu_1=1$, $\mu_2(C,O)=1$; $q_2(C,O)=q_1(s_1)$. $\square$

Note that in the AA game $s_1=\varnothing$, hence $q_1(s_1)=q_1(\varnothing)=p$. Therefore, Proposition 5.1 follows directly from Lemma 5.1.

Proof of Lemma 5.1.
For $q_1(s_1)\notin\{0,a^2,a,1\}$ the lemma follows directly from Propositions 1 and 2 in Potters and Van Winden (1990). So, here we just consider the four cases left. Because the lemma considers the continuation game after message s_1 which starts from updated belief $q_1(s_1)$, we write $q_2(q_1(s_1),x_1,y_1)$ instead of $q_2(p,s_1,x_1,y_1)$. The posterior belief $q_2(q_1(s_1),x_1,y_1)$ that the interest group is of type t_2 after history (x_1,y_1), and given initial (but already updated) beliefs $q_1(s_1)$, is determined by Bayes' rule whenever this rule applies:[44]

$$q_2(q_1(s_1),N,O) = \frac{q_1(s_1)(1-\pi_1(t_2))}{q_1(s_1)(1-\pi_1(t_2))+(1-q_1(s_1))(1-\pi_1(t_1))} \qquad (5.6)$$

$$q_2(q_1(s_1),N,E) = \frac{q_1(s_1)\pi_1(t_2)}{q_1(s_1)\pi_1(t_2)+(1-q_1(s_1))\pi_1(t_1)} \qquad (5.7)$$

$$q_2(q_1(s_1),C,O) = q_1(s_1) \qquad (5.8)$$

In case the denominators in (5.6) and (5.7) vanish, Bayes' rule does not apply. Restrictions on out-of-equilibrium beliefs are given by (5.4) and (5.5) above (with $q_2(q_1(s_1),x_1,s)$ written instead of $q_2(p,s_1,x_1,s)$), where $c((O,\varnothing)|N;t_1)=1$, $c((E,\varnothing)|N;t_1)=2-g$, $c((O,\varnothing)|N;t_2)=2$, and $c((E,\varnothing)|N;t_2)=2-g$. Note that in these costs the first period outcome $(x_1,y_1)=(C,O)$ is taken as reference point, that is, $c((O,\varnothing)|C;t_1)=c((O,\varnothing)|C;t_2)=0$.

First, let $q_1(s_1)=0$. Suppose the policymaker chooses $x_1=N$ and the weak type chooses $y_1=E$ with positive probability, i.e. $\pi_1(t_1)>0$. We get $q_2(0,N,E)=0$ by (5.7), and given (5.1) the weak type strictly prefers $y_1=O$. Hence the weak type always chooses $y_1=O$ after $x_1=N$. In that case the policymaker indeed chooses $x_1=N$. To secure that the weak type does not want to deviate to $y_1=E$ we must have $\mu_2(N,E)\leq1-g$. By refinement (5.5) we must have $q_2(0,N,E)\geq0$, imposing no further restrictions. Note that applying refinement (5.4) rather than (5.5) would let us to conclude $q_2(0,N,E)=1$, upsetting equilibrium AAE4, and no equilibrium satisfying (5.4) exists.

From Potters and Van Winden (1990) follows that when $0<q_1(s_1)<1$ the interest group's equilibrium reaction strategy in any plausible equilibrium is given by $\pi_1(t_1)=\min\{(1-a)q_1(s_1)/(1-q_1(s_1))a,1\}$ and $\pi_1(t_2)=1$. Given this reaction strategy, when $q_1(s_1)>a^2$ the policymaker chooses $x_1=C$, when $q_1(s_1)<a^2$ she chooses $x_1=N$. When $q_1(s_1)=a^2$ the policymaker is indifferent. Together with the reaction strategy after $x_1=N$ this yields AAE5. In case $q_1(s_1)=a$ the policymaker chooses $x_1=C$ for sure, and $\mu_2(C,O)$ arbitrarily follows from (5.1) and (5.8).

Lastly, let $q_1(s_1)=1$. From (5.1), (5.6) and (5.7) follows that the strong type uses a pure reaction strategy to $x_1=N$ in equilibrium. Always choosing $y_1=E$

induces $\mu_1=1$, $q_2(C,O)=1$ by (5.8), and $\mu_2(C,O)=1$ by (5.1). This yields AAE3. When the strong type always chooses $y_1=O$ after $x_1=N$, the policymaker prefers to choose $x_1=N$ and we get the supposed equilibrium path $\mu_1=0$, $\pi_1(t_1)=0$, $\pi_1(t_2)=0$, $\mu_2(N,O)=1$ and $\mu_2(N,E)\leq1-g/(2-g)$. The latter inequality is required to secure that the strong type does not want to deviate to $\pi_1(t_2)=1$. Applying the refinement in (5.5) we get from $R_1((N,E))=\varnothing$ and $R_2((N,E))=[1-g/(2-g),1]$ that $q_2(N,E)=1$, and thus $\mu_2(N,E)=1$, upsetting the supposed equilibrium path.   *QED.*

Lemma 5.2
In models AL and LL the plausible equilibrium paths after $(x_1,y_1)=(C,O)$ are given by:

$\underline{q_1(s_1)=0}$:   C1: $\lambda_2(t_1)=0$, $\lambda_2(t_2)=0(\in[0,1])[=1]$ if $\mu_2(c)<(=)[>]c/(2-g)$, $\mu_2(0)=0$,
                     $\mu_2(c)\leq c$; $q_2(0)=0$, $q_2(c)\leq a$.

$\underline{0<q_1(s_1)<a}$:   C2: $\lambda_2(t_1)=q_1(s_1)(1-a)/a(1-q_1(s_1))$, $\lambda_2(t_2)=1$, $\mu_2(0)=0$, $\mu_2(c)=c$;
                     $q_2(0)=0$, $q_2(c)=a$.

$\underline{q_1(s_1)=a}$:   C3: $\lambda_2(t_1)=\lambda_2(t_2)=0$, $\mu_2(0)\in[1-c/(2-g),1]$; $q_2(0)=a$.
                     C4: $\lambda_2(t_1)=\lambda_2(t_2)=1$, $\mu_2(c)\in[c,1]$; $q_2(c)=a$.

$\underline{a<q_1(s_1)<1}$:   C5: $\lambda_2(t_1)=\lambda_2(t_2)=0$, $\mu_2(0)=1$; $q_2(0)=q_1(s_1)$.
                     C6: $\lambda_2(t_1)=\lambda_2(t_2)=1$, $\mu_2(c)=1$; $q_2(c)=q_1(s_1)$.
                     C7: $\lambda_2(t_1)=0$, $\lambda_2(t_2)=(q_1(s_1)-a)/q_1(s_1)(1-a)$, $\mu_2(0)=1-c/(2-g)$, $\mu_2(c)=1$;
                     $q_2(0)=a$, $q_2(c)=1$.

$\underline{q_1(s_1)=1}$:   C5.
                     C8: $\lambda_2(t_1)=0(\in[0,1])[=1]$ if $\mu_2(0)>(=)[<]1-c$, $\lambda_2(t_2)=1$, $\mu_2(0)\leq1-$
                     $c/(2-g)$, $\mu_2(c)=1$; $q_2(0)\leq a$, $q_2(c)=1$.   □

Proof of Lemma 5.2.
When $x_1=C$ and $y_1=O$ we have $q_2(p,s_1,C,O)\equiv q_2(q_1(s_1),C,O)=q_1(s_1)$ by (5.8). The proof for $q_1(s_1)\notin\{0,a,1\}$ is given in Potters and Van Winden (1992) (but also follows directly from the proof of Proposition 4.5), so here we only consider the three cases left. The posterior belief $q_2(s_2)=q_2(q_1(s_1),C,O,s_2)$ that the interest group is of type t_2 after receiving signal s_2, and given beliefs $q_1(s_1)$ $(\equiv q_2(q_1(s_1),C,O))$ at the beginning of period 2, is determined by Bayes' rule whenever this rule applies:

$$q_2(0) = \frac{q_1(s_1)(1-\lambda_2(t_2))}{q_1(s_1)(1-\lambda_2(t_2))+(1-q_1(s_1))(1-\lambda_2(t_1))} \qquad (5.9)$$

$$q_2(c) = \frac{q_1(s_1)\lambda_2(t_2)}{q_1(s_1)\lambda_2(t_2)+(1-q_1(s_1))\lambda_2(t_1)} \qquad (5.10)$$

In case the denominators in (5.9) and (5.10) vanish, Bayes' rule does not apply. Additional restrictions on out-of-equilibrium beliefs follow from (5.4) and (5.5) in the STABAC procedure. For the continuation game after $x_1=C$ follows $c((0,0)|C;t_j)=0$ and $c((0,c)|C;t_j)=c$ for $j=1,2$.

First, let $q_1(s_1)=0$. When the weak type mixes we get $q_2(0)=q_2(c)=0$ by (5.9) and (5.10), and given (5.1), the weak type does not want to send $s_2=c$. Hence the weak type chooses $s_2=0$ for sure; $q_2(0)=0$, and by (5.1), $\mu_2(0)=0$. To secure that the weak type does not want to deviate to $s_2=c$ we must have $\mu_2(c)\leq c$. By refinement (5.5) and $R_1((0,c))=[c,1]\subset[c/(2-g),1]=R_2((0,c))$ follows $q_2(c)\geq0$. This gives no additional restrictions and yields C1. Note that equilibrium C1 does not satisfy (5.4).

Next, let $q_1(s_1)=a$. From the sorting condition $2-g>1$ follows $\lambda_2(t_1)>0 \Rightarrow \lambda_2(t_2)=1$. Using (5.1) this implies in turn that when $q_1(s_1)=a$ the types must necessarily pool on a pure strategy. First, suppose they pool on $s_2=0$. In case $\mu_2(0)\geq1-c/(2-g)$ we get $R_1((0,c))=R_2((0,c))=\varnothing$, and (5.4) does not restrict $q_2(c)$; pooling on $s_2=0$ can be sustained with any out-of-equilibrium belief $q_2(c)$ (C3). When $\mu_2(0)<1-c/(2-g)$ we get $R_1((0,c))=[\mu_2(0)+c,1]$ and $R_2((0,c))=[\mu_2(0)+c/(2-g),1]$;[45] from $g<1$ follows $R_1((0,c))\subset R_2((0,c))$, and by (5.4) then $q_2(c)=1$. But when $q_2(c)=1$ (at least) the strong type has an incentive to deviate, upsetting pooling on $s_2=0$ as an equilibrium when $\mu_2(0)<1-c/(2-g)$. Second, consider the case of pooling on $s_2=c$. We get $q_2(c)=a$ by (5.10). Note that $\mu_2(c)\geq c$ is required to let both types choose $s_2=c$. In that case $R_1((0,0))=[\mu_2(c)-c,1]$ and $R_2((0,0))=[\mu_2(c)-c/(2-g),1]$; $q_2(0)=0$ by (5.4), and thus $\mu_2(0)=0$ by (5.1) (C4).[46]

Lastly, when $q_1(s_1)=1$ the strong type must play a pure strategy, and chooses either $s_2=0$ or $s_2=c$ for sure. In the first case $R_1((0,c))=R_2((0,c))=\varnothing$, and even out-of-equilibrium belief $q_2(c)=1$ sustains C5. In the latter case $R_2((0,0))=[1-c/(2-g),1]\subset[1-c,1]=R_1((0,0))$, and (5.5) requires $q_2(0)\leq1$. Only choices $q_2(0)\leq a$ and $\mu_2(0)\leq1-c/(2-g)$ sustain the choice of $s_2=c$ by the strong type, to avoid deviation to $s_2=0$. This yields C8.[47]   QED.

Lemma 5.3
In models AL and LL the plausible equilibrium paths after $x_1=N$ are given by:

$\underline{q_1(s_1)=0}$:

N1: $\pi_1(t_1)=0$, $\pi_1(t_2)=1$, $\lambda_2(t_1;O)=0$, $\lambda_2(t_2;E)=0(\in[0,1])[=1]$ if $\{\mu_2(E,c)-\mu_2(E,0)\}$ $<(=)[>]$ $c/(2-g)$, $\mu_2(O,O)=0$, $\mu_2(E,0)\leq1-g$, $\mu_2(E,c)\leq1-g+c$; $q_2(O,O)=0$, $q_2(E,0)\leq a$, $q_2(E,c)\leq a$ when $c<g$.

$\underline{0<q_1(s_1)<a}$:

c<g   N2: $\pi_1(t_1)=(1-a)q_1(s_1)/(1-q_1(s_1))a$, $\pi_1(t_2)=1$, $\lambda_2(t_1;O)=0$, $\lambda_2(t_1;E)=1$, $\lambda_2(t_2;E)=1$, $\mu_2(O,O)=0$, $\mu_2(E,c)=1-g+c$; $q_2(O,O)=0$, $q_2(E,c)=a$.

g<c<(2-g)g   N3: $\pi_1(t_1)=0$, $\pi_1(t_2)=1$, $\lambda_2(t_1;O)=0$, $\lambda_2(t_2;E)=1$, $\mu_2(O,O)=0$, $\mu_2(E,c)=1$; $q_2(O,O)=0$, $q_2(E,c)=1$.

(2-g)g<c   N4: $\pi_1(t_1)=(1-a)q_1(s_1)/(1-q_1(s_1))a$, $\pi_1(t_2)=1$, $\lambda_2(t_1;O)=\lambda_2(t_1;E)=0$, $\lambda_2(t_2;E)=0$, $\mu_2(O,O)=0$, $\mu_2(E,0)=1-g$; $q_2(O,O)=0$, $q_2(E,0)=a$.

$q_1(s_1)=a$:

N5: $\pi_1(t_1)=1$, $\pi_1(t_2)=1$, $\lambda_2(t_1;E)=0$, $\lambda_2(t_2;E)=0$, $\mu_2(E,0)\in[1-\min\{g,c/(2-g)\},1]$; $q_2(E,0)=a$.

$c<g$ N6: $\pi_1(t_1)=1$, $\pi_1(t_2)=1$, $\lambda_2(t_1;E)=1$, $\lambda_2(t_2;E)=1$, $\mu_2(E,c)\in[1-g+c,1]$; $q_2(E,c)=a$.

$g<c<(2-g)g$ N3.

$a<q_1(s_1)<1$:

N7: $\pi_1(t_1)=1$, $\pi_1(t_2)=1$, $\lambda_2(t_1;E)=0$, $\lambda_2(t_2;E)=0$, $\mu_2(E,0)=1$; $q_2(E,0)=q_1(s_1)$.

$c<g$ N8: $\pi_1(t_1)=1$, $\pi_1(t_2)=1$, $\lambda_2(t_1;E)=1$, $\lambda_2(t_2;E)=1$, $\mu_2(E,c)=1$; $q_2(E,c)=q_1(s_1)$.

$c<(2-g)g$ N9: $\pi_1(t_1)=1$, $\pi_2(t_2)=1$, $\lambda_2(t_1;E)=0$, $\lambda_2(t_2;E)=(q_1(s_1)-a)/(1-a)q_1(s_1)$, $\mu_2(E,0)=1-c/(2-g)$, $\mu_2(E,c)=1$; $q_2(E,0)=a$, $q_2(E,c)=1$.

$g<c<(2-g)g$ N3.

$q_1(s_1)=1$:

N10: $\pi_1(t_2)=1$, $\lambda_2(t_2;E)=0$, $\mu_2(E,0)=1$; $q_2(E,0)=1$.

$c<(2-g)g$ N11: $\pi_1(t_2)=1$, $\lambda_2(t_2;E)=1$, $\mu_2(E,c)=1$, $\mu_2(E,0)\leq1-c/(2-g)$, $\mu_2(O,0)\leq1-(g-c)/(2-g)$; $q_2(E,c)=1$, $q_2(E,0)\leq a$, $q_2(O,0)\leq a$. (In N10 and N11 the strategy of the weak type is left implicit.)   □

Proof of Lemma 5.3.
The following table gives for each signal $s=(y_1,s_2)$ the costs $c(s|N;t_j)$ of sending the signal for each type $(j=1,2)$ after $x_1=N$ (note that again the first period outcome $(x_1,y_1)=(C,O)$ serves as reference point):

Table 5.2 Costs of sending a signal S after $x_1=N$

s	$c(s|N;t_1)$	$c(s|N;t_2)$
(E,c)	2-g+c	2-g+c
(E,0)	2-g	2-g
(O,0)	1	2
(O,c)	1+c	2+c

The maximal gain from sending (y_1,s_2) is 2-g for the strong type, and equals 1 for the weak type (cf. Table 5.1; the maximal gain is the additional payoff the interest group gets in period 2 by persuading the policymaker to choose $x_2=C$ rather than $x_2=N$). By assumption the sorting condition 2-g>1 holds.

We first consider the degenerate belief cases $q_1(s_1) \in \{0,1\}$. When $q_1(s_1)=0$ the policymaker chooses $\mu_2(y_1,s_2)=0$ for sure after any signal (y_1,s_2) sent by the weak type in equilibrium. Consequently, the weak type chooses its cheapest message $(y_1,s_2)=(O,0)$ for sure. We must have $\mu_2(O,c) \leq c$, $\mu_2(E,0) \leq 1-g$, and $\mu_2(E,c) \leq 1-g+c$ to ensure that the weak type does not want to deviate. By (5.5) $q_2(E,0) \geq 0$, $q_2(E,c) \geq 0$ and $q_2(O,c) \geq 0$ is required. By choosing all these out-of-equilibrium beliefs $q_2(\cdot) \leq a$ path N1 is sustained. Note that N1 does not satisfy (5.4), and no such equilibrium exists when $q_1(s_1)=0$.

When $q_1(s_1)=1$ after any signal (y_1,s_2) sent by the strong type the policymaker chooses $\mu_2(y_1,s_2)=1$ for sure. Consequently, the strong type does not mix and chooses a specific signal for sure (the knife-edge case $g=c$ in which signals (E,c) and $(O,0)$ are cost equivalent is excluded). When the strong type chooses $(E,0)$ for sure, we get by (5.5) $q_2(O,0) \leq 1$, and $q_2(O,c) \leq 1$ when $c<1-g$, implying no further restrictions. The equilibrium strategy of the weak type may take several forms depending on how the policymaker reacts to out-of-equilibrium signals, and here we just choose $q_2(y_1,s_2)=\mu_2(y_1,s_2)=0$ when $(y_1,s_2) \neq (E,0)$ to induce the weak type to choose $(y_1,s_2)=(O,0)$ for sure. This yields N10. When the strong type chooses (E,c) for sure we need $\mu_2(E,0) \leq 1-c/(2-g)$ and $\mu_2(O,0) \leq 1-(g-c)/(2-g)$ to secure that the strong type does want to deviate. We get $q_2(O,c) \leq 1$ by (5.5), for the strong type can never gain by deviating to (O,c). Note that, since the weak type can always get a total payoff of at least 2 (not incorporating any expenditures made on s_1) by choosing $(O,0)$, when it does not choose $(E,0)$ $R_1((E,0)) \subseteq [1-g,1]$, and when it does choose $(E,0)$ $\mu_2(E,0) \geq 1-g$. Since $R_2((E,0))=[1-c/(2-g),1]$ we necessarily must have $1-c/(2-g)>1-g$, i.e. $c<(2-g)g$. In that case the path is sustainable with several specifications for the equilibrium strategy of the weak type and the policymaker, and here we just choose one satisfying (5.5); $q_2(y_1,s_2)=\mu_2(y_1,s_2)=0$ when $(y_1,s_2) \neq (E,c)$. So, when $c<g$ the weak type chooses (E,c) as well, and indeed by $R_2((E,0)) \subset [1-c,1]=R_1((E,0))$ follows $q_2(E,0)=0$. Similarly, $R_2((O,0))=\emptyset \subset [g-c,1]=R_1((O,0))$ and $R_2((O,c))=\emptyset \subset [g,1]=R_1((O,c))$, implying $q_2(O,0)=q_2(O,c)=0$ by (5.5). When $c>g$ the weak type chooses $(O,0)$. In that case $R_2((E,0)) \subset [1-g,1]=R_1((E,0))$ when $c<(2-g)g$ (N11).

Now suppose the strong type chooses $(O,0)$ for sure. Hence the weak type can get its maximum payoff by choosing its cheapest signal. We have $R_1((E,0))=\emptyset \subset [1-g/(2-g),1]=R_2((E,0))$, and thus $q_2(E,0)=1$ by (5.5). But then the strong type wants to deviate. Lastly, when the strong type chooses (O,c) for sure, we get $q_2(O,c)=\mu_2(O,c)=1$ and the weak type can get at least $2-c$ by choosing (O,c). Hence $R_1((E,c))=\emptyset \subset [1-g/(2-g),1]=R_2((E,c))$, and thus $q_2(E,c)=1$ by (5.5). In that case the strong type has an incentive to deviate.

Next, let $0<q_1(s_1)<1$. We first argue that signal (O,c) is not sent in an equilibrium satisfying refinement (5.4). Suppose to the contrary that the strong type chooses (O,c) with positive probability. We necessarily have $\mu_2(O,c)>\mu_2(E,c)$, implying that the weak type does not choose (E,c) in equilibrium (cf. Table 5.2). In turn, (E,c) cannot be chosen in equilibrium by the strong type, for otherwise $q_2(E,c)=\mu_2(E,c)=1$. Hence (E,c) is an out-of-equilibrium signal. Now suppose (O,c)

is not chosen by the weak type. Then we get $\mu_2(O,c)=1$, and $R_2((E,c))=[1-g/(2-g),1]$. Since the weak type does not choose (O,c), it gets at least 3-c, and therefore never wants to deviate to (E,c); $R_1((E,c))=\varnothing$. By (5.4) then $q_2(E,c)=1$, implying $\mu_2(E,c)=1$, a contradiction. So, (O,c) must chosen by the weak type as well with positive probability. In this case $R_1((E,c))=[\mu_2(O,c)+(1-g),1]$ and $R_2((E,c))=[\mu_2(O,c)-g/(2-g),1]\cap[0,1]$. By (5.4) $q_2(E,c)=1$ and the strong type wants to deviate, a contradiction. Consequently, the strong type never chooses (O,c) in equilibrium. In turn, the weak type will not choose (O,c) in equilibrium as well.

Now suppose the strong type sends $(O,0)$ in equilibrium. Using Table 5.2 this is only possible when $\mu_2(O,0)>\mu_2(E,0)$. In that case the weak type strictly prefers signal $(O,0)$. Hence, $(E,0)$ cannot be sent by the strong type, for otherwise $\mu_2(E,0)=1$ by Bayes' rule and (5.1). Therefore, $(E,0)$ must be an out-of-equilibrium signal. Now suppose $(O,0)$ is not chosen by the weak type. Then we get $\mu_2(O,0)=1$. Using Table 5.2, the weak type strictly prefers $(O,0)$, a contradiction. So, $(O,0)$ must be chosen by the weak type with positive probability as well. We have $R_1((E,0))=[\mu_2(O,0)+1-g,1]$ and $R_2((E,0))=[\mu_2(O,0)-g/(2-g),1]\cap[0,1]$; by (5.4) $q_2(E,0)=1$ implying $\mu_2(E,0)=1$, and the strong type does not want to send $(O,0)$, a contradiction.

In summary, the strong type only chooses between (E,c) and $(E,0)$ in equilibrium, the weak type never chooses (O,c) in equilibrium. Below we consider the three possible types of equilibria — separating, hybrid and pooling — separately.

(i) separating equilibria. When g<c signal (E,c) is dominated for the weak type. In this case separation is possible, with the strong type choosing (E,c) and the weak type choosing $(O,0)$. Separation is sustainable only when $q_2(E,0)\leq a$. We have $R_1((E,0))=[1-g,1]$ and $R_2((E,0))=[1-c/(2-g),1]$. We get $q_2(E,0)=0$ by (5.4) when 1-g<1-c/(2-g), i.e. when c<(2-g)g (N3). When c>(2-g)g we get $q_2(E,0)=1$ and separation is not possible. Likewise, we get $q_2(O,c)=0$, because $R_2((O,c))=\varnothing\subset[c,1]=R_1((O,c))$. When g>c neither signal is dominated for the weak type and separation is not possible.

(ii) hybrid equilibria. In a hybrid equilibrium, at least one type uses a mixed strategy. When the strong type mixes between (E,c) and $(E,0)$ we must have $\mu_2(E,c)=\mu_2(E,0)+c/(2-g)$, and the weak type strictly prefers $(E,0)$ over (E,c). Hence $q_2(E,c)=\mu_2(E,c)=1$, $q_2(E,0)=a$, and $\mu_2(E,0)=1-c/(2-g)$. $q_2(E,0)=a$ requires that $(E,0)$ is also chosen by the weak type. The weak type strictly prefers $(E,0)$ over $(O,0)$ when 1-c/(2-g)>1-g, i.e. when c<(2-g)g (the knife-edge case c=(2-g)g is ignored). $q_2(E,0)=a$ gives by Bayes' rule $\lambda_2(t_2)=(q_1(s_1)-a)/(1-a)q_1(s_1)$. This is only possible when $q_1(s_1)>a$. Now $R_1((O,0))=[g-c/(2-g),1]\cap[0,1]$ and $R_2((O,0))=\varnothing$, implying $q_2(O,0)=0$ by (5.4). Similarly, $R_1((O,c))=[g+c(1-g)/(2-g),1]$ and $R_2((O,c))=\varnothing$, i.e. $q_2(O,c)=0$ or $q_2(O,c)$ unrestricted. In the latter case just choose $q_2(O,c)=0$. This yields N9.

When the weak type mixes between $(E,0)$ and (E,c) we must have $\mu_2(E,c)=\mu_2(E,0)+c$, and the strong type chooses (E,c) for sure. But then signal $(E,0)$

is a sure sign of being weak, and the weak type prefers (O,0) instead of (E,0). Hence the weak type does not mix between (E,c) and (E,0). In case the weak type mixes between (E,c) and (O,0) we get $\mu_2(E,c)=1-g+c$, i.e. $g>c$ is required. The strong type must send signal (E,c) for sure, and we must have $q_2(E,c)=a$; this is only possible when $q_1(s_1)<a$. Mixing between (E,c) and (O,0) for the weak type is equivalent with mixing between E and O with the same probability, and choosing lobbying message c (0) for sure after E (O). Together with $q_2(E,c)=a$ and Bayes' rule the strategies in N2 follow. Note that $R_1((E,0))=[1-g,1]$ and $R_2((E,0))=[1-g+c(1-g)/(2-g),1]$. Hence $q_2(E,0)=0$ by (5.4). Likewise, $R_1((O,c))=[c,1]$ and $R_2((O,c))=[1+c-g(1-g)/(2-g),1]$, implying $q_2(O,c)=0$. Finally, when the weak type mixes between (E,0) and (O,0) we get $\mu_2(E,0)=1-g$. Ignoring the knife-edge case $c=(2-g)g$ the strong type must choose (E,0) for sure. It follows that $R_1((O,c))=[c,1]$ and $R_2((O,c))=[1-g+(g+c)/(2-g),1]$, implying $q_2(O,c)=0$. We also get $R_1((E,c))=[1-g+c,1]$ and $R_2((E,c))=[1-g+c/(2-g),1]$. So, when $c<(2-g)g$ it holds that $R_2((E,c))\neq\varnothing$ and $q_2(E,c)=1$ follows from (5.4). In that case the strong type wants to deviate, upsetting the supposed equilibrium strategies. When $c>(2-g)g$ $q_2(E,c)$ is unrestricted, and even $q_2(E,c)=1$ sustains the supposed equilibrium path N4. $\pi_1(t_1)$ directly follows from $q_2(E,0)=a$ and Bayes' rule.

(iii) pooling equilibria. In a pooling equilibrium the two types choose exactly the same strategy. From the above follows that the strong type may only mix between (E,c) and (E,0), and in that case the weak type prefers (E,0). Thus, in a pooling equilibrium both types necessarily choose the same pure strategy. First, consider pooling on (E,c). We have $R_1((E,0))=[\mu_2(E,c)-c,1]\cap[0,1]$ and $R_2((E,0))=[\mu_2(E,c)-c/(2-g),1]\cap[0,1]$. Hence by (5.4) $q_2(E,0)=0$ (or $q_2(E,0)$ unrestricted). Similarly, universal divinity yields $q_2(O,0)=q_2(O,c)=0$. For the weak type not to deviate to (O,0) it is required that $\mu_2(E,c)\geq1-g+c$, and thus $q_1(s_1)\geq a$ and $g>c$ (N6 and N8). When both types pool on (E,0), we get $R_1((E,c))=[\mu_2(E,0)+c,1]$ and $R_2((E,c))=[\mu_2(E,0)+c/(2-g),1]$. Hence by (5.4) $q_2(E,c)=\mu_2(E,c)=1$ when $\mu_2(E,0)<1-c/(2-g)$, and in that case the strong type wants to deviate. Similarly, by universal divinity $q_2(O,0)=0$, and $q_2(O,c)=0$ or $q_2(O,c)$ not restricted. For the strong type not to deviate to (E,c) it is required that $\mu_2(E,0)\geq1-c/(2-g)$, for the weak type not to deviate to (O,0) we must have $\mu_2(E,0)\geq1-g$. Therefore, $\mu_2(E,0)\geq1-\min\{g,c/(2-g)\}$ and $q_1(s_1)\geq a$ is required. When $q_1(s_1)>a$ we get $q_2(E,0)>a$ and $\mu_2(E,0)=1$ by (5.1) (N5 and N7).   *QED.*

Proof of Proposition 5.2.
Let $U_P^{sub}(x_1|q_1(s_1))$ denote the expected payoffs of the policymaker in equilibrium of the continuation game after choice x_1, given her updated belief $q_1(s_1)$ after the first lobbying stage. With Lemmas 5.2 and 5.3, we are able to determine the first period strategy of the policymaker. The following table can be derived:

Table 5.3 Expected equilibrium payoffs for the policymaker

choice of x_1 and continuation path	$U_P^{sub}(x_1 \mid q_1(s_1))$
$x_1=C$ and all paths C1 through C8	$\max\{0, a-q_1(s_1)\}$
$x_1=N$ and all paths, except N3	$\max\{a-1, 2a-q_1(s_1)-q_1(s_1)/a\}$
$x_1=N$ and path N3	$2a-q_1(s_1)-aq_1(s_1)$

In the game AL necessarily $s_1=\varnothing$, and thus $q_1(s_1)=q_1(\varnothing)=p$. Note that by assumption $0<p<1$, and thus paths after $q_1(s_1)\in\{0,1\}$ in Lemmas 5.2 and 5.3 do not have to be taken into account. When $c\notin[g,(2-g)g]$ path N3 is not possible. Now $\max\{a-1, 2a-p-p/a\}>\max\{0, a-p\}$ only when $p<a^2$. It follows that the policymaker chooses $x_1=N$ only when $p<a^2$, otherwise she chooses $x_1=C$ (note that the knife-edge case $p=a^2$ is ignored in Proposition 5.2). When $p>a$ either C5, C6 or C7 follows after $x_1=C$. With Lemmas 5.2 and 5.3 this yields all the equilibria for the case $c\notin[g,(2-g)g]$. Next, consider the case $c\in[g,(2-g)g]$. Suppose always continuation path N3 follows after $x_1=N$. By $2a-p-ap>\max\{0, a-p\}$ when $p<2a/(a+1)$ we get ALE2. When $p>a$ N3 is not the only possible continuation path. When N7 or N9 follows, $x_1=N$ yields $a-1$ and the policymaker prefers $x_1=C$; this yields ALE5, ALE6 and ALE7 for the case $c\in[g,(2-g)g]$, based on C5, C6 and C7.

$$QED.$$

Proof of Proposition 5.3.
Let $U_G^{sub}(t_j \mid q_1(s_1))$ denote the expected payoffs of the interest group of type t_j in the continuation game that starts after the first lobbying stage (i.e. not including any expenditures on first period lobbying), given the beliefs of the policymaker $q_1(s_1)$. From Lemma 5.1 we derive the following table:

Table 5.4 Expected continuation payoffs of G after first stage lobbying

$q_1(s_1)$	$U_G^{sub}(t_1 \mid q_1(s_1))$	$U_G^{sub}(t_2 \mid q_1(s_1))$
$q_1(s_1) = 0$	2	$2g+\mu_2(s_1,N,E)(2-g)$
$0 < q_1(s_1) < a^2$	2	$2-g+g^2$
$q_1(s_1) = a^2$	$2+\mu_1(s_1)$	$\mu_1(s_1)[2+g]+(1-\mu_1(s_1))[2-g+g^2]$
$a^2 < q_1(s_1) < a$	3	$2+g$
$q_1(s_1) = a$	$3+\mu_2(s_1,C,O)$	$2+g+\mu_2(s_1,C,O)(2-g)$
$a < q_1(s_1) \leq 1$	4	4

Now $q_1(s_1)$ is determined by Bayes' rule whenever this rule applies for s_1, that is, along the equilibrium path. Bayes' formula for $q_1(s_1)$ follows from a straightforward adaption of (5.9) and (5.10) to first period strategies.

It follows easily from Table 5.4 and the assumption c<1 that separation at the first lobbying stage cannot occur in equilibrium. In order to determine the first period lobbying strategy of the interest group we distinguish three cases: (i) $p<a^2$, (ii) $a^2<p<a$ and (iii) $p>a$. For each case we verify whether pooling on no-message ($s_1=0$), pooling on a lobbying message ($s_1=c$), or a hybrid lobbying stage (at least one type strictly mixes between $s_1=0$ and $s_1=c$) is possible. In a hybrid lobbying equilibrium Bayes' rule applies for both $q_1(0)$ and $q_1(c)$. From Table 5.4 follows that $U_G^{sub}(t_j|q_1(s_1))$ is weakly increasing in $q_1(s_1)$ for j=1,2 (from the proof of Lemma 5.1 it is obtained that necessarily $\mu_2(s_1,N,E)\leq 1-g$ for $q_1(s_1)=0$ when AAE4 applies, and thus $U_G^{sub}(t_2|0)\leq 2-g+g^2$). Hence, because message $s_1=c$ ($s_1=0$) bears costs c (0), it follows that $q_1(0)<q_1(c)$ is required for at least one type to mix. In turn, this requires $\lambda_1(t_1)<\lambda_1(t_2)\leq 1$ and yields $q_1(0)<p<q_1(c)$ by Bayes' rule.

(i) First, let $p<a^2$. When the types pool on $s_1=0$ we get $q_1(0)=p$ by Bayes' rule. The weak type has an incentive to deviate to $s_1=c$ whenever this leads to $q_1(c)>a^2$. The strong type does not always gain when $q_1(c)>a^2$ (e.g. when $a^2<q_1(c)<a$ and c>(2-g)g). When $q_1(c)=a^2$ the weak type gains from deviating when $\mu_1(c)>c$, the strong type only when $\mu_1(c)>c/(2-g)g$. By (2-g)g<1 the weak type has a stronger incentive to deviate. Both types do not have an incentive to deviate when $q_1(c)<a^2$. We get by universal divinity $q_1(c)=0$. This sustains pooling on no-lobbying and yields LAE1. Pooling on $s_1=c$ is certainly not possible, for this yields the weak type 2-c and it wants to deviate to $s_1=0$.

Now consider the possibility of a hybrid lobbying stage. When $q_1(c)>a^2$ the weak type wants to send $s_1=c$ for sure and we do not get $\lambda_1(t_1)<1$. Similarly, $q_1(c)<a^2$ is not possible, for the weak type will not send message $s_1=c$. A hybrid lobbying equilibrium then requires the strong type to mix between $s_1=c$ and $s_1=0$, but this is incompatible with $q_1(c)<a^2$. Thus, necessarily, $q_1(c)=a^2$. Now suppose both the weak and the strong type play a mixed lobbying strategy. We get $0<q_1(0)<p$ and $q_1(c)=a^2$. For the weak type to mix it is required that $\mu_1(c)=c$, whereas for the strong type to mix it is required that $\mu_1(c)=c/(2-g)g$. By g<1 these requirements are incompatible. Hence one of the types must play a pure strategy, and this must be message $s_1=c$ sent by the strong type (otherwise not $q_1(c)=a^2$). Hence, we have $q_1(0)=0$ and $q_1(c)=a^2$, with the weak type mixing and $\mu_1(c)=c$. The strong type will indeed want to send $s_1=c$ in case $\mu_1(c)=c$ and $\mu_2(0,N,E)\leq(1-g)[1-c(1-g)/(2-g)]$. $\lambda_1(t_1)$ follows from $q_1(c)=a^2$, $\lambda_1(t_2)=1$ and Bayes' rule (LAE2).

(ii) Next, let $a^2<p<a$. Pooling on $s_1=0$ yields $q_1(0)=p$ by Bayes' rule. Both types want to deviate to $s_1=c$ whenever this leads to $q_1(c)>a$, and have no incentive when this leads to $q_1(c)<a$. When $q_1(c)=a$ the weak type gains from deviating when $\mu_2(c,C,O)>c$, the strong type when $\mu_2(c,C,O)>c/(2-g)$. Hence the strong type has a stronger incentive to deviate, and thus $q_1(c)=1$ by universal divinity. But then both types want to deviate, and pooling on $s_1=0$ is not possible.

Pooling on $s_1=c$ leads to $q_1(c)=p$. Both types have an incentive to deviate to $s_1=0$ whenever $q_1(0)>a^2$. The weak type has no incentive to deviate when $q_1(0)<a^2$, although the strong type has when $c>(2-g)g$. In case $q_1(0)=a^2$ the weak type gains from deviating when $\mu_1(0)>1-c$, the strong type when $\mu_1(0)>1-c/(2-g)g$. Hence the strong type has a stronger incentive to deviate, and thus $q_1(0)=1$ by universal divinity. But then pooling on $s_1=c$ is not possible, for both types want to deviate.[48]

Now suppose the weak or the strong type plays a mixed lobbying strategy (or both). As noted we get $q_1(0)<p<q_1(c)$. When $q_1(c)>a$ neither the weak nor the strong type wants to mix, they both prefer to send message $s_1=c$. So $q_1(c)\leq a$ and message c must be sent by both the weak and the strong type. Hence $0<\lambda_1(t_1)<\lambda_1(t_2)\leq 1$ and the weak type must be indifferent between $s_1=0$ and $s_1=c$. In turn, this requires $q_1(0)\geq a^2$, otherwise the weak type strictly prefers $s_1=c$. Therefore, both types must mix. First suppose $q_1(0)>a^2$. From Table 5.4 it follows that for the weak type to mix it is then required that $q_1(c)=a$ and $\mu_2(c,C,O)=c$. But, when $q_1(c)=a$ and $\mu_2(c,C,O)=c$ the strong type prefers $s_1=c$, and we get $q_1(0)=0$, a contradiction. Next, suppose $q_1(0)=a^2$. When $q_1(c)<a$ the weak type wants to mix when $\mu_1(0)=1-c$, the strong type when $\mu_1(0)=1-c/(2-g)g$. By $g<1$ these two requirements are incompatible. Finally, when $q_1(c)=a$ the weak type wants to mix when $\mu_2(c,C,O)=\mu_1(0)+c-1$, the strong type when $\mu_2(c,C,O)=g\mu_1(0)+c/(2-g)-g$. These requirements lead to $\mu_1(0)=1-c/(2-g)$ and $\mu_2(C,O)=c-c/(2-g)$. Moreover, $q_1(0)=a^2$ and $q_1(c)=a$ together imply $\lambda_1(t_1)=(p-a^2)/a(1-p)$ and $\lambda_1(t_2)=(p-a^2)/p(1-a)$. This yields LAE3.

(iii) Lastly, let $p>a$. In case both types send $s_1=0$ for sure we get $q_1(0)=p$ by Bayes' rule, the policymaker concedes in both periods, and neither type has an incentive to deviate to $s_1=c$ (LAE4). Pooling on lobbying in the first period leads to $q_1(c)=p$. Both types have an incentive to deviate to $s_1=0$ whenever $q_1(0)>a$, and no incentive when $q_1(0)<a$. When $q_1(0)=a$ the weak type gains from deviating when $\mu_2(0,C,O)>1-c$, the strong type when $\mu_2(0,C,O)>1-c/(2-g)$. Hence the weak type has a stronger incentive to deviate, and thus $q_1(0)=0$ by universal divinity. This yields LAE5.

From the above, in any hybrid lobbying equilibrium $\lambda_1(t_1)<\lambda_1(t_2)\leq 1$ and $q_1(0)<p<q_1(c)$. When $q_1(0)>a$ neither type wants to mix, in case $q_1(0)<a$ the weak type strictly prefers $s_1=c$ and we cannot have $\lambda_1(t_1)<1$. Hence we necessarily have $q_1(0)=a$. By $\lambda_1(t_1)<1$, $q_1(0)=a$ and $q_1(c)>p$ it follows that $\lambda_1(t_2)<1$. Hence the strong type must mix. The strong type is indifferent between $s_1=c$ and $s_1=0$ when $\mu_2(0,C,O)=1-c/(2-g)$. In that case the weak type strictly prefers $s_1=0$ (LAE6).

$QED.$

Proof of Proposition 5.4.
We have to consider three cases for the first lobbying stage: (i) pooling on $s_1=c$, (ii) pooling on $s_1=0$ and (iii) a hybrid first lobbying stage. Firstly, when the interest group types pool on $s_1=c$ we get $q_1(c)=p<a^2$. Using Table 5.3 we get $x_1=N$. In every continuation path after $x_1=N$ the weak type gets its lowest possible

expected utility of 2. (See Table 5.5 below. Note that after the first lobbying stage in LL the continuation game equals the AL game, and for this game the equilibrium payoffs are given in Table 5.5. For instance, the payoffs obtained on the continuation paths N2, N3 and N4 of Lemma 5.3 correspond to the payoffs in ALE1, ALE2 and ALE3, respectively.) Hence, including the cost of lobbying c, choosing $s_1 = c$ yields 2-c. The weak type thus wants to deviate to $s_1 = 0$, which yields at least 2, and pooling on $s_1 = c$ is not possible.

Secondly, when the types pool on $s_1 = 0$, we get $q_1(0) = p < a^2$ and hence $x_1 = N$. The continuation paths after $s_1 = 0$ and $x_1 = N$ are specified in Lemma 5.3. Rather than deriving the full implications of applying universal divinity when either N2 or N3 follows, we show that universal divinity never excludes $q_1(c) = 0$ in these cases. By choosing $q_1(c) = 0$ then, pooling on $s_1 = 0$ is sustained, for neither type has an incentive to deviate. This yields LLE1 and LLE3. Concerning LLE6 the full implications of universal divinity are derived.

When $c < g$ the supposed equilibrium continuation path N2 yields the strong type $2g + (1-g+c)(2-g) - c$, the weak type 2 (cf. Table 5.5, ALE1). In case out-of-equilibrium message $s_1 = c$ leads the policymaker to belief $q_1(c) = a^2$, she is indifferent between choosing $x_1 = C$ and following continuation path C2, and choosing $x_1 = N$ and following path N2. The choice of $x_1 = C$ followed by the continuation path C2 yields the weak type 3, the strong type $2 + g + c(1-g)$ (cf. Table 5.5 below, ALE4). Consequently, taking account of the costs of first period lobbying the strong type has an incentive to deviate when $\mu_1(c) > c/(2-g)g$, the weak type when $\mu_1(c) > c$. With $(2-g)g < 1$ follows that the weak type has a stronger incentive to deviate to $s_1 = c$ in case this leads the policymaker to belief $q_1(c) = a^2$. Universal divinity thus never leads us to conclude $q_1(c) = 1$, and does not exclude $q_1(c) = 0$.

Next, let $g < c < (2-g)g$. The path N3 yields the strong type $2 + g - c$, the weak type 2 (cf. Table 5.5, ALE2). Suppose out-of-equilibrium message $s_1 = c$ leads the policymaker to belief $q_1(c) = 2a/(a+1)$. In a possible equilibrium continuation path, C7 follows after $x_1 = C$ and path N3 after $x_1 = N$. In that case, the policymaker is indifferent between $x_1 = C$ and $x_1 = N$, and $\mu_1(c) \in [0,1]$. Now the weak type gains from deviating when $\mu_1(c) > (2-g)c/(4-2g-c)$, the strong type when $\mu_1(c) > c/(2-g)$ (payoffs in continuation path C7 equal the payoffs in ALE7, see Table 5.5). By $(2-g)c/(4-2g-c) < c/(2-g)$ when $c < (2-g)g$ the weak type has a stronger incentive to deviate. Again, universal divinity either requires $q_1(c) = 0$ or does not put restrictions on $q_1(c)$. LLE3 follows.

Finally, consider the case $c > (2-g)g$. The supposed equilibrium path N4 yields the strong type $2 - g + g^2$, the weak type 2 (cf. Table 5.5 below, ALE3). Both types want to deviate when $q_1(c) > a^2$, neither type when $q_1(c) < a^2$. When $q_1(c) = a^2$ the strong type wants to deviate when $\mu_1(c) > c/[(2-g)g + c(1-g)]$, the weak type when $\mu_1(c) > c$. When $c > 1-g$ the strong type has a stronger incentive to deviate, upsetting the supposed equilibrium. When $c < 1-g$ $q_1(c) = 0$ by universal divinity, sustaining pooling on $s_1 = 0$ (LLE6).

Thirdly, we consider the possibility of a hybrid first lobbying stage. Suppose $q_1(c)<p<q_1(0)$. Then the weak type gets $2-c$ when choosing $s_1=c$, and at least 2 when choosing $s_1=0$ (cf. Table 5.5 below). The weak type will not choose $s_1=c$, thus $q_1(c)=1$, a contradiction. Hence $q_1(0)<p<q_1(c)$, and the weak type must choose $s_1=0$ with a larger probability. Suppose $q_1(c)=1$, that is, only the strong type chooses $s_1=c$. Then the policymaker chooses $x_1=C$ after $s_1=c$, and the weak type prefers $s_1=c$ over $s_1=0$. So, necessarily $q_1(c)<1$.

When $c\notin[g,(2-g)g]$, $q_1(c)>a^2$ induces the policymaker to choose $x_1=C$ (cf. Table 5.3). But then the weak type gets at least $3-c>2$ by choosing $s_1=c$, and does not want to choose $s_1=0$ (by $q_1(0)<p<a^2$, $s_1=0$ yields the weak type 2). Hence the strong type must mix, leading to $q_1(0)=1$, a contradiction. So, we must have that $q_1(c)=a^2$. The weak type must choose both $s_1=0$ and $s_1=c$ with positive probability, and this requires $\mu_1(c)=c$. When $q_1(0)>0$ the strong type will prefer to send a lobbying message at cost c only when $\mu_1(c)\geq c/(2-g)g$. By $(2-g)g<1$ we have $c<c/(2-g)g$, and no such hybrid equilibrium exists. So, $q_1(0)=0$. Taking $\mu_2(y_1,s_2)=0$ in N1 for all (y_1,s_2), the strong type strictly prefers $s_1=c$. We get LLE2 and LLE7.

When $c\in[g,(2-g)g]$ $q_1(c)>2a/(a+1)$ induces $x_1=C$ for sure, and by the reasoning in the previous paragraph $q_1(c)\leq2a/(a+1)$. So, the weak type must choose $s_1=c$ as well. Since necessarily N3 follows after $x_1=N$ when $q_1(c)<a$ the policymaker certainly chooses $x_1=N$ when $q_1(c)<a$, yielding the weak type $2-c$. The weak type strictly prefers $s_1=0$. Hence necessarily $q_1(c)\geq a$. For the same reason, $a<q_1(c)<2a/(a+1)$ followed by $x_1=N$ and N3 is not possible. $a<q_1(c)<2a/(a+1)$ followed by $x_1=C$ is not possible as well, for then the weak type strictly prefers $s_1=c$. Since when $a<q_1(c)<2a/(a+1)$ either $x_1=N$ or $x_1=C$ for sure (i.e. no mixing), necessarily $q_1(c)=a$ or $q_1(c)=2a/(a+1)$. When $q_1(c)=a$ and N3 follows after $x_1=N$ the policymaker chooses $x_1=N$ for sure after $s_1=c$ and the weak type does not want to mix. When either N5 or N6 follows, the policymaker chooses $x_1=C$ for sure. In that case, by $c<1$, the weak type strictly prefers $s_1=c$. So, necessarily $q_1(c)=2a/(a+1)$. For the weak type to mix it is required that $\mu_1(c)=\frac{1}{2}c$ for C5, $\mu_1(c)=c/(2-c)$ for C6 and $\mu_1(c)=c(2-g)/(4-2g-c)$ for C7 (again, the continuation payoffs in C5, C6 and C7 equal the payoffs in ALE5, ALE6 and ALE7, respectively, cf. Table 5.5). For the strong type to weakly prefer $s_1=c$ it is required that $\mu_1(c)\geq c/(2-g+c)$ when C5 follows after $x_1=C$, and $\mu_1(c)\geq c/(2-g)$ when C6 or C7 (the strong type only wants to mix when these inequalities become equalities, hence it immediately follows that the strong type uses a pure strategy). When C5 or C6 follows the respective inequalities hold, yielding LLE4 and LLE5. By $c(2-g)/(4-2g-c)<c/(2-g)$ when $c<(2-g)g$, path C7 cannot follow after $x_1=C$. *QED.*

Proof of Proposition 5.5.
Given the strategy of the policymaker (cf. Table 5.3), pressure can only occur when $q_1(s_1)\leq a^2$ for some $s_1\in\{0,c\}$ sent in equilibrium, and $c\notin[g,(2-g)g]$ (note that in N3 pressure does not occur). Hence necessarily $s_1=0$ and $s_1=c$ both sent in equilibrium. Suppose $q_1(c)\leq a^2$. Choosing $s_1=c$ yields the weak type at most $3-c$, choosing $s_1=0$ at least 3 (cf. Table 5.5, payoffs on continuation path equal payoffs in AL game). The weak type does not want to choose $s_1=c$, a contradiction. Hence,

necessarily $q_1(0) \leq a^2$, and for pressure to occur the strong type must choose $s_1 = 0$ as well. Now when $q_1(0) < a^2$ $s_1 = 0$ yields the weak type 2 and $s_1 = c$ at least $4 - 2c$ (note that $q_1(c) > p > a$), i.e. the weak type does not want to choose $s_1 = 0$. So, necessarily $q_1(0) = a^2$. Both types mixing only may occur in knife-edge cases, which we exclude, so $q_1(c) = 1$.

$s_1 = 0$ yields the weak type $2 + \mu_1(0)$. $s_1 = c$ followed by $x_1 = C$ and C5 (Lemma 5.2) after $q_1(c) = 1$ would yield $4 - c$, and the weak type wants to deviate. Note that N10 and N11 after $s_1 = c$ and $q_1(c) = 1$ are not possible, due to the strategy of the policymaker (cf. Table 5.3). So, necessarily C8 follows, yielding both types $4 - 2c$. For the weak type to prefer $s_1 = 0$ $2 + \mu_1(0) \geq 4 - 2c$ is required, i.e. at least $c > \frac{1}{2}$.

(i) $c < g$. For the strong type to be indifferent we need $\mu_1(0)[2 + g + c(1-g)] + (1 - \mu_1(0))[2 - g + g^2 + c(1-g)] = 4 - 2c$. (When $s_1 = 0$ path N2 follows after $x_1 = N$ and path C2 follows after $x_1 = C$, the equilibrium payoff for these continuation paths follow from Table 5.5 below from ALE1 and ALE4, respectively). This yields $\mu_1(0) = [2 - 2c + (g - c)(1-g)]/(2-g)g$. From $(2-g)g < 1$ and $c < g$ follows $\mu_1(0) > 2 - 2c$. In order to secure $\mu_1(0) \leq 1$ it is required that $g \geq (2-3c)/(1-c)$ (LLE8).

(ii) $c > (2-g)g$. After $x_1 = N$ path N4 follows. For the strong type to be indifferent we need $\mu_1(0)[2 + g + c(1-g)] + (1 - \mu_1(0))[2 - g + g^2] = 4 - 2c$ (cf. Table 5.5). This yields $\mu_1(0) = [2 - 2c + g(1-g)]/[(2-g)g + c(1-g)]$. In order to secure $\mu_1(0) \leq 1$ it is required that $g \geq (2-3c)/(1-c)$. The latter inequality can only hold when $c > \frac{1}{2}$. When $c > \frac{1}{2}$, we certainly have $\mu_1(0) \geq 2 - 2c$, for this is satisfied when $2(1-c)^2 \geq g(1-2c)$. This gives LLE9.   QED.

Table 5.5 Expected equilibrium payoffs

Eq.	$U_G^*(t_1)$	$U_G^*(t_2)$	U_P^*
AAE1	2	$2-g+g^2$	$2a-p-p/a$
AAE2	3	$2+g$	$a-p$
AAE3	4	4	0
****			
ALE1	2	$2-g+g^2+c(1-g)$	$2a-p-p/a$
ALE2	2	$2+g-c$	$2a-p(1+a)$
ALE3	2	$2-g+g^2$	$2a-p-p/a$
ALE4	3	$2+g+c(1-g)$	$a-p$
ALE5	4	4	0
ALE6	$4-c$	$4-c$	0
ALE7	$4-c/(2-g)$	$4-c$	0
****			
LAE1	2	$2-g+g^2$	$2a-p-p/a$
LAE2	2	$2-(1-g)[c(1-g)+g]$	$2a-p-p/a$
LAE3	$3-c/(2-g)$	$2+g(1-c)$	$a-p$
LAE4	4	4	0
LAE5	$4-c$	$4-c$	0
LAE6	$4-c/(2-g)$	$4-c$	0
****			
LLE2	2	$2-g(1-g)(1-c)$	$2a-p-p/a$
LLE4	2	$2+g-\frac{1}{2}c(2+g-c)$	$2a-p(1+a)$
LLE5	2	$2+g-c(2+g-2c)/(2-c)$	$2a-p(1+a)$
LLE7	2	$2-(1-c)(1-g)(g+c)$	$2a-p-p/a$
LLE8	$2+[2-2c+(g-c)(1-g)]/(2-g)g$	$4-2c$	$(1-p)a/(a+1)$
LLE9	$2+[2-2c+g(1-g)]/[(2-g)g+c(1-g)]$	$4-2c$	$(1-p)a/(a+1)$

Note: the payoffs for LLE1, LLE3 and LLE6 equal those for ALE1, ALE2 and ALE3, respectively.

APPENDIX 5.C ADDITIONAL REFINEMENTS: GLOBAL PLAUSIBILITY AND SUPPORT RESTRICTIONS

Of the four models discussed, only the LA and LL model are effectively repeated signaling games. In these models there are two subsequent rounds of signaling; first in the initial lobbying stage through s_1, and then, after the policymaker's choice of x_1, through the choice of (multi-dimensional) signal (y_1, s_2). Note that in the LA model $s_2 = \emptyset$ necessarily. Our STABAC refinement concept applies the universal divinity refinement in a local manner. Out-of-equilibrium signals (y_1, s_2) are only considered in the context of the continuation game that starts after x_1, and out-of-equilibrium signals s_1 only by assuming that after s_1 an equilibrium path for the ensuing continuation game is followed. A drawback of proceeding in this way is that not all out-of-equilibrium histories are assessed in a universal divinity like perspective. Consider for instance LAE2. In this equilibrium out-of-equilibrium history $h_1 = (0, N, E)$ is not covered by a universal divinity type of argument, because after $s_1 = 0$ the continuation game starts from a degenerate belief $q_1(0) = 0$, and applying universal divinity refinement (5.4) would lead to non-existence of an equilibrium for the continuation game (cf. the proof of Lemma 5.1). For that reason, the weaker refinement of divinity (5.5) is applied in that case, only leading to the (non-)restriction $q_2(0, N, E) \geq 0$ (as before, p and s_2 are left out as an argument). A global application of universal divinity, however, would force us to conclude that $q_2(0, N, E) = 1$. This would upset LAE2 as a plausible equilibrium because both types want to deviate.

The argument that yields this conclusion runs as follows. Following the prescribed equilibrium path of LAE2 the strong type gets an expected payoff of $2 - (1-g)[c(1-g) + g]$. In case the strong type deviates to $s_1 = 0$, the policymaker has no reason to belief that the interest group deviated since $s_1 = 0$ occurs with positive probability in LAE2. Hence the policymaker follows her equilibrium strategy and chooses $x_1 = N$. Compared with its equilibrium payoff in LAE2 the strong type now gains from choosing $y_1 = E$ — yielding out-of-equilibrium history $h_1 = (0, N, E)$ — when $\mu_2(0, N, E) > 1 - g - c(1-g)^2/(2-g)$. Similarly, the weak type gains from deviating to $\lambda_1(t_1) = 0$ and $y_1 = E$, assuming that the policymaker sticks to her first period equilibrium strategy $\mu_1(0) = 0$, when $\mu_2(0, N, E) > 1 - g$. Because the first inequality holds whenever the latter does, but not vice versa, the strong type is more easily induced to "produce" out-of-equilibrium history $h_1 = (0, N, E)$, and a universal divinity like argument requires us to conclude that $q_2(0, N, E) = 1$. This upsets LAE2 as a *globally* plausible equilibrium.

Likewise, in the proof of Proposition 5.3 we claim by local universal divinity in the STABAC procedure that $q_1(c) = 0$ in LAE1, and that the policymaker reacts with $x_1 = N$, assuming that equilibrium continuation path N1 follows (cf. lemma 5.3). But, when choosing $s_1 = c$ leads to $q_1(c) = 0$ and $x_1 = N$, the strong type could reason as follows. When I choose $s_1 = c$, the policymaker chooses $x_1 = N$ and beliefs $q_1(c) = 0$. In case I react, again out-of-equilibrium, with $y_1 = E$ what should the policymaker conclude? For me, the strong type, this sequence is profitable, compared with the equilibrium payoff of LAE1, when $\mu_2(c,N,E)\geq1-g+c/(2-g)$. For the weak type this is only profitable in case $\mu_2(c,N,E)\geq1-g+c$. Hence when $c<(2-g)g$ the policymaker must conclude, on the basis of a global universal divinity argument, that $q_2(c,N,E)=1$, upsetting LAE1 with out-of-equilibrium belief $q_1(c)=0$. However, because equilibrium path LAE1 is sustainable with other out-of-equilibrium beliefs, like e.g. $q_1(c)=p$, the global argument does not delete equilibrium path LAE1. In short, contrary to the STABAC procedure, a global argument also considers multiple deviations simultaneously, that is, a deviation at the first lobbying stage followed by a deviation from the ensuing (supposed) equilibrium path at the second signaling stage.[49]

The above reasoning indicates that, although the STABAC procedure as used in deriving our propositions is a reasonable refinement procedure (cf. Cho, 1993), it may be considered not strong enough to take full account of multiple deviations. We briefly investigate therefore which of the plausible equilibria of the LA and LL model (cf. Propositions 5.3, 5.4, and 5.5) are still reasonable when such multiple deviations are fully taken into account. To that purpose we check whether the plausible equilibria of these two models satisfy certain additional restrictions on out-of-equilibrium beliefs. Those that do are labelled *globally plausible*. With the additional restrictions each out-of-equilibrium history is assessed in a universal divinity like perspective. Thus, the somewhat inconvenient method of applying the two different refinements (5.4) and (5.5) disappears. (In a sense, the restrictions fill the gap when local divinity (cf. expression (5.5)) does not completely determine $q_2(s_1,x_1,y_1,s_2)$.) The additional restrictions are *global*, because now also out-of-equilibrium signals (y_1,s_2) are assessed in view of the equilibrium payoffs obtained in the equilibrium of the complete game under consideration. In the STABAC procedure they were just assessed in the context of the payoffs obtained in an equilibrium of the continuation game *after* the policymaker's choice of a specific fist period action x_1.

The additional restrictions come down to the following procedure. Take a plausible equilibrium of either the LA or LL model, and consider out-of-equilibrium history $h^\#=(s_1^\#,x_1^\#,y_1^\#,s_2^\#)$. First, when $(s_1^\#,x_1^\#)$ occurs in the specific equilibrium with positive probability, multi-dimensional signal $(y_1^\#,s_2^\#)$ is the first observable deviation from the equilibrium path. Then proceed as follows. When $s_1^\#$ is sent by both types of interest group, no additional restrictions are needed.[50] When $s_1^\#$ is sent by only one type, the policymaker reacts with choosing $x_1^\#$ for sure along the equilibrium path (i.e. because $q_1(s_1^\#)\in\{0,1\}$ she does not play a mixed strategy). Given that the policymaker chooses $x_1^\#$ after $s_1^\#$, determine the set of second period best responses $R_j(h^\#)\subseteq[0,1]$ for which the interest group of type t_j has a weak incentive to deviate, compared with the equilibrium payoff $U_G^*(t_j|E)$ it obtains in the plausible equilibrium E of the complete game under consideration (and thus not just the payoff it would get in the continuation game after the policymaker's choice of $x_1^\#$). Hence $R_j(h^\#)=\{\mu_2\in[0,1]|$ type t_j gets a higher expected payoff after history $h^\#$ than $U_G^*(t_j|E)$ when history $h^\#$ induces the policymaker to choose $x_2=C$ with probability $\mu_2\}$. Now when $R_1(h^\#)\subset R_2(h^\#)$ we must have $q_2(s_1^\#,x_1^\#,y_1^\#,s_2^\#)=1$, when $R_2(h^\#)\subset R_1(h^\#)$ we restrict the second period belief of the policymaker to $q_2(s_1^\#, x_1^\#, y_1^\#, s_2^\#) = 0$.[51] If neither applies, the out-of-equilibrium belief is unrestricted.

Second, when $s_1^\#$ is sent in the particular equilibrium but is never followed by $x_1^\#$, $x_1^\#$ is the first observable deviation from the equilibrium path. Such a deviation is initialized by the policymaker, and cannot be interpreted as a rational deviation by the interest group meant to signal information to the policymaker. Global universal divinity has thus nothing to say about such a deviation, and hence when $x_1^\#$ is the first observable deviation, $q_2(s_1^\#, x_1^\#, y_1^\#, s_2^\#)$ is not further restricted by global plausibility.

Third, when $s_1^\#$ is sent by neither type, we also want to reason using the potentially best responses of the policymaker. These potentially best responses are determined by the policymaker's belief $q_1(s_1^\#)$ after observing $s_1^\#$, and what to expect after her own choice of $x_1^\#$. These latter expectations can only be rationally given by the plausible equilibrium paths for the continuation game starting after $x_1^\#$, given belief $q_1(s_1^\#)$. STABAC interprets $s_1^\#$ as a single deviation because the procedure assumes that after $x_1^\#$ an equilibrium in the continuation path is followed, and draws a conclusion from such an interpretation. We get either $q_1(s_1^\#) = 0$, $q_1(s_1^\#) = 1$, or $q_1(s_1^\#)$ unrestricted. Here we just add the restriction that the belief $q_1(s_1^\#)$ that is consistent with STABAC, is not inconsistent when the deviation to $s_1^\#$ is in fact interpreted as part of a two-stage deviation and evaluated in the context of universal divinity (cf. the example given above with respect to LAE1). That is, given the out-of-equilibrium belief $q_1(s_1^\#)$ that is consistent with STABAC, a specific equilibrium continuation path must follow after $s_1^\#$ (in particular, one of the equilibrium paths specified in Proposition 5.2, where p in this case has to be replaced by $q_1(s_1^\#)$). Deviations of this particular path at the second signaling stage (y_1, s_2) can then be assessed using universal divinity as described in the first case above. Again, $R_j(h^\#) = \{\mu_2 \in [0,1] \mid$ type t_j gets a higher expected payoff after history $h^\#$ than $U_G^*(t_j \mid E)$ when history $h^\#$ induces the policymaker to choose $x_2 = C$ with probability $\mu_2\}$. In case this leads to a conclusion opposed to the conclusion initially obtained from STABAC, the original conclusion is dropped and the out-of-equilibrium belief obtained from STABAC cannot be used to sustain the path.

In summary, the idea behind global plausibility is that a firm conclusion about out-of-equilibrium beliefs can only be reached when both single-stage and two-stage deviations are considered. To that purpose a sequential procedure is proposed. First single-stage deviations are considered, yielding the plausible equilibria satisfying STABAC. Subsequently, for the set of plausible equilibria the validity of the conclusions obtained by STABAC is verified by also taking two-stage deviations into account. Lemma 5.4 below indicates which of the plausible equilibria are globally plausible as well.

Lemma 5.4

(a) In Proposition 5.3 only path LAE2 is not globally plausible.

(b) In Proposition 5.4 only paths LLE2, and LLE7 when $c<1-g$ are not globally plausible.

(c) In Proposition 5.5 both LLE8 and LLE9 are globally plausible. □

Proof of Lemma 5.4.

(a) LAE2 is deleted as globally plausible equilibrium by the reasoning in the text above. Similarly, from the text above follows that $q_1(c)=0$ cannot be concluded when $c<(2-g)g$, but that $q_1(c)=p$ still sustains LAE1 for that case. In LAE3 $q_1(0)=a^2$ and $q_1(c)=a$, and the additional restrictions have no bite. LAE4 yields both types their highest possible utility, so neither type wants to deviate. In LAE5 $q_1(0)=0$ by STABAC, followed by $x_1=N$ and path N1 (cf. Lemma 5.3). A multiple deviation, first from LAE5 and then from N1, to (0,N,E) is profitable for neither type. Lastly, In LAE6 $q_1(c)=1$, and $x_1=C$ for sure after $s_1=c$. A multiple deviation for the weak type is thus not possible (after $x_1=C$ necessarily $y_1=O$), and no additional restrictions follow.

(b) First, consider the no-lobbying equilibria LLE1, LLE3 and LLE6. STABAC either requires $q_1(c)=0$ or does not put restrictions on $q_1(c)$ (cf. the proof of Proposition 5.4). Now $q_1(c)=p$ sustains all three equilibrium paths, so when $q_1(c)=0$ from STABAC is inconsistent with a global argument, we just can choose $q_1(c)=p$. (Note that $q_1(c)=p$ is robust against the two-stage deviations considerations, because when $q_1(c)=p$ after the first deviation to $s_1=c$ the equilibrium continuation path equals the one after equilibrium signal $s_1=0$, and this latter continuation path is robust against deviations at the second stage (y_1,s_2) because it is part of a plausible equilibrium (cf. Proposition 5.2).) Next consider LLE2 where $q_1(0)=0$. The equilibrium path prescribes $x_1=N$, $y_1=O$ and $s_2=0$ after $s_1=0$, and history $h=(0,N,E,c)$ constitutes an out-of-equilibrium history. We get $R_1((0,N,E,c))=[1-g+c,1]\subset R_2((0,N,E,c))=[1-g+c(1+g-g^2)/(2-g),1]$ (cf. Table 5.5, the inclusion follows from $1+g-g^2<2-g$). Hence we must have $\mu_2(0,N,E,c)=1$, upsetting equilibrium LLE2. In LLE4 and LLE5 we also have $q_1(0)=0$, with $x_1=N$ after $s_1=0$. The strong type never has an incentive to deviate to (0,N,E,c) or (0,N,O,c), and setting belief $q_2(\cdot)=0$ for these two histories sustains the two paths. In addition, for LLE4 holds $R_2((0,N,E,0))=[1-\frac{1}{2}c(2+g-c)/(2-g),1]\subset R_1((0,N,E,0))=[1-g,1]$, whereas for LLE5 we have $R_2((0,N,E,0))=[1-c(2+g-2c)/(2-g)(2-c),1]\subset R_1((0,N,E,0))=[1-g,1]$ (inclusions follow from $g<c<(2-g)g$). In both cases $q_2(0,N,E,0)=0$, sustaining the two paths. Finally, consider LLE7. For this path out-of-equilibrium histories $(0,N,E,s_2)$ have to be considered (neither type wants to deviate to (0,N,O,c)). We obtain $R_1((0,N,E,0))=[1-g,1]$ and $R_2((0,N,E,0))=[(1-g)(2-(1-c)(g+c))/(2-g),1]$. When $c<1-g$ we have $R_1(\cdot)\subset R_2(\cdot)$, when $c>1-g$ we get $R_2(\cdot)\subset R_1(\cdot)$. Hence LLE7 is certainly deleted when $c<1-g$. Similarly, $R_1((0,N,E,c))=[1-g+c,1]=\varnothing$ (by $g<c$) and $R_2((0,N,E,c))=[1-\{(g-c)+(1-c)(1-g)(g+c)\}/(2-g),1]$. Now, it holds that $R_2((0,N,E,c))\neq\varnothing$ when $\{(g-c)+(1-c)(1-g)(g+c)\}>0$. This inequality can be rewritten to $(1-c)(2-g)g>c^2(1-g)$ (*). We show that (*) does not add any restrictions to $c>1-g$. In case (*) holds $\mu_2(0,N,E,c)=1$ and LLE7 is deleted. We consider two cases. First, let $(1-g)<(2-g)g$, i.e. $1-3g+g^2<0$. We verify whether (*) holds for $c=(2-g)g$; $(1-g)^2g(2-g)>(2-g)^2g^2(1-g)$ $\Leftrightarrow$ $(1-g)>(2-g)g$, contradicting our starting assumption. Hence (*) does not hold for $c=(2-g)g$, and by the l.h.s. decreasing and the r.h.s. of (*) increasing in c, the inequality does not hold for any value of $c>(2-g)g$. Next, let $(2-g)g<(1-g)$, i.e. $1-3g+g^2>0$. For $c=(1-g)$ inequality (*) reduces to $g^2(2-g)>(1-g)^3$ $\Leftrightarrow$ $0>1-3g+g^2$, contradicting the starting assumption. Hence (*) does not hold for $c=(1-g)$, and by the reasoning above not for any value of $c>(1-g)$. In summary, history $(0,N,E,c)$ does not add any restrictions to $c<1-g$.

(c) The only history to consider is $(c,C,O,0)$. In both LLE8 and LLE9 we have $R_2((c,C,O,0))=[1-c/(2-g),1]$. In LLE8 $R_1((c,C,O,0))=[(1/g)(1-c+gc-c/(2-g)),1]$. Now $(1/g)(1-c+gc-c/(2-g))<1-c/(2-g)$ $\Leftrightarrow$ $(1-g)(1-c)<(1-g)c/(2-g)$. So, when $g>(2-3c)/(1-c)$, which holds, $R_2((c,C,O,0))\subset R_1((c,C,O,0))$ and $q_2(c,C,O,0)=0$ by global universal divinity. In LLE9 $[c+(2-g-c(3-g))/(2-g)g,1]\subset R_1((c,C,O,0))$. Now $c+(2-g-c(3-g))/(2-g)g<1-c/(2-g)$ $\Leftrightarrow$ $c>(2-g)/(3-g)$, which is equivalent to $g>(2-3c)/(1-c)$. So, $q_2(c,C,O,0)=0$ by global universal divinity.   *QED.*

From the proof of Lemma 5.4 it follows that global plausibility in fact only adds restrictions to STABAC with respect to the cases where $q_1(s_1^\#)\in\{0,1\}$, that is, when $q_1(s_1^\#)$ is degenerate. An alternative way of dealing with degenerate beliefs is imposing a *support* restriction. Some refinement concepts, like the Perfect Sequential Equilibrium (PSE) concept of Grossman and Perry (1986), incorporate such a restriction. Therefore, we briefly explore the consequences of imposing support restrictions besides STABAC. In general, a support restriction requires that a belief which assigns zero probability to a certain type of interest group at some point in the game cannot later on be updated to assign positive probability to this type (in the words of Harrington, 1993, "never dissuaded once convinced"). Specifically, in the models LA and LL with repeated possibilities for signaling, a support restriction entails that the following implication must hold:

$$q_1(s_1)=0\ (=1)\ \Rightarrow\ q_2(s_1,x_1,y_1,s_2)=0\ (=1)\ \text{for all } x_1, y_1 \text{ and } s_2 \tag{5.11}$$

Restriction (5.11) implies that once the policymaker is convinced that the interest group is of the weak (strong) type after the lobbying stage in the first period, subsequent reactions of the interest group — like choosing $(y_1,s_2)=(E,c)$ after $x_1=N$ — will not alter her conviction. Note that along the equilibrium path (5.11) is necessarily satisfied through the requirement of Bayesian updating. Hence the restriction is just an additional restriction on out-of-equilibrium beliefs. Also note that from $0<p<1$ and $q_1(\varnothing)=p$ follows that $q_1(\varnothing)\notin\{0,1\}$, and thus that (5.11) trivially holds for the equilibria of models AA and AL with just a single round of signaling.

It should be realized that imposing the support restriction does not interfere with our application of divinity in the STABAC procedure when

$q_1(s_1) \in \{0,1\}$. The refinement in expression (5.5) namely never excludes $q_2(q_1,s_1,x_1,y_1,s_2) = q_1(s_1)$. The support restriction is really a far stronger refinement than expression (5.5). The consequences of imposing (5.11) besides STABAC for the LA and LL game are summarized in the following lemma:

Lemma 5.5
(a)   In Proposition 5.3 all paths survive imposing support restriction (5.11).
(b)   In Proposition 5.4 all paths survive imposing support restriction (5.11).
(c)   In Proposition 5.5 both paths LLE8 and LLE9 are deleted by imposing support restriction (5.11).   □

Proof of Lemma 5.5.
Because Lemma 5.1 is not affected by imposing (5.11), part (a) follows directly. Note that the restriction deletes continuation path C8 in lemma 5.2, and continuation path N11 in Lemma 5.3. Part (c) follows directly from the fact that C8 is deleted, and the observation in the proof of Proposition 5.5 that for the type of equilibria considered, after $x_1=C$ necessarily C8 must follow. However, part (c) can also directly be seen from the fact that in both LLE8 and LLE9 (5.11) requires $q_2(c,C,O,0)=1$, and in that case both types want to deviate to $s_1=c$ and $s_2=0$. Finally, the proof of Proposition 5.4 does not make use of the existence of continuation paths C8 and N11, and still applies when C8 and N11 are deleted in Lemmas 5.2 and 5.3, respectively. This proves part (b).   *QED.*

The result that restriction (5.11) deletes equilibria LLE8 and LLE9, equilibria which we find very plausible (cf. Lemma 5.4), renders the condition suspect. In fact, although support restrictions are widely used in applications, they have received considerable critique in the literature. For instance, Harrington (1993) argues on intuitive grounds that the restriction is not plausible because it prevents in the game he considers certain, according to him implausible, equilibria from being destabilized. Madrigal et al. (1987), on the other hand, reason that the requirement may be too strong in certain cases since it may lead to non-existence of a sequential (perfect Bayesian) equilibrium. The critique typically applies to games where detection of out-of-equilibrium behavior is possible only after a specific player has taken two subsequent actions and, in case observed outcomes indicate deviation for only one of the two actions, it is not possible to verify whether this player deviated in his first or in his second action. In that case a support restriction is rather strong because it forces the conclusion that the player deviated in his second action.

To illustrate the restrictiveness of a support restriction, consider again LAE2. Above we observed that history $h_1=(0,N,E)$ does not occur in equilibrium. When observing history $(0,N,E)$, though, restriction (5.11) would require the policymaker to conclude that a deviation took place at the enforcement stage; the weak type deviated from its equilibrium reaction strategy $\pi_1(t_1;0)=0$. However, it would also be possible to conclude that a deviation took place at the first signaling stage, viz. the lobbying stage. In that case the strong type deviated from its lobbying strategy $\lambda_1(t_2)=1$, but subsequently followed its enforcement strategy $\pi_1(t_2;0)=1$. Applying universal divinity in a global manner, as we saw above, would indeed lead to such a conclusion. This would upset the equilibrium, since both types would want to deviate (cf. Lemma 5.4). The criticism Harrington (1993) has on the support restriction precisely entails that the requirement excludes such a type of (global plausibility) reasoning.

The idea behind a support restriction can also be extended to the multi-dimensional signal (y_1,s_2) in models AL and LL when we separate the enforcement stage from the second lobbying stage. Such an extension entails that, in case y_1 is only used by the weak (strong) type in equilibrium, s_2 should not alter the belief of the policymaker. A rationale for such a restriction seems to be lacking, because only after observing the combined signal (y_1,s_2) the policymaker has a reason to determine and update her belief. There is no real reason to already update beliefs after just observing y_1, for the interest group has to make another decision yet before it is the policymaker's turn to move. On the other hand, in the model the decisions on y_1 and s_2 are really separated in time, and just after observing y_1 the policymaker already has some additional information to update her beliefs. Formally such a restriction entails:

$$q_2(s_1,x_1,y_1)=0\ (=1)\ \Rightarrow\ q_2(s_1,x_1,y_1,s_2)=0\ (=1)\ \text{for}\ s_2\in\{0,c\} \tag{5.12}$$

Imposing (5.12) besides STABAC has substantial consequences, as the following lemma shows:

Lemma 5.6[52]
(a)   In Proposition 5.2 path ALE2 is deleted by imposing restriction (5.12).
(b)   In Proposition 5.4 paths LLE3, LLE4 and LLE5 are deleted by imposing restriction (5.12).
(c)   In Proposition 5.5 paths LLE8 and LLE9 survive imposing restriction (5.12).   □

Proof of Lemma 5.6.
In Lemma 5.3 path N3 is deleted because we must have $q_2(s_1,N,E,0)=1$ by (5.12), implying $\mu_2(s_1,N,E,0)=1$ by (5.1). But then both types want to deviate to signal $(y_1,s_2)=(E,0)$. Consequently, all equilibria in which continuation path N3 is followed in equilibrium with positive probability are deleted. This proves part (a) and (b), part (c) follows from simple verification.   *QED.*

As a consequence, when $g<c<(2-g)g$ requiring (5.12) to hold together with STABAC leads to non-existence of an equilibrium when $p<a^2$. This is caused by the fact that under (5.12) full separation after $x_1=N$ through $(y_1,s_2)=(E,c)$ is not possible. Consider for instance ALE2. This separating equilibrium is deleted by imposing (5.12) for the following reason. After observing that the interest group enforced the threat in the first period ($y_1=E$), the policymaker should conclude — using Bayes' rule along the equilibrium path — that the interest group is of the

strong type for sure ($q_2(\emptyset,N,E)=1$). Starting from this degenerate belief second period lobbying cannot alter the conviction of the policymaker under restriction (5.12). So, even if the policymaker does not get a lobbying message in the second period, she should still believe that the interest group is of the strong type and thus choose $x_2=C$. In that case the strong type has an incentive to deviate from the equilibrium by not lobbying in the second period. Acknowledging this, the policymaker employs strategy $\mu_2(N,E,0)=1$. But then the weak type also wants to play $y_1=E$ after $x_1=N$, upsetting the equilibrium. In respect to part (c) it must be noted that the result is in fact of little interest, because there seems to be no rationale for imposing (5.12) without imposing (5.11).

We conclude that imposing additional refinements like global plausibility or support restrictions has some consequences, but they do not alter our main conclusions (the Results 5.1, 5.2 and 5.3 in the main text). However, as follows from the results in this appendix, support restrictions are less desirable because they delete the paths LLE8 and LLE9, and because they may exclude the possibility of full separation in cases where this seems quite reasonable (cf. Lemma 5.6).

NOTES

1. This chapter draws heavily from Sloof and Van Winden (1998).

2. As an example, consider the following situation. A union of public sector workers (the interest group) wants the government (the policymaker) to concede (C) to its wage claim. If the government does not concede (N), the union may either enforce (E) its threat to strike, or abstain from striking (O). Perhaps because of a commitment to its members, one type of union — the "strong type" — would prefer to strike if the government does not concede. A "weak type" of union, on the other hand, would prefer not to strike if the government does not concede.

        Another example concerns a firm which has to decide on a specific investment and approaches the government for support through a subsidy or tax exemption. The policymaker prefers the firm to invest, giving a boost to employment. The position of a "strong type" of firm would be such that it prefers not to invest when the government does not give financial support (because of better investment opportunities abroad, for example). A "weak type" of firm would prefer to invest even without governmental aid.

3. Also in a setting where the game described is repeated, a situation which is analyzed in this chapter, the content of the lobbying message is not influential. That is, two different equilibrium lobbying messages with the same fixed cost cannot induce different continuation paths. The content of the lobbying message is, therefore, *essentially uninformative*.

4. These modeling assumptions are in line with Potters and Van Winden (1990), and they contrast with the assumptions made in Lohmann (1995). In the latter study, although the policymaker may care about the (deadweight) costs of political pressure, she cannot avoid pressure, and thereby the associated costs, by taking a specific course of action.

5. This last assumption is in line with Kreps and Wilson (1982), but contrasts with e.g. Alt et al. (1988) and Calvert (1987). In the latter studies the type of leader (comparable to our interest group) may vary over the two periods. The leader knows its *expected* type, and becomes informed about its period specific type at the beginning of each period. Followers (policymakers) have and update beliefs about the leader's *expected* type.

6. By assumption the interest group cannot enforce its threat in case the policymaker gives in (cf. Kreps and Wilson, 1982, and Potters and Van Winden, 1990). Potters et al. (1991) investigate the consequences of this assumption by extending the Potters and Van Winden (1990) model with the possibility of enforcing the threat (choice of E) even after the government gives in (choice of C). In equilibrium the strong type indeed sometimes chooses to enforce its threat after a concession by the policymaker, even though this is costly in the short run. Because the policymaker anticipates such behavior, she is less likely to give in. An extension of the models analyzed in this chapter in a similar vein as in Potters et al. (1991) would allow an assessment of whether the second alleged difference between lobbying and pressure — that the opportunities to exert pressure are dependent of the decisionmaker's action — is really needed for obtaining our results. We conjecture that, under reasonable assumptions about the relative costs of lobbying and threat enforcement, our results will carry over.

7. Of course, one could incorporate the enforcement of a threat by the strong type into the definition of pressure as well. Such an alternative definition would not change our results in a qualitative sense. In our view, it seems more appropriate to speak of *structural coercion* in case the strong type enforces its threat (Van Winden, 1983).

8. With $x_1 \equiv N$ and $x_2 \equiv C$ the preferences given in Table 5.1 fit in the setup of the basic signaling game of Section 3.1 in the following way: $d_1 \equiv a$, $d_2 \equiv 1-a$, $e_1 \equiv 1$ and $e_2 \equiv 2-g$, where the l.h.s. of each identity refers to the notation used in Table 3.1.

9. Because in (a plausible, cf. Section 5.1.3) equilibrium the strong type always chooses to enforce after no-concession, the potential gain follows from comparing the payoff after (N,E), equal to g, with the payoff after (C,O), equal to 2.

10. In the light of this interpretation model AA may correspond to a situation in which there is no opportunity to get access due to a high turnover of policymakers. Admittedly, the interpretation of gaining access is somewhat loose because in the AL model access will be "earned" in the second period regardless of the interest group's action in the first period. Such an interpretation would therefore be more appropriate if the decisionsmaker's choice to grant access for lobbying activities was actually made endogenous in the model (cf. Austen-Smith, 1995).

11. In this case policy is settled more precisely after elections. The same model could apply for a different type of policy which is determined before elections. For instance, during the election campaign the platform of the policymaker concerning certain policies might still be susceptible to lobbying. Once the policymaker is (re)elected on the basis of this platform, it may be only amendable after strong agitation by the interest group. With respect to such a platform, one could also think of the agreements among coalition partners in multi-party systems at the start of a new cabinet.

12. Note that in this chapter subscripts typically refer to the period considered. Only in t_j (j=1,2) subscripts refer to the type of interest group under consideration. Although initially may be somewhat confusing, in this way we can avoid the use of superscripts. Numerical superscripts have the disadvantage that it is difficult to tell them apart from raising to a power.

13. For an application to the Kreps and Wilson (1982) chain store game see Van Damme (1991), and see Noldeke and Van Damme (1990) for an application to a dynamic version of the Spence labor market signaling model.

    The closely related Perfect Sequential Equilibrium (PSE) concept of Grossman and Perry (1986) imposes an additional *support* restriction on out-of-equilibrium beliefs. Such a restriction entails that, once the policymaker is convinced that the interest group is of a certain type (i.e. $q_1(s_1) \in \{0,1\}$), she will never alter these beliefs. Support restrictions are not innocuous, and may even lead to non-existence of an equilibrium (Madrigal et al., 1987). In addition, they have received some critique on intuitive grounds (cf. Harrington, 1993). Therefore, we do not impose the support restriction here. The main consequences of imposing support restrictions in our models are that they delete equilibria in which the period 2 decision of the policymaker x_2 is always based on complete knowledge concerning the interest group's type, and lead to non-existence of a plausible equilibrium in case these separating equilibria are the only ones surviving STABAC (cf. Appendix 5.C).

14. When $q_1(p,s_1) \in \{0,1\}$ we use the somewhat weaker *divinity* equilibrium refinement, in order to avoid non-existence of an equilibrium of the continuation game. This remark also applies to the other models considered. For a complete description and justification of the refinement procedure, see Appendix 5.B.

15. Proposition 5.1 follows from Potters and Van Winden (1990). Proofs of the other propositions are relegated to an appendix (cf. Appendix 5.B). Here, and in the sequel, the knife-edge cases $p=a^2$ and $p=a$ are disregarded.

16. The first period equilibrium strategy of the interest group is given by $\pi_1(t_1)=\min\{(1-a)p/(1-p)a,1\}$ and $\pi_1(t_2)=1$. Note that in both AAE2 and AAE3 we have $\mu_1=1$, thus $x_1=C$ for sure.

17. The maximal possible gain is given by the gain the weak type gets from inducing the policymaker to choose $x_2=C$ rather than $x_2=N$.

18. See remark E on page 266 in Kreps and Wilson (1982), and Proposition 4 in Calvert (1987).

19. Specifically, the single adaption that has to be made is that ALE2 only exists when $p<a$. Of course, with two short-run policymakers the period 2 policymaker must observe the history of play in period 1 for the equilibria of Proposition 5.2 to apply. Interestingly, when we identify social welfare with the sum of the payoffs of the period 1 and the period 2 policymaker, society is better off having one long-run policymaker than with having two short-run policymakers (cf. Table 5.5 in Appendix 5.B, ALE2 yields the policymaker the highest expected payoff). In contrast to the commonly held view that "cozy" long-term relationships between policymakers and interest groups are detrimental to society, this result suggests that, like with ordinary government investments, society may sometimes be better off in case policymakers take a long-run view when "investing" in their relationship with interest groups.

20. If only lobbying leads the policymaker to belief that the interest group is strong, and thus to choose $x_2=C$ rather than $x_2=N$, also the weak type interest group wants to lobby. This invalidates the belief, after observing a lobbying message, that the interest group is strong. A similar reasoning holds for playing $y_1=E$.

21. Recall that only separating equilibrium ALE2 of Proposition 5.2 not immediately applies for the case with two short-run policymakers. All other, non-separating, equilibrium paths of Proposition 5.2 do directly apply to that case. Proposition 5.3 does not contain any separating equilibrium, and it easily verified that it directly applies to the case with two short-run policymakers.

22. See e.g. Kreps and Wilson (1982). This feature does not apply to the models of Alt et al. (1988) and Calvert (1987), because there it is assumed, contrary to the setup in this chapter, that the type of the reputation building leader (interest group) may vary over the two periods. Reputation building in their models in fact occurs with respect to the *expected* type of the leader.

23. These results also hold for the AA model. In AAE2, ALE2 and ALE4 pressure does not occur. In AAE1, ALE1, ALE3 and LAE1 pressure occurs with probability $(1-a)p/a$, in LAE2 with probability $(1-c)(1-a)p/a$. In LAE3 the expected occurrence of pressure equals $c(a-p)/(2-g)$.

24. Lobbying occurs with probability p/a in ALE1 and ALE4, with probability p in ALE2, and does not occur in ALE3 and LAE1. The expected occurrence of lobbying equals p/a^2 in LAE2 and $(p-a^2)/a(1-a)$ in LAE3. When we focus solely on lobbying by the weak type, the same qualitative conclusion can be drawn with respect to the dependence on p.

25. Admittedly, this conclusion is somewhat weakened by LAE2 because in this path pressure has to be preceded by initial lobbying in order to build up a high reputation. However, a *global universal divinity* argument in the spirit of Harrington (1993) questions the plausibility of equilibrium LAE2 (cf. Appendix 5.C). In case one endorses this argument, LAE2 is deleted as a plausible equilibrium, and the suggestion that pressure is mainly used to build up a reputation is (even) more stronger.

26. In Proposition 5.4 these equilibria correspond to LLE1, LLE3 and LLE6, respectively. The STABAC refinement adds the additional restriction $c<1-g$ to the original restriction $(2-g)g<c$ of ALE3 when continuation path ALE3 follows after $s_1=0$ (cf. LLE6).

27. In class (i) the continuation path after $x_1=N$ follows from ALE1, and $c<g$ is required. In class (ii) the continuation path after $x_1=N$ follows from ALE2, and $g<c<(2-g)g$ is required. Finally, in class (iii) the continuation path after $x_1=N$ follows ALE3, and thus $c>(2-g)g$ is required. In Proposition 5.4 of

Lobbying or pressure?   179

Appendix 5.A class (i) consists of LLE2, class (ii) of LLE4 and LLE5, and class (iii) just contains equilibrium LLE7.

28. The two equilibria of class (ii), viz. LLE4 and LLE5, in fact exist whenever $p<2a/(a+1)$. Propositions 5.4 and 5.5 immediately apply for the case with two short-run policymakers. However, with two short-run policymakers LLE4 and LLE5 do not constitute an equilibrium when $p>a$.

29. After $s_1=c$ we have $q_1(c)=2a/(a+1)$ in equilibria LLE4 and LLE5, and only $q_1(c)=a^2$ in LLE2 and LLE7.

30. Only when $c>\max\{(2-g)g,1-g\}$ a plausible no-lobbying equilibrium does not exist and the interest group must necessarily lobby in the first period to build up a reputation.

31. Hicks (1963, p. 146) refers to the following explanation for the occurrence of strikes: "The most able Trade Union leadership will embark on strikes occasionally, not so much to secure greater gains upon that occasion (which are not very likely to result) but in order to keep their weapon burnished for future use, and to keep employers thoroughly conscious of the Union's power." The following quote from Rees (1977, p. 33) holds a similar contention: "The strike is part of a long-range strategy for both parties. Union gains won without a strike are usually won through the threat of a strike, stated or implied. Such threats cannot retain much force if they are never carried out."

32. When $p<a^2$ this follows directly from Table 5.5 in Appendix 5.B. For the case $p>a^2$ the plausible equilibria of the LL model are not presented. The statement, however, follows directly from the following two observations. Firstly, when $p>a$ in both the AA and the LA model the policymaker gets an equilibrium payoff of 0. By choosing $x_1=x_2=C$ in model LL the policymaker can always secure herself the same payoff of 0, and any equilibrium in LL will therefore yield the policymaker at least 0 in expected utility. Secondly, when $a^2<p<a$ the policymaker can always ignore the information from the lobbying process and secure herself an expected payoff of $a-p$ by choosing $x_1=C$ and $x_2=N$. In models AA and LA an expected payoff of $a-p$ is obtained in equilibrium when $a^2<p<a$. Rather intuitively, in our setting the policymaker cannot lose from additional possibilities for information transfer.

33. For any value of p the policymaker can get at least the same expected payoff in some equilibrium of model LL as can be obtained in any equilibrium of model AL. However, for high values of p a comparison between the two models is more problematic due to the existence of multiple equilibria in both models.

34. Of course, when the costs of establishing lobbying channels are extremely high model AA is preferred for sure. In that case also second period lobbying possibilities should not be provided.

35. This suggestion hinges on our assumption that $0<c<1$. In case the policymaker, besides choosing the institutional setup, could completely determine the costs of lobbying c, she would prefer model LA with $2<c<4-2g$. In that case a separating equilibrium exists in which the interest group types completely separate themselves out in the first lobbying stage. However, by establishing the possibility to lobby, i.e. by providing routes for exerting influence with "words", in a way the policymaker already sets the costs of lobbying at a low level. The assumption $0<c<1$ seems reasonable then, especially in comparison with the costs of exerting pressure $(1-g)$.

36. An observation made by Van Riemsdijk (1996, p. 78) in a recent popular publication on conflicts between interest groups and private enterprises is also suggestive in this respect: "The parties should recognize each other as legitimate parties concerned, and they should accept that without each others collaboration they cannot find their way out. That has to be the actual situation. For really new

problems, however, the situation is not that clear straight away. Therefore, generally this [collaboration] is a strategy which only comes into the picture after a successful action. Then the balance of powers is clear. [..] *Collaborating* will be under discussion only after a given problem is settled that far such that legitimate representatives of different points of view to the problem are recognized as such by all parties involved, and thus there is no need for actions anymore."

37. Support for this impression is given by the following passage taken from a Dutch newspaper (de Volkskrant, November 23, 1996, Soft image, hard guilders):

> Also Shell intensively confers, even daily, with all sorts of pressure groups. That was already the case before the Brent Spar and Ogoni land made the front pages. "However, nowadays we take a different attitude towards these consultations", says Mr Berkhout [managing director Oil of Shell Nederland]. "Before we mainly thought these were very annoying. We just heard their opinion and then made a decision on our own. In the future we will *share* important decisions more often. For we want to prevent that pressure groups once again will call: *Shell, and now you do this or that, damn it, or we'll boycott you.*"

38. In a commentary in a Dutch newspaper (NRC Handelsblad, January 6, 1996) it is argued that: "..they [interest groups] have experienced that codifying does not work until they have struggled and won their battle."

39. See, for instance, the extensions of the basic signaling game discussed in Section 3.3, which can be easily transferred and interpreted in the context of the model of this chapter.

40. It was already observed that, when a political newcomer is *completely* informed on the interest group's record, the equilibrium analysis for the case with two short-run (one period) policymakers almost equals the analysis for the case with one long-run (two period) policymaker. In models AA and LA the equilibrium outcomes in the two cases are the same. Only in respect to models AL and LL some minor adaptions have to be made, namely the range of prior belief p for which equilibria exist in which separation may occur (ALE2, LLE4, LLE5) should be curtailed (from $p<2a/(a+1)$ to $p<a$) in case of two short-run policymakers. In comparison with other models of reputation building, e.g. Kreps and Wilson (1982) and Calvert (1987), the result that the two cases are not completely equivalent in terms of equilibrium outcomes is somewhat surprising.

41. Alternatively, one could build a model in which the policymaker has non-perfect recall concerning the interest group's past actions and those of her own. Especially since the theory of games with non-perfect recall is not yet very well developed (cf. Piccione and Rubinstein, 1997a, 1997b), such an extension may be difficult to tackle, though.

42. The superscript [sub] refers to subgame. Although the game after x_1 is strictly speaking a continuation game rather than a subgame, the superscript [con] could lead to confusion in case one has still the contributions model of Chapter 4 in mind.

43. Consider the following example of a standard signaling game. A privately informed sender with type $t \in \{t_1, t_2\}$, with prior probability $p=P(t=t_2)$, chooses a message from $\{m_1, m_2\}$. After receiving a message from the sender, a receiver chooses between x_1 and x_2. Payoffs for message-type-action combinations are given by the following tables (with the first number referring to the payoff of the sender):

m_1	x_1	x_2
t_1	0,2	2,3
t_2	1,4	3,3

m_2	x_1	x_2
t_1	1,0	3,1
t_2	0,2	2,1

Let the prior belief be degenerate, specifically let p=1. Suppose type t_2 mixes between m_1 and m_2. We get $q(m_1)=q(m_2)=1$ by Bayes' rule, and the receiver chooses x_1 after both messages ($q(m)$ denotes the receiver's posterior belief that $t=t_2$ after having received message m). In that case type t_2 does not want to choose m_2, a contradiction. Hence type t_2 necessarily plays a pure strategy in equilibrium, and always chooses m_1, followed by the receiver choosing x_1. Note that $q(m_2)$ cannot be determined by Bayes' rule. In order to sustain these strategies as a PBE we must have $q(m_2)\geq\frac{1}{2}$ (in a sequential equilibrium necessarily $q(m_2)=1$, cf. Fudenberg and Tirole, 1991, remark on p. 243). Applying universal divinity to this game yields $q(m_2)=0$; type t_1 has an incentive to deviate when $\mu(m_2)\geq0$ (with $\mu(m)$ the probability that the receiver chooses $x=x_2$ after message m), type t_2 when $\mu(m_2)\geq\frac{1}{2}$. Note that the receiver's best response to m_2 is given by $\mu(m_2)=0(\in[0,1])[=1]$ when $q(m_2)>(=)[<]\frac{1}{2}$. The single PBE of this game does not satisfy universal divinity extended to degenerate priors.

44. Expressions (5.6), (5.7) and (5.8) also apply when s_1 does not occur in equilibrium. That is, in case s_1 leads to out-of-equilibrium beliefs $q_1(s_1)$, out-of-equilibrium beliefs $q_2(q_1(s_1),x_1,y_1)$ must be based on $q_1(s_1)$, the equilibrium strategies of the two players, and Bayes' rule (when it applies). Similarly, when x_1 is an out-of-equilibrium action of the policymaker, (5.6), (5.7) and (5.8) still apply. These restrictions on out-of-equilibrium beliefs follow from the definition of a PBE as given by Fudenberg and Tirole (1991), which is used here. In a weaker version of PBE, after any out-of-equilibrium history the choice of beliefs is arbitrary.

45. When $1-c<\mu_2(0)<1-c/(2-g)$ we in fact have $R_1((O,c))=\varnothing$. In the sequel, intervals [a,b] with a>b always refer to the empty set $\varnothing$.

46. The CFIEP refinement discussed in Chapter 3 puts the same restrictions on C3, but contrary to universal divinity deletes C4. Interestingly, the equilibria of C3 with $\mu_2(0)<1-c/(2-g)$ can only be deleted by perfect Bayesian equilibria — viz. C4 with $\mu_2(c)>\mu_2(0)+c/(2-g)$ — which themselves are deleted (by e.g. C3 with $\mu_2(0)=1$). Hence, implausible (in the context of CFIEP) equilibria are used to argue the implausibility of other equilibria. For an argument why such a, at first glance suspicious, procedure may still be justified see Mailath et al. (1993).

47. In relation to the previous note, the CFIEP refinement concept deletes C4, C6 and C8 as plausible equilibria of the continuation game after $h_1=(s_1,C,O)$. As a consequence, when the CFIEP refinement concept is used rather than (universal) divinity in the STABAC procedure, our result that lobbying is typically used to maintain a reputation for being strong (through pooling on lobbying) vanishes. This shows that the use of specific equilibrium refinements may have important consequences for the final results obtained (cf. Umbhauer, 1994). Therefore, though typically having a rather formal and abstract character, a discussion of the consequences of using different refinements in the context of a specific application is useful. Our choice for using the (universal) divinity refinement in Chapter 5 is guided by the fact that (i) existence of a universally divine equilibrium is guaranteed, at least as long as non-degenerate (prior) beliefs are considered, (ii) the concept is theoretically related to strategic stability, and (iii) it is widely used in applications. CFIEP lacks these characteristics. A strict formal argument of why universal divinity may be preferred over CFIEP does not exist, though.

48. This reasoning seems at first sight a bit odd because the strong type only gains more than the weak type in case a deviant message is interpreted as being more likely to come from the weak type; the cases in which the strong type gains from deviating and the weak type does not, are cases in which the deviant message is, in a sense, interpreted in a wrong way! It must be noted, though, that the same oddity applies when using (universal) divinity in a one-shot signaling game. To illustrate, consider the following example taken from Banks and Sobel (1987). Take the setup of the signaling game described in an earlier note of this appendix, with $0<p<1$ and the following payoffs:

m_1	x_1	x_2
t_1	-3,3	-6,0
t_2	-3,3	-11,5

m_2	x_1	x_2
t_1	-5,5	-6,0
t_2	-5,5	-11,5

For this game pooling on m_2 constitutes a perfect Bayesian equilibrium, with the receiver choosing x_1 after m_2. Such a pooling equilibrium, however, is not (universally) divine. With $\mu(m_1)$ the probability that the receiver chooses x_2 after message m_1, t_1 wants to deviate when $\mu(m_1)\leq\frac{2}{3}$, type t_2 when $\mu(m_1)\leq\frac{1}{4}$. Concerning the best-responses of the receiver holds that $\mu(m_1)=1(\in[0,1])[=0]$ when $q(m_1)>(=)[<]3/5$. Hence, for $p<3/5$ the pooling on m_2 equilibrium is not (universally) divine (Banks and Sobel, 1987, p. 654). However, for $p<3/5$ type t_1 only gains more than t_2 whenever $q(m_1)=3/5$ and $\mu(m_1)\in(\frac{1}{4},\frac{2}{3})$.

49. In that way, the global argument considers situations in which forward induction arguments might be partly incompatible with backwards induction arguments (cf. Myerson, 1991). In the STABAC procedure forward induction arguments have a very short (one-move) reach, because STABAC assumes that after each (out-of-equilibrium) move a plausible equilibrium is played in the continuation game. In this way, STABAC lets backward induction arguments prevail over forward induction arguments. For instance, in the example of Myerson (1991, Figure 4.15, p. 192) STABAC selects the equilibrium predicted by the backwards induction logic, and not the ones predicted by the forward induction argument. The global argument to be discussed in the main text extends the forward induction logic to multiple moves (deviations), and is therefore more sympathetic to forward induction arguments than STABAC is.

50. In that case applying universal divinity to $(y_1^{\#},s_2^{\#})$ in the local manner (STABAC) is equivalent to a global universal divinity argument to history $h^{\#}$. This follows because, when both types send $s_1^{\#}$ with positive probability, both their expected equilibrium payoffs (as needed for global universal divinity) can be determined by considering only the path after this message $s_1^{\#}$. That is exactly what local universal divinity does.

51. The conclusion drawn on the basis of the global argument cannot be inconsistent with the conclusion obtained from (5.5) in the STABAC procedure. This can be seen as follows. An inconsistency may only arise when $q_2(h^{\#})=0$ or $q_2(h^{\#})=1$ is concluded from (5.5). Suppose $q_2(h^{\#})=0$ by (5.5) (the reasoning when $q_2(h^{\#})=1$ is symmetric). The equilibrium payoff $U_G^{*}(t_1|E)$ can be obtained when considering the path after $s_1^{\#}$ alone. Let $U_G(t_2|$ send $s_1^{\#}$, then behave according to equilibrium path for continuation game after $(s_1^{\#},x_1^{\#})$) be the expected payoff of type t_2 when it deviates to $s_1^{\#}$, but then follows its equilibrium strategy in the continuation game. When determining the set of best responses for which the types have an incentive to deviate, local divinity (5.5) uses $U_G^{*}(t_1|E)$ and $U_G(t_2|$ send $s_1^{\#},..)$ as reference points, and concludes that type t_1 has a larger incentive to deviate. Global universal divinity uses $U_G^{*}(t_1|E)$ and $U_G^{*}(t_2|E)$ as reference points. By $U_G^{*}(t_2|E)\geq U_G(t_2|$ send $s_1^{\#},..)$, for otherwise the strong type would indeed deviate to $s_1^{\#}$ and it would not be an equilibrium, the strong type has even less incentive to deviate and a global comparison leads to the same conclusion that $q_2(h^{\#})=0$.

52. In lemma 5.6 we apply restriction (5.12) after applying STABAC, i.e. we just check whether the plausible equilibria given in Propositions 5.2, 5.4 and 5.5 satisfy (5.12). In that way, (5.12) may "overrule" a conclusion drawn on the basis of universal divinity in the STABAC procedure (restriction (5.4)). E.g., in path N3 of lemma 5.3 $q_2(s_1,N,E,0)=0$ by universal divinity when $c<(2-g)g$. By applying (5.12) afterwards, however, we revise this conclusion to $q_2(s_1,N,E,0)=1$. Changing the order, by first deriving the equilibria that satisfy (5.12), and then applying universal divinity as in STABAC may possibly lead to a different set of equilibria. However, the equilibria deleted in lemma 5.6 would also not survive the reverse order of application. Here lemma 5.6 is only presented to show that intuitive, separating equilibria are deleted by (5.12).

REFERENCES

Ainsworth, S., 1993, Regulating lobbyists and interest group influence, *Journal of Politics* 55, 41-56.
Austen-Smith, D., 1995, Campaign contributions and access, *American Political Science Review* 89, 566-581.
Alt, J.E., Calvert, R.L., and B.D. Humes, 1988, Reputation and hegemonic stability, *American Political Science Review* 82, 445-466.
Banks, J.S. and J. Sobel, 1987, Equilibrium selection in signaling games, *Econometrica* 55, 647-661.
Brinig, M.F., Holcombe, R.G. and L. Schwartzstein, 1993, The regulation of lobbyists, *Public Choice* 77, 377-384.
Calvert, R.L., 1987, Reputation and legislative leadership, *Public Choice* 55, 81-119.
Card, D., and C.A. Olson, 1995, Bargaining power, strike durations, and wage outcomes: An analysis of strikes in the 1880s, *Journal of Labor Economics* 13, 32-61.
Carlsen, F., 1994, Asymmetric information, reputation building, and bureaucratic inefficiency, *Public Finance/Finances Publiques* 49, 350-357.
Cho, I.K., 1993, Strategic stability in repeated signaling games, *International Journal of Game Theory* 22, 107-121.
Cho, I.K. and D.M. Kreps, 1987, Signaling games and stable equilibria, *Quarterly Journal of Economics* 102, 179-221.
Cho, I.K. and J. Sobel, 1990, Strategic stability and uniqueness in signaling games, *Journal of Economic Theory* 50, 381-413.
Fudenberg, D. and J. Tirole, 1991, Perfect Bayesian equilibrium and sequential equilibrium, *Journal of Economic Theory* 53, 236-260.
Grossman, S.J. and M. Perry, 1986, Perfect sequential equilibrium, *Journal of Economic Theory* 39, 97-119.
Harrington, J.E., 1993, The impact of reelection pressures on the fulfillment of campaign promises, *Games and Economic Behavior* 5, 71-97.
Hicks, J.R., 1963, The Theory of Wages. 2nd edition (Macmillan, London).
Konrad, K.A. and G. Torsvik, 1997, Dynamic incentives and term limits in bureaucracy regulation, *European Journal of Political Economy* 13, 261-279.
Kreps, D.M. and R. Wilson, 1982, Reputation and imperfect information, *Journal of Economic Theory* 27, 253-279.
Lohmann, S., 1995, A signaling model of competitive political pressures, *Economics and Politics* 7, 181-206.
Lowery, D. and Gray, V., 1997, How some rules just don't matter: The regulation of lobbyists, *Public Choice* 91, 139-147.
Madrigal, V., Tan, T.C.C. and S. Ribeiro da Costa Werlang, 1987, Support restrictions and sequential equilibria, *Journal of Economic Theory* 43, 329-334.
Mailath, G.J., Okuno-Fujiwara, M. and A. Postlewaite, 1993, Belief-based refinements in signaling games, *Journal of Economic Theory* 60, 241-276.
Massó, J., 1996, A note on reputation: More on the chain-store paradox, *Games and Economic Behavior* 15, 55-81.
Myerson, R.B., 1991, Game theory. Analysis of conflict (Harvard University Press, Cambridge).
Noldeke, G. and E.E.C. Van Damme, 1990, Signaling in a dynamic labor market, *Review of Economic Studies* 57, 1-24.
Piccione, M., and A. Rubinstein, 1997a, On the interpretation of decision problems with imperfect recall, *Games and Economic Behavior* 20, 3-24.
Piccione, M., and A. Rubinstein, 1997b, The absent-minded driver's paradox: Synthesis and responses, *Games and Economic Behavior* 20, 121-130.
Potters, J., 1992, Lobbying and pressure. Theory and experiments (Thesis Publishers, Amsterdam).
Potters, J. and F. van Winden, 1990, Modelling political pressure as transmission of information, *European Journal of Political Economy* 6, 61-88.
Potters, J. and F. van Winden, 1992, Lobbying and asymmetric information, *Public Choice* 74, 269-92.

Potters, J., Van Winden, F. and M. Mitzkewitz, 1991, Does concession always prevent pressure?, in: R. Selten, ed., Game equilibrium models IV. Social and political interaction (Springer Verlag, Berlin) 41-63.

Rees, A., 1977, The economics of trade unions. Revised Edition (The University of Chicago Press, Chicago).

Sloof, R. and F. van Winden, 1998, Show them your teeth first! A game-theoretic analysis of lobbying and pressure, revised version submitted to: *Public Choice*.

Sobel, J., 1985, A theory of credibility, *Review of Economic Studies* 52, 557-573.

Umbhauer, G., 1994, Information transmission in signaling games: Confrontation of different forward induction criteria, in: B. Munier and M.J. Machina, eds., Models and experiments in risk and rationality (Kluwer, Dordrecht) 413-438.

Van Damme, E.E.C. 1991, Stability and perfection of Nash equilibria. 2nd Edition (Springer Verlag, Berlin).

Van Riemsdijk, M.J., 1996, Actie, dialoog en onderhandeling. Een plaatsbepaling, in: H.J. Tieleman, red., Conflicten tussen actiegroepen en ondernemingen. De democratisering van het moreel gezag (Stichting Maatschappij en Onderneming, Den Haag) 68-96.

Van Winden, F., 1983, On the interaction between state and private sector (North Holland, Amsterdam).

6 LOBBYING POLITICIANS OR BUREAUCRATS?[1]

According to Laffont and Tirole (1991, p. 1090) one methodological limitation of the theoretical literature on interest groups is that most models mainly focus "...on the 'demand side' in their study of political and regulatory decision-making, in that all the action takes place on the side of interest groups. By 'blackboxing' the 'supply side' (the political and regulatory institutions), they have ignored a crucial agency relationship between politicians and their delegates in the bureaucracy."[2] In this chapter the relationship between politicians and bureaucrats is explicitly taken into account when studying the choice an interest group has between lobbying politicians or lobbying bureaucrats. On the one hand, the internal organization of government has an impact on the way in which interest groups may try to influence political decision-making. On the other hand, when deciding on whether to delegate policy authority to bureaucrats or not, politicians may take the potential influence of interest groups on bureaucrats into account. From this perspective the internal organization of government and (the potential for) interest group influence are interrelated. This chapter presents a model that analyzes this latter relationship.

Specifically, in this chapter the following questions with respect to the relationship between interest group influence and the delegation of policy authority are addressed. Firstly, when do politicians, acknowledging the potential influence of interest groups on bureaucrats, prefer to delegate policy authority to the bureaucracy? Secondly, do politicians always prefer an unbiased bureaucrat over a biased one? Thirdly, under what circumstances do interest groups try to influence the delegation decision of politicians, and when the policy decision of bureaucrats? Lastly, do interest groups typically pursue delegation of policy authority, or do they rather prefer the politician not to delegate?

The model presented in this chapter that is used to tackle these questions fits into two strands of literature. Firstly, it contributes to the theoretical research on the choice of an interest group between various means of influence. Secondly, it adds to the formal literature on delegation of policy authority from politicians to bureaucrats. The first line of research has been briefly reviewed in Section 2.2. Therefore, Section 6.1 below starts with a short overview of some recent studies belonging to the second strand of literature. Then the specific game-theoretic model that is analyzed in this chapter is described, and related to the delegation studies reviewed. Moreover, it is indicated how the model is connected to the basic signaling game of Chapter 3. Section 6.2 presents the equilibrium analysis of our game model, giving some answers to the questions posed above. Since the discussion in this section is rather lengthy, the substantive equilibrium results obtained are summarized at the end of the section in a separate subsection. Section 6.3 elaborates on a number of generalizations of the model. Finally, Section 6.4 concludes.

6.1   DESCRIPTION OF THE MODEL AND ITS RELATIONSHIP TO RECENT DELEGATION STUDIES

Short overview of some recent delegation studies. One of the main reasons often mentioned for delegation of policy authority from legislators to bureaucrats is that the latter are better informed and have more expertise than the former (cf. Niskanen, 1971, Breton, 1995). In his study of repeated delegation Legros (1993) takes the advantages of delegation as given and focuses on the choice of a (female) politician between delegates (males) in a signaling game setting. Through repeated interaction the politician extracts information about the, to her unknown, preferences of the delegate (his type).[3] The delegate has an incentive to establish a reputation for having similar preferences as the politician in order to increase his probability of reappointment. The main result Legros obtains is that, when no strict concavity assumptions are made regarding the utility functions of the players, there exist equilibrium outcomes in which the politician reappoints some extremely biased types with a higher probability than less biased, moderate ones. Although the politician strictly prefers unbiased delegates over biased ones, in these equilibria extremely biased types exactly mimic the behavior of (almost) unbiased types. Only moderately biased types of delegates take a different behavior. As a result, the politician can only distinguish between, on the one hand, moderately biased bureaucrats, and, on the other hand, the pool of (almost) unbiased and extremely biased bureaucrats.[4] In short, biased bureaucrats are appointed not because the politician likes them, but because she cannot identify them.

Epstein and O'Halloran (1994) analyze the decision problem of a politician how much discretion to give to a specific bureaucrat of a known type. Bureaucrats are better informed about how public policy maps into consequences. Delegation therefore leads to a gross "informational gain" to the politician because a more informed policy decision can be made. On the other hand, bureaucrats also have different policy preferences, and therefore propose a policy different from the one the politician would prefer were she to have the same information (expertise). This bias leads to a gross "distributional loss" to the politician. When deciding on the level of discretion the politician weighs the informational gains against the distributional losses. In case of an ex post veto of the politician there exists a "discretionary floor" in equilibrium; always a certain minimum amount of discretion is given to the bureaucrat, irrespective of his preferences (type).[5] When the politician has no ex post veto the amount of discretion given declines monotonically in the degree of conflict of interests among the bureaucrat and the politician. Also, less discretion is given in case policy (technical) uncertainty decreases.

Bawn (1995) considers the choice of a bureaucrat and the level of his discretion simultaneously.[6] In her model the politician chooses the (mean) preferences of the bureaucrat and his independence. It is assumed that, on the one hand, independence increases the policy expertise of the bureaucrat and thus lowers technical (policy) uncertainty. On the other hand, independence is supposed to increase the variance in the political effect of administrative procedures. This is labeled procedural uncertainty. As in Epstein and O'Halloran (1994) the politician then faces a trade-off between control, to curb procedural uncertainty, and expertise, to reduce technical uncertainty.[7] It appears that the optimal bureaucrat has (mean) preferences equal to the preferences of the politician. The optimal level of independence depends on whether technical or procedural uncertainty is the largest problem.

The papers discussed up till now analyzed the problem of delegation independently of the potential role of interest groups. As already touched upon in Section 3.3, however, a number of scholars argue that interest groups may help politicians control the negative effects of delegation. They may provide politicians with information about the performance of the bureaucracy, for instance. Rationally acting politicians benefit from this watchdog role of interest groups, even when different interests induce these groups to strategically transmit their information. In these "fire alarms" models interest groups do not directly affect the behavior of the bureaucrats, but the latter are indirectly constrained in their choices by the monitoring behavior of interest groups. Only a few theoretical studies do take into consideration the direct influence interest groups have on the bureaucracy (cf. Laffont and Tirole, 1991, Mazza and Van Winden, 1998, and Spiller, 1990). This influence is typically modeled as a (conditional) monetary transfer from the interest group to the bureaucrat. In these studies politicians take into account this potential for collusion between bureaucrats and interest groups when interacting with them. Only in Spiller (1990), however, the politician actually controls the potential for collusion directly. He finds, inter alia, that politicians sometimes explicitly allow for collusion in order to extract rents from bureaucrats competing for office.

The model presented in this chapter fits best in the last category of models that study the three-tier relationship between politicians, bureaucrats and interest groups. However, it differs from these models through a combination of the following three aspects. First, interest groups try to affect governmental decision-making through the strategic transmission of information, rather than through monetary transfers. Second, the question whether to delegate policy authority or not is studied within the context of the potential for interest group influence. Third, interest groups may try to affect the decisions of bureaucrats, as well as those of politicians. In the description of the model given below these three aspects are made more explicit.

Description of the game. In our model there are three players, a female politician P, a male bureaucrat B, and an interest group G. Government thus consists of two agents, namely P and B. The government has to decide on which policy to implement, for instance the level of a subsidy to a firm. The final policy decision is represented by x. The authority to design policy is initially given to P. She may either be a (non-) benevolent dictator, or her preferences may be induced by electoral considerations. In any case, her ideal policy is assumed to depend on the actual state of the world t. In the subsidy example t could for instance represent the economic "health" of the firm. The preferred policy of the politician is assumed to be increasing in the state of the world. In the example, when the economic prospects of the firm deteriorate (higher t), a higher subsidy (higher x) is preferred by the politician in order to forestall forced layoffs. Because the politician also takes other interests into account (e.g. those of the taxpayers) she does not prefer policy x to be as high as technically feasible.

Interest group G is really a "special" interest group, because when it comes to government policies it is assumed to be solely concerned with policy x. Due to this narrow focus, the interest group's preference *ordering* over public policy is independent of the actual state of the world. That is, G always prefers policy x as "high" as possible. However, the preference *intensity* of the group is assumed to depend on the actual state of the world t. In the running example, the firm always prefers the subsidy to be as high as possible. But, the more severe the economic prospects of the firm, the more urgent a high subsidy becomes.

The bureaucrat is assumed to be solely concerned with his career opportunities, either within or outside government. Due to the first concern, the bureaucrat to a certain extent wants to serve the interests of the politician (Peacock, 1994, Tirole, 1994). On the other hand, the existence of "revolving doors" — that is, career opportunities within the private sector — induces the bureaucrat to care somewhat about the objectives of the interest group.[8] This second concern may be amplified in case the bureaucrat obtains special favors (as for instance explicit bribes) from interest groups. In the model the two main concerns of the bureaucrat are reflected in the assumption that B's (reduced form) preferences are a weighted sum of the utilities of the politician and the interest group. The bias parameter $b \geq 0$ represents the relative weight the bureaucrat attaches to the interests of the interest group. Note that by specifying the preferences of the bureaucrat in this way, the relationship between the politician and the bureaucrat, as well as the relationship between the bureaucrat and the interest group, is analyzed partly in reduced form.

The following specific functional forms of the players' preferences used in the model fit the more general description given above:[9]

$$U_P(x;t) = -(x-t)^2 \tag{6.1}$$
$$U_G(x;t,g) = gtx \tag{6.2}$$
$$U_B(x;t,g) = -(x-t)^2 + bgtx \tag{6.3}$$

Recall that the final policy decision of the government is represented by x (with $x \geq 0$), and that the actual state of the world is denoted with t ($t \geq 0$). The parameter $g > 0$ represents the general characteristics of the interest group, that is, the general stake it has in public policy. Ceteris paribus, for a fixed value of the state of the world t, it holds that the larger g, the larger the incentive for the interest group to convince the government that "really" a large x is needed.

Next, the interactions between the three players are discussed. The interest group is assumed to become better informed than both the bureaucrat and the politician about the "need" for policy x. That is, only G is assumed to become informed about the actual state of the world t. In the subsidy example, for instance, the firm obtains information about its own state of economic health from a market and internal analysis. In contrast to the politician, the bureaucrat has the expertise (and the time) to assess the information coming from the interest group about the specific policy needed. Thus the bureaucrat is assumed to be the only one within government that is able to understand technical information about t coming from the interest group.[10] Moreover, due to a long term relationship with the interest group the bureaucrat knows its general characteristics, i.e. the general stake g it has in persuading government, whereas the politician does not.[11] The politician may therefore want to delegate policy authority to the bureaucrat in order to benefit from the potential for information transmission between the interest group and the bureaucracy ("informational gain"). On the other hand, when bureaucrats have different interests they will implement a policy that differs from the optimal policy of the politician ("distributional loss"). In case policy authority is delegated to the bureaucrat, G may send B a lobbying message concerning the actual state of the world. In line with previous chapters it is assumed that sending a lobbying message bears a fixed cost $c_B > 0$, and that not lobbying B bears no cost.

When deciding whether to delegate policy authority or not, the politician takes the potential for information transfer between the bureaucrat and the interest group into account, as well as the policy bias that might result from having the bureaucrat implementing policy x. Contrary to Bawn (1995) and Epstein and O'Halloran (1994), here the benchmark case is studied in which the politician either gives the bureaucrat full discretion or, otherwise, decides on policy completely by herself.[12] Besides, like in Legros (1993) it is assumed that the politician cannot block the implementation of a policy with which she disagrees once authority is delegated. Before the politician decides whether to delegate or not, the interest group may try to affect this decision by lobbying her. Lobbying P comes down to transmitting information about the general stake g the interest group has in persuading government. The value of g namely co-determines the scope for information transmission between B and G about the actual state of the world t, and thus determines the relative attractiveness of delegation to the politician. As with lobbying B, it is assumed that lobbying P bears a fixed cost $c_P > 0$. The specific order of moves in the game can now be summarized as follows:

(i)   Nature draws G's general stake g and reveals it to both G and B.

(ii)   G sends a costly lobbying message to P at cost c_P, or does not send P a lobbying message (signal 0).

(iii)   P decides whether to delegate policy authority or to choose policy x herself. In the latter case P chooses policy x.

(iv)   Nature draws the actual state of the world t, and reveals t only to G.

(v)   In case P decided to delegate at stage (iii), G sends a costly message to B at cost c_B, or does not send such a message.

(vi)   In case P decided to delegate at stage (iii), B chooses policy x.

(vii)   The actual values of t and g are revealed to all players and payoffs are obtained.[13]

The random variables g and t are drawn independently of each other. With respect to the distribution of g it is assumed that $g \in \{g_1, g_2\}$, with $0 < g_1 \leq g_2$, and that $P(g=g_2)=p$.[14] The state of the world t is assumed to be uniformly distributed on the interval $T=[t_m-z, t_m+z]$, with $t_m \geq z \geq 0$. Hence $E(t)=t_m$ and $var(t)=z^2/3$. Policy x can be chosen from $[0,\infty)$. The setup of the game — that is, the order of moves and the prior distributions from which g and t are drawn — is common knowledge.

The assumption that the interest group decides on lobbying the politician before the state of the world t becomes known is made for two reasons.[15] First, it seems fairly plausible that in practice an interest group takes a long run perspective when approaching politicians for (no-)delegation. In such a case the group will reason that, although the particular future circumstances are not known yet, these may become such that on average the group is better off having the bureaucrat (not) deciding on policy. From this perspective, it seems reasonable that the interest group does not know the actual state of the world yet when it decides on lobbying the politician. Second, the assumption is used to reflect the idea that the group cannot transmit specific technical information to the politician because she would not (have the time to) understand. By assuming that G is not informed on t when lobbying P, lobbying her cannot be interpreted as transmitting information about t, and exclusively refers to G's general stake g.

Relationship with the basic signaling game. We end this section by briefly connecting the model analyzed in this chapter to the basic signaling game discussed in Chapter 3. The general structure of the basic signaling game emerges in two instances in the present model. Firstly, the lobbying (continuation) game between B and G is a straightforward extension of the basic signaling game to continuous type and action spaces. The actual state of the world t is drawn from an interval rather than from just two possible values, and instead of only two actions, the decisionmaker (bureaucrat) now has a continuum of actions to choose from. In Potters (1992) it is shown that many results obtained for the basic signaling game carry over to this continuous version. For instance, because G's preference ordering over policy x is independent of t cheap talk cannot be influential. As a result, the content of a lobbying message cannot contain relevant information (cf. Result 3.1(a)), and hence the identification of a lobbying message to B solely with its costs c_B is justified. Besides, for information transfer to be possible in equilibrium a sorting condition on the preferences of the interest group has to be met (cf. Result 3.1(b)). By the assumption that G's preference intensity is increasing in t (cf. expression (6.2)), this condition holds from the outset.

Secondly, the lobbying game between P and G (stages (ii) and (iii)) is a direct application of the basic signaling game when we first solve for the equilibria of the two possible continuation games afterwards. This is explained in more detail in Subsection 6.2.5 below.

6.2   EQUILIBRIUM ANALYSIS

In this section the equilibrium analysis of the model is presented and answers to various questions posed in the introduction to this chapter are given. The equilibrium analysis begins with the determination of the equilibria of the two continuation games that start after the delegation decision of the politician. First the continuation game after no-delegation is analyzed in Subsection 6.2.1. Subsequently, Subsection 6.2.2 considers the continuation game in case the politician decides to delegate policy authority. In the latter case a signaling game is played between the interest group and the bureaucrat. In order to select the more plausible perfect Bayesian equilibria of this signaling continuation game, again the universal divinity refinement concept is used.[16] With the equilibria of the two continuation games at hand, the politician's equilibrium delegation strategy is determined in Subsection 6.2.3. In a similar vein, Subsection 6.2.4 derives the induced equilibrium preferences of the interest group over no-delegation and delegation, respectively. Then, in Subsection 6.2.5 the signaling game between the interest group and the politician is solved again using the universal divinity refinement, assuming that after this first lobbying stage an (universally divine) equilibrium is played in the continuation games.[17] The final subsection provides a summary of the results obtained from the equilibrium analysis.

A formal definition of the equilibrium notion used in this chapter is given in Appendix 6.A. For expositional reasons, all proofs of the propositions are relegated to this appendix as well.

6.2.1   The continuation game when the politician does not delegate

In case P decides not to delegate, equilibrium policy x_P is easily determined from her preferences reflected in (6.1) and the prior distribution of t. P implements policy $x_P = t_m$ in equilibrium (cf. the proof of Proposition 6.1 below). Note that, because $U_P(x;t)$ is independent of g, this optimal policy is independent of P's expectations concerning the general stake g of the interest group and its stage (ii) lobbying behavior. As a result, the expected equilibrium payoffs of the politician after no-delegation do not depend on the lobbying signal $s_P \in \{0, c_P\}$ received. Denoting the expected equilibrium payoffs after No-delegation for the politician as $E_N(U_P)$, and for the interest group with general characteristics $g \in \{g_1, g_2\}$ as $E_N(U_G;g)$, the following expressions are obtained:

$$E_N(U_P) = -\tfrac{1}{3}z^2 \tag{6.4}$$

$$E_N(U_G;g) = gt_m^2 \tag{6.5}$$

(Because $E_N(U_G;g)$ denotes G's expected payoff *after* No-delegation at stage (iii), the term does not include any outlays made by the interest group on lobbying P at stage (ii).) The expected payoffs of the bureaucrat after no-delegation directly follow from $E_N(U_B;g) \equiv E_N(U_P) + bE_N(U_G;g)$. The equilibrium policy and expected payoffs when P does delegate the policy decision to B are less straightforwardly determined. The signaling continuation game between B and G that follows after delegation is analyzed in the next subsection.

6.2.2 The continuation game between the bureaucrat and the interest group

Equilibrium analysis. When P delegates policy authority at stage (iii), the bureaucrat makes the final decision concerning policy x. Before the bureaucrat decides on which policy to implement (stage (vi)), however, the interest group becomes informed on the state of the world t (stage (iv)) and may lobby him (stage (v)). Lobbying B serves the purpose of transmitting information about the actual state of the world. From Potters (1992) it follows that any non-pooling perfect Bayesian equilibrium of the continuation game between B and G is a partition equilibrium of size two over states of the world t in T. Such a two-partition equilibrium is most easily illustrated with the use of the following figure of the type space T:

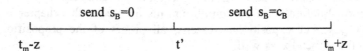

Figure 6.1 Partition of the type space T in equilibrium

As before, lobbying signals are associated with their costs. So, when the interest group does not send a lobbying message to B we have $s_B = 0$, and when G does lobby the bureaucrat we have $s_B = c_B$.

In a two-partition equilibrium high types $(t > t')$ send a costly lobbying message to the bureaucrat, whereas low types $(t < t')$ abstain. A lobbying message is thus a clear signal that $t > t'$, and no-lobbying a clear signal that $t < t'$. Because his optimal policy is increasing in t in equilibrium, the bureaucrat chooses a higher policy after being lobbied than after not being lobbied. Although a costly lobbying message induces the bureaucrat to implement a "better" policy from the viewpoint of G (cf. Result 3.1(c)), not all types are prepared to make the costly lobbying effort. Cut-off type t' is indifferent between lobbying and no-lobbying, because the additional benefit the better policy equals the costs of lobbying c_B. In equilibrium cut-off type t' randomizes between $s_B=0$ and $s_B=c_B$ with arbitrary probability. Its actual randomizing behavior is inessential for the analysis of the complete game.

A non-pooling equilibrium does not always exist (cf. Proposition 6.1 below). In a pooling equilibrium either all types $t \in T$ send a costly lobbying message, or all types abstain from lobbying. In the first case the pooling equilibrium is equivalent with the degenerate two-partition equilibrium in which $t'=t_m-z$ and t' chooses $s_B=c_B$ for sure. In the second case the pooling equilibrium corresponds with the degenerate two-partition equilibrium in which $t'=t_m+z$ and t' chooses $s_B=0$ for sure. Therefore, all pooling and non-pooling equilibria can be represented with the use of Figure 6.1, and are completely characterized by the value of t'. In Proposition 6.1 presented below all the universally divine equilibrium paths of the continuation game between B and G are given using this characterization.

Proposition 6.1
In any universally divine PBE of the continuation game between B and G the equilibrium strategy of the interest group is characterized by the unique value of t' (cf. Figure 6.1):

$\frac{1}{2}g(2+bg)(t_m+z)z \le c_B$ U1: $t'=t_m+z$.

$\frac{1}{2}g(2+bg)(t_m-z)z < c_B < \frac{1}{2}g(2+bg)(t_m+z)z$ I: $t'=2c_B/[g(2+bg)z]$.

$c_B \le \frac{1}{2}g(2+bg)(t_m-z)z$ U2: $t'=t_m-z$.

Let $x_B(s_B;g)$ denote the equilibrium policy B chooses after having received signal $s_B \in \{0,c_B\}$ from an interest group with general stake g. It holds that $x_B(s_B;g)= E(t|s_B)+\frac{1}{2}bgE(t|s_B)$. After $s_B=0$ B's posterior belief on t is uniform on $[t_m-z,t']$, while after $s_B=c_B$ his posterior belief on t is uniform on $[t',t_m+z]$. Hence, $E(t|s_B=0)=E(t|t<t')=\frac{1}{2}(t_m-z+t')$ and $E(t|s_B=c_B)=E(t|t>t')=\frac{1}{2}(t_m+z+t')$. □

Proposition 6.1 states that for all parameter values there exists a unique universally divine equilibrium of the continuation game between B and G.[18] In regimes U1 and U2 only Uninformative pooling equilibria exist, and the interest group t-types either pool on no-lobbying (U1) or on lobbying the bureaucrat (U2). Under regime U1 the costs of lobbying c_B are prohibitive for all types of interest group. Even if a costly lobbying message would induce B to believe that $t=t_m+z$ and act accordingly, the costs of lobbying exceed the benefits. When regime U2 applies the costs of lobbying are so low that the interest group is prepared to lobby B irrespective of the actual state of the world, because no-lobbying is credibly interpreted as indicating that $t=t_m-z$. Under regime U2 the expenditures G makes on lobbying constitute a pure social waste to society because a costly lobbying message does not alter B's belief on the actual state of the world. In the non-pooling Informative regime I the costs of lobbying are in between, such that some types ($t>t'$) lobby whereas other types ($t<t'$) do not.

Comparative statics. We focus on the effect of changes in the basic parameters c_B, b, g (that is, g_1 and g_2), z and t_m on (i) the amount of lobbying, and (ii) on the impact of (no-)lobbying on the equilibrium policy chosen by B. The amount of lobbying is measured by the relative number of t-types that send a lobbying message ($s_B=c_B$), that is, by the probability $(t_m+z-t')/2z$ that lobbying occurs in equilibrium (cf. the *expected occurrence* of a costly message in Chapter 3). The impact of (no-)lobbying is measured by the equilibrium strategy of the bureaucrat, viz. $x_B(s_B;g)$. The discussion is based in part on Table 6.1 presented below, which lists the exact comparative statics results obtained when informative regime I applies. This table can easily be derived from Proposition 6.1. (In each cell of the table the first order partial derivative with respect to the basic parameter is reflected, together with its sign in brackets.)

Table 6.1 Comparative statics results in the informative regime I

	$(t_m+z-t')/2z$	$x_B(0;g)$	$x_B(c_B;g)$
c_B	$-t'/2zc_B$ (-)	$1/2gz$ (+)	$1/2gz$ (+)
b	$t'g/2(2+bg)z$ (+)	$\frac{1}{4}g(t_m-z)$ (+)	$\frac{1}{4}g(t_m+z)$ (+)
g	$t'(1+bg)/g(2+bg)z$ (+)	$\frac{1}{4}[b(t_m-z)-2c_B/g^2z]$ (-)	$\frac{1}{4}[b(t_m+z)-2c_B/g^2z]$ (±)
z	$2t'-t_m)/2z^2$ (±)	$-\frac{1}{4}(2+bg)(z+t')/z$ (-)	$\frac{1}{4}(2+bg)(z-t')/z$ (±)
t_m	$1/2z$ (+)	$\frac{1}{4}(2+bg)$ (+)	$\frac{1}{4}(2+bg)$ (+)

Trivially, the amount of lobbying in the two uninformative regimes is independent of the basic parameters, as long as the regimes apply. Under the informative regime more t-types send a costly lobbying message if the costs c_B of sending such a message decrease, the bureaucrat takes larger care of the interests of the interest group (higher b), the group's general stake g in persuading government increases, or the state of the world is more "extreme" in general (t_m larger). All these changes increase the potential gain that can be obtained from lobbying, and therefore cause t' to decrease relative to the lowest possible type t_m-z. As was the case for the basic signaling game presented in Chapter 3, it is obtained here that costly lobbying is more likely to occur when the costs are lower and the interest group's stakes are higher (cf. Result 3.1(d)). When z increases not only t' decreases, but also the set T of possible types becomes larger. The overall effect of an increase in this measure of uncertainty on the amount of lobbying is ambiguous, and depends on the relative difference between cut-off point t' and mean value t_m (cf. Table 6.1).

A change in one of the basic parameters may affect B's equilibrium policy both directly through an altered reaction to the same kind of information, and indirectly through its influence on the kind of information received.[19] Only changes in b and g have a direct impact. An increase in either one of these parameters increases the policy level the bureaucrat prefers given the value of t, and induces him to react more favorably, from G's point of view, to the same information. When either regime U1 or U2 applies, no information is transmitted and $E(t|s_B)=t_m$ for s_B sent in equilibrium. Hence in these two regimes only changes in t_m have an indirect influence on the impact of (no-)lobbying. Taking the direct and indirect influence together, it follows that in the uninformative regimes the impact of (no-)lobbying is increasing in b, g and t_m, and independent of c_B and z.

Under informative regime I all basic parameters have an indirect effect on equilibrium policy, either through their influence on cut-off type t', or via their impact on the boundaries of T (or both). As already noted, the parameters b and g have a direct effect as well. First, consider the three variables c_B, t_m and z that only have an indirect effect. An increase in c_B or t_m leads to a larger impact of both a lobbying message ($E(t|c_B)$ increases) and of no such message ($E(t|0)$ increases). An increase in z lowers the impact of no-lobbying because $E(t|0)$ decreases, whereas the change in the impact of lobbying depends on the value of t' relative to z (cf. Table 6.1). Second, with respect to both b and g the direct effect and the indirect effect work in opposite directions. The direct effect of an increase in b or g leads an increase in equilibrium policy $x_B(s_B;g)$. The indirect effect, on the other hand, leads to a reduction of $x_B(s_B;g)$ irrespective of whether the signal $s_B=0$ or $s_B=c_B$ is obtained in equilibrium. A higher b or g induces more t-types to send a lobbying message (t' decreases), and therefore lowers the impact of both a lobbying message and of no such message ($E(t|s_B)$ decreases). Overall, with respect to b the direct effect dominates. This does not hold in regard to g. An increase in g always leads to a decrease in $x_B(0;g)$, whereas the total effect of an increase of g on the impact of lobbying is ambiguous (cf. Table 6.1).[20]

One could argue that perhaps a better measure of the effect of lobbying is given by the additional impact it has on the equilibrium policy chosen, viz. by $x_B(c_B;g)-x_B(0;g)$ (cf. Chapter 3). It follows from Proposition 6.1 that when the informative regime applies this additional impact equals $\frac{1}{2}(2+bg)z$. Surprisingly, the additional impact of lobbying is independent of its costs. Although when c_B increases less types lobby B and the (indirect) impact of lobbying increases, also the impact of no-lobbying increases. It appears that both effects cancel each other out, such that the additional impact of lobbying stays the same.

Welfare implications. Proposition 6.1 reveals that the equilibrium policy chosen by the bureaucrat is given by $x_B(s_B;g)=E(t|s_B)+\frac{1}{2}bgE(t|s_B)$.[21] The first term represents the policy the politician would have chosen were she to have the same information as B has in equilibrium, that is, in case she could observe and interpret lobbying signal s_B. The second term reflects the policy bias caused by the different interests of the bureaucrat. Note the direct presence of bias parameter b in this term. On the one hand delegation is attractive to P in the sense that B can base his policy decision on more information about t, namely on $E(t|s_B)$ rather than $E(t)=t_m$. On the other hand delegation is unattractive to P because the different interests of B introduce a policy bias. It is exactly this trade-off that governs the delegation decision of the politician in our model.

Of course, the ultimate decision of P whether to delegate policy authority or not depends on the expected payoffs she gets in the continuation game after delegation. (A similar remark applies to the decision of G whether to lobby the politician for (no-) delegation.) Because P neither directly observes g nor directly gets to know the true value of t, we have to take expectations both with respect to g and with respect to t to obtain her expected payoffs. Specifically, we have to calculate these expected payoffs in two steps. First we have to take expectations with respect to t, taking the value of g as given. The following proposition specifies the expected equilibrium payoffs P and G obtain in the continuation game after delegation, *conditional* on the true value of g. In this proposition $E_D(U_P|g)$ denotes the expected equilibrium payoff to P of Delegation, assuming that P knows the true value of g. Because G in fact knows the true value of g, we use $E_D(U_G;g)$ to denote G's payoffs conditional on g.

Proposition 6.2
Given the value of g, the expected equilibrium payoffs obtained by P and G in the equilibria of the continuation game between the bureaucrat and the interest group are given by:

$$E_D(U_P|g) = -(b^2g^2/16)[3t_m^2+z^2+t'(2t_m-t')] - (1/12)[3t_m^2+z^2-3t'(2t_m-t')] \quad \text{and}$$
$$E_D(U_G;g) = \tfrac{1}{8}g(2+bg)[3t_m^2+z^2+t'(2t_m-t')] - c_B(t_m+z-t')/2z \qquad \Box$$

In order to arrive at the expected payoffs of the politician in the continuation game between B and G we need another step. In this second step we have to take expectations with respect to g. Specifically, we have to take expectations based on P's posterior belief about g after stage (ii). This posterior belief depends on the "lobbying-the-politician" strategy of G at this stage. Therefore, to determine these expected payoffs we have to consider the complete game rather than just the continuation game between B and G. In order not to confuse matters, these issues, which are not solely connected to the continuation game between B and G, are discussed in the next subsection.

6.2.3   The induced preferences of the politician: to delegate or not?

In this subsection the induced preferences of the politician with respect to delegation and no-delegation are considered. Recall that the expected equilibrium payoffs of No-delegation are denoted as $E_N(U_P)$ (cf. expression (6.4)). These expected payoffs are independent of the politician's beliefs about the value of the general stake g the interest group has in persuading government. They are therefore independent of the lobbying signal s_P received at stage (ii). In contrast, as observed in the previous subsection the expected payoffs to the politician of delegation do depend on her expectations about the actual value of g. Now, at the time she has to decide whether to delegate or not (stage (iii)) these expectations are determined by her posterior belief about the value of g at this decision stage. This posterior belief depends on the lobbying signal $s_P \in \{0,s_P\}$ she has received at stage (ii). Specifically, the posterior belief after having received signal s_P is endogenously determined within the model through the stage (ii) lobbying strategy of the interest group, together with Bayes' rule and the prior belief p. As a result, the politician's expected equilibrium payoffs of delegation depend on the signal s_P she receives. These conditional expected payoffs are denoted as $E_D(U_P|s_P)$. With Proposition 6.2 at hand these payoffs can be obtained from $E_D(U_P|s_P) \equiv E(E_D(U_P|g)|s_P)$ and the stage (ii) lobbying strategy of the interest group. That is, the expected value of $E_D(U_P|g)$ is determined using P's posterior belief on g after having received message s_P.

Taking the equilibrium lobbying strategy of the interest group at stage (ii) as implicit, expression (6.6) below can easily be obtained from Proposition 6.2. The expression relates $E_D(U_P|s_P)$ to $E_N(U_P)$ in a convenient way. Of course, the politician prefers to delegate policy authority after having received lobbying signal $s_P \in \{0,c_P\}$ when $E_D(U_P|s_P)>E_N(U_P)$, and prefers to determine policy herself in case $E_D(U_P|s_P)<E_N(U_P)$.

$$E_D(U_P|s_P) = -\tfrac{1}{3}z^2 - \tfrac{1}{4}b^2 t_m^2 E(g^2|s_P) + E(\tfrac{1}{4}[z^2-(t_m-t')^2]|s_P) - \tfrac{1}{4}b^2 E(g^2\tfrac{1}{4}[z^2-(t_m-t')^2]|s_P)$$

$$= E_N(U_P) \quad - \quad J \quad + \quad K \quad - \quad Q \qquad (6.6)$$

Here $E(\cdot|s_P)$ is used to denote expectations with respect to g based on the posterior belief of P after having received signal s_P. The subterm $[z^2-(t_m-t')^2]$ equals 0 when $t'=t_m-z$ or $t'=t_m+z$, and is strictly positive in case $t' \in (t_m-z,t_m+z)$. Hence terms K and Q are both non-negative, just like term J. Note that when b=0 both J and Q reduce to 0, implying $E_D(U_P|s_P) \geq E_N(U_P)$. When the bureaucrat is unbiased, the politician never loses from delegation. The terms J, K and Q can be interpreted in the context of the two different effects of delegation. To that purpose, we make use of the following table:

Table 6.2 Decomposition of the expected payoff P obtains in equilibrium

	decision based on t_m	decision based on $E(t\|s_B)$
P decides on policy	$E_N(U_P)$	$E_N(U_P) + K$
B decides on policy	$E_N(U_P) - J$	$E_N(U_P) - J + K - Q$

In Table 6.2 the top row describes the situation in which the politician decides on policy herself, the bottom row refers to the case in which policy authority is delegated to the bureaucrat. The two columns refer to the amount of information available to the decisionmaker when deciding on policy. In the left column the decisionmaker has to decide on the basis of the prior belief about t, in the right column on the basis of the same amount of information as is revealed in equilibrium of the continuation game between B and G. For instance, in case the politician would have obtained the same information as B obtains in equilibrium after delegation — that is, were P to know the exact value of g and able to interpret and observe signal s_B — she would have chosen policy $x_P(s_B;g)=E(t|s_B)$ (with $E(t|s_B)$ from Proposition 6.1). The expression $E_N(U_P)+K$ in the top right cell gives the expected payoff the politician would get in such a case.

Two types of effects of delegation can be disentangled. Firstly, delegation of authority leads to "bureaucratic drift". When the interests of the bureaucrat differ from those of the politician, B implements a different policy than the one P would have chosen on the basis of the same information. Through delegation B is able to bias policy from P's ideal point towards his own ideal point. The gain to B (and, thus, G) comes at the expense of the loss to P. This type of effect is therefore defined as a *distributional effect*. Distributional effects are a consequence of the different interests of the agents. Purely distributional effects follow from Table 6.2 by comparing for each column, thus for a fixed amount of information about the state of the world, the top row with the bottom row. In Table 6.2 only term J reflects a pure distributional effect, because for each column it holds that going from top to bottom the term J is subtracted.

Secondly, delegation may entail that the government gets additional information from G about the actual state of the world. The pure *informational effect* refers to changes in payoffs caused by the fact that the same decisionmaker can base his (her) decision on more information. In other words, holding constant distributional effects (thus who decides on policy), the pure informational effect is given by the effect of additional information on payoffs, irrespective of who the actual decisionmaker is. In Table 6.2 the pure informational effect is reflected by term K. Going from left to right term K is added in each row.

The third term Q is a combination of distributional and informational effects. It represents changes in payoffs caused by the fact that additional information about t is used in a different way by B than P would do. The combined effect only occurs when a different decisionmaker, B instead of P, bases his policy decision on different (more) information.[22] Though arbitrary, with respect to term Q we let verbally the informational effect prevail. In the sequel we will refer to the difference between the terms K and Q as the impure informational effect, and to term J as the pure distributional effect.

The pure distributional effect is always non-positive, so that term J can be said to represent the *distributional loss* from delegation. Only when the bureaucrat has the same preferences as the politician (b=0), term J reduces to 0. The impure informational effect K-Q, on the other hand, may be either positive, equal to zero, or negative. In the first case term K-Q represents the (impure) *informational gain* from delegation, in the last case the (impure) *informational loss*.[23] Comparing the distributional loss with the (impure) informational effect, the following proposition is obtained:

Proposition 6.3
The politician prefers to delegate policy authority to the bureaucrat only when the (impure) informational effect K-Q exceeds the distributional loss J. In equilibrium the politician strictly prefers no-delegation when either (a) or (b) holds:
(a) $b>0$, and $c_B \geq \frac{1}{2}g_2(2+bg_2)(t_m+z)z$ or $c_B \leq \frac{1}{2}g_1(2+bg_1)(t_m-z)z$.
(b) $(bg_1)^2 > 4z^2/[4t_m^2+z^2]$. $\qquad\qquad\qquad\qquad\qquad\qquad\qquad$ □

Proposition 6.3 provides the formal answer to the first question posed in the introduction to this chapter. It reflects the by now familiar (cf. Section 6.1) trade-off between informational gains and distributional losses that governs the delegation decision of the politician. However, contrary to these existing models the model presented in this chapter explicitly relates this trade-off to, and derives it from, interest group influence on bureaucrats.

In case the bureaucrat is biased (b>0), the politician is only willing to delegate authority when at least some information is revealed in equilibrium of the continuation game between the bureaucrat and the interest group. Only in that case it is possible that the distributional loss J is compensated for by an informational gain K-Q>0. Therefore, when the costs of lobbying the bureaucrat c_B are either very high or very low, such that either uninformative regime U1 or U2 applies in the continuation game between B and G and K=Q=0, the politician does not delegate authority. Proposition 6.3(a) thus reveals that for delegation to occur interest groups must have some (c_B not too high), but not too easy (c_B not too low) lobbying access to the bureaucracy.

Part (b) of Proposition 6.3 shows that when bg becomes too large, the losses invariably exceed the gains and the politician does not want to delegate. In that event the preferences of P and B are so diverse that the additional information B (potentially) obtains is used in an unprofitable way from P's point of view. Loosely put, when the bureaucrat cares too much about his career opportunities outside government (high b), or the general stake g of the interest group is too high, delegation is not profitable for the politician. This result corresponds with the result obtained by Epstein and O'Halloran in a setting without interest group influence, which entails that the level of discretion given to a bureaucrat declines when the conflict of interests between the politician and the bureaucrat increases (cf. Section 6.1). In line with Laffont and Tirole (1991) the results so far also suggest that the organizational response to the possibility of "capture" of B by G may be to try to reduce the stakes the interest group has in public policy (see below, however). Of course, another response is simply not to delegate policy authority, but then the politician forgoes the potential informational gains that can be obtained from delegation.

For delegation to occur the preferences of the politician, on the one hand, and those of the bureaucrat and the interest group, on the other hand, must be sufficiently aligned. However, an increase in the bias parameter b or an increase in the general stake of the interest group g does not always harm the politician. We first consider changes in the bias parameter b. Contrary to what might be conjectured the most preferred situation for P is not necessarily a completely loyal bureaucracy (b=0), because P may benefit from some disloyalty of B. This assertion is reflected in the following proposition, which is proved by means of an explicit numerical example (cf. Appendix 6.A).

Proposition 6.4
The net gain of delegation to the politician $E_D(U_P|s_P)-E_N(U_P)=-J+K-Q$ may both be non-negative and increasing in b over some interval, and the politician may prefer a biased (b>0) bureaucrat over an unbiased (b=0) one.   □

Proposition 6.4 gives the answer to the second question posed in this chapter by indicating that politicians do not always prefer an unbiased bureaucrat over a biased one. That is, contrary to the delegation model of Legros (and Bawn) reviewed in Section 6.1, in the present model the politician may actually prefer to appoint a biased bureaucrat. Loosely put, Proposition 6.4 states that it is sometimes profitable for the politician when the bureaucrat would care somewhat more about his outside opportunities. Although this leads to a larger distributional loss of delegation, it may lead to an even larger informational gain because "more" information is transmitted between the interest group and the (biased) bureaucrat. More information transmission is possible in that case because the interests of the bureaucrat and the interest group are more aligned.

The result that a principal (here the politician) may prefer biased delegates over unbiased ones is in line with results obtained in a number of other studies. Firstly, some scholars find that politicians may prefer biased political advisors. In Calvert (1985), for instance, this result is obtained with respect to a non-strategic advisor for whom the bias in his messages is given exogenously. For purely statistical reasons it appears that the politician may prefer a biased advisor in order to gain from the possibility of an unexpected advice from a biased source. Other studies confirm the "biased-advisor-preferred" result in a setting with strategically acting advisors. In Epstein and O'Halloran (1995) and Milner and Rosendorff (1996) a biased advisor disciplines the behavior of the bureaucracy because it needs the advisor's endorsement (the watchdog role of interest groups, cf. Section

3.3 and 6.1). Letterie and Swank (1997) find that a biased advisor is chosen in order to support political legitimacy in parliament or among voters. Secondly, the preference for a biased delegate is also observed in models of monetary policy institutions. In these models it is typically obtained that politicians prefer to appoint a central banker that is more "conservative" than themselves in order to gain credibility (cf. Rogoff, 1985). Thirdly, also in certain bargaining situations the principal may prefer a biased delegate over an unbiased one. In the international negotiations model analyzed by Mo (1995), for instance, negotiators sometimes prefer to grant veto power to biased agents in order to curtail the range of proposals the other party can successfully make.

The result that the politician sometimes prefers somewhat biased bureaucrats firstly suggests that the politician may benefit from interest groups having some *indirect* influence on bureaucrats, for instance through the existence of revolving doors. Che (1995) and Salant (1995) draw the same conclusion from their formal models, albeit for different reasons. Che (1995) argues that the existence of revolving doors provides incentives for bureaucrats to acquire more expertise in order to qualify for the post-government job. In turn, this increase in human capital may have a positive spill-over effect on the bureaucrats' performance while still in public office. In Salant (1995) revolving doors solve the hold-up investment problem of a regulated industry. Through revolving doors regulators have a stake in the future profitability of the regulated firm, and therefore they will credibly not engage in regulation which reduces the industries investment incentives. Secondly, the potential preference for somewhat biased bureaucrats also suggests that politicians may prefer to allow interest groups to have some *direct* influence on the bureaucracy. Such an interpretation is in line with a result obtained by Spiller (1990). He finds that Congress may prefer to allow direct transfers from the industry to the regulator. Contrary to Spiller (1990), however, here the informational gains are emphasized, rather than the gains from the increase in the regulators' rents which are appropriated by politicians.

Having discussed the effect of changes in bias parameter b on P's potential benefit from B's expertise, we next consider the effect of changes in the general stake the interest group has in persuading government on this potential gain. To that purpose (only) the case $g_1=g_2=g$ is considered, such that there is no uncertainty about the value of g. Before additional information about t is known, the optimal policy of the politician equals $x_P=t_m$. The interest group, on the other hand, prefers policy x to be as large as possible. An increase in policy from x_P to $x_P+\Delta x$ yields the interest group with general stake g an additional expected payoff of $gt_m\Delta x$. Hence the larger g, the larger the incentive the interest group has in persuading the politician to go beyond her a priori optimal policy. From this point of view, stake parameter g can be interpreted as a measure of conflict of interests. When g increases, the interests of the politician and the interest group become more opposed.

As suggested, it appears that the politician may prefer the interest group to have a larger general stake g. Put differently, she may prefer a situation in which the interests of G are more opposed to her own interests. This assertion is summarized in the proposition below, and proved with an example in which an increase in g induces the politician to delegate more often (cf. Appendix 6.A).

Proposition 6.5
Consider the case $g_1=g_2=g$. The net gain of delegation to the politician $E_D(U_P)-E_N(U_P)=-J+K-Q$ may both be non-negative and increasing in g over some interval, and the politician may prefer the interest group to have interests that are more opposed to her own.   □

In Proposition 6.5 we write $E_D(U_P)$ instead of $E_D(U_P|s_P)$ because, when $g_1=g_2=g$, the value of g is known to P and her expected payoff is independent of the lobbying signal s_P. The result reflected in the proposition is driven by the informational effect. An increase in the general stake of the interest group may facilitate an increased scope for information transmission between the group and the bureaucracy. The proposition indicates that the resulting increase in the informational gain may outweigh the increase in the distributional loss. Consequently, an increase in the stake an interest group has in public policy may in fact be beneficial to the politician, and induce her to delegate more often. This conclusion is in line with results obtained in Spiller (1990, Corollary 7) and by Austen-Smith and Wright (1992).[24] The conclusion qualifies the one drawn by Laffont and Tirole (1991) discussed above, because it shows that reducing the stake the interest group has in persuading government is not always the "right" organizational response to its potential influence on the bureaucrat.[25]

6.2.4   The induced preferences of the interest group: to pursue delegation or no-delegation?

The expected equilibrium payoffs of the interest group after delegation can be analyzed in the same way as we did for the politician in the previous subsection. As before, let $E_D(U_G;g)$ denote the expected equilibrium payoff of Delegation for an interest group with general characteristics g. Rewriting the expression for $E_D(U_G;g)$ given in Proposition 6.2 the following expression is obtained for the induced preferences of G:

$$E_D(U_G;g) = gt_m^2 + \tfrac{1}{2}bg^2t_m^2 + [\tfrac{1}{4}g[z^2-(t_m-t')^2] - c_B(t_m+z-t')/2z] + \tfrac{1}{8}bg^2[z^2-(t_m-t')^2]$$

$$= E_N(U_G;g) + X + \qquad [\,Y1 - Y2\,] \qquad\qquad + \quad Z \qquad (6.7)$$

Define $Y\equiv[Y1-Y2]$. Recall that the subterm $[z^2-(t_m-t')^2]$ equals 0 when $t'=t_m-z$ or $t'=t_m+z$, and is strictly positive when $t'\in(t_m-z,t_m+z)$. Hence, the terms X, Y1, Y2, and Z are all non-negative. When b=0 both X and Z reduce to 0. In that case $E_D(U_G;g)=E_N(U_G;g)+Y\leq E_N(U_G;g)$, because as will be shown below, term Y appears to be always non-positive.

As in case of the induced preferences of the politician, the expected equilibrium payoff the interest group obtains after delegation differs from its expected payoff after no-delegation due to the presence of both a distributional and an informational effect. The terms X, Y and Z can be interpreted in the context of these two effects in exactly the same way as was done for the induced preferences of the politician. To that purpose the following table is helpful:

Table 6.3 Decomposition of the expected payoff G obtains in equilibrium

	decision based on t_m	decision based on $E(t \mid s_B)$
P decides on policy	$E_N(U_G;g)$	$E_N(U_G;g) + Y$
B decides on policy	$E_N(U_G;g) + X$	$E_N(U_G;g) + X + Y + Z$

The left column in Table 6.3 describes the case in which no information is available about the actual state of the world t except for its prior distribution. The right column refers to the situation in which the decisionmaker obtains the same information as B gets in equilibrium of the continuation game between B and G. The rows refer to which governmental agent, either P or B, takes the final policy decision. In each cell the expected payoff of the interest group is presented, given the actual decisionmaker (row) and the amount of information this decisionmaker has (column) when deciding on which policy to implement. From the definitions given in the previous subsection it follows that X represents a pure distributional effect, Y a pure informational effect, and Z a combined effect.[26] Once more, with some abuse of terminology, the terms Y and Z are summed to represent the impure informational effect.

The pure informational effect Y can be decomposed into two components. The first component Y1 represents the gain to the interest group from having a better informed decisionmaker, who now bases the policy decision on $E(t \mid s_B)$ instead of just $E(t) \equiv t_m$. From Y1≥0 it follows that the additional information the decision-maker gets is, in expected utility terms, beneficial to the interest group. The second component Y2 reflects the expected costs $P(t > t')c_B$ of supplying the additional information. These costs are a consequence of the potential informativeness of the lobbying process, and they are therefore incorporated in the informational effect. In case the interest group would not lobby, the decisionmaker would believe that t<t' and act accordingly.

The pure distributional effect is always non-negative, and term X represents the *distributional gain* to the interest group of delegation. Only when the bureaucrat is unbiased (b=0), term X reduces to 0. The impure informational effect Y+Z, on the other hand, is always non-positive (see Proposition 6.6 below). The term Y+Z therefore represents the interest group's (impure) *informational loss* from delegation. (An elaborated discussion of the modeling assumptions that drive the result that the informational effect is always non-positive is given in Subsection 6.3.1 below.) This (impure) informational loss equals 0 only when $t'=t_m+z$, because only in that case no expenditures are made on lobbying the bureaucrat. Comparing the distributional gain with the (impure) informational loss, Proposition 6.6 is obtained:

Proposition 6.6
It holds that $Y+Z\leq0$. Therefore, the interest group prefers delegation of policy authority only when the distributional gain X exceeds the (impure) informational loss $Y+Z$. This certainly occurs in equilibrium when $g>2/3b$.   □

Proposition 6.6 shows that when the general stake g of the interest group is rather high, the interest group will, were it to lobby the politician, typically pursue delegation of policy authority. In the next subsection it is derived exactly when the interest group will lobby the politician in equilibrium.

Proposition 6.6 also reveals that it is not the fact that information about the actual state of the world can be transmitted that induces G to prefer delegation, but rather the distributional effect. Although the additional information B receives from G is in itself beneficial to the interest group (in expected payoff terms), the fact that credible information transmission can only occur at a certain cost yields that G's net expected gain of information transmission is negative. (The potential restrictiveness of this result is discussed in Section 6.3.1 below.) Indeed, when $g<2/3b$ it may occur that the informational loss exceeds the distributional gain, such that the interest group in fact loses from delegation. Put differently, although from the perspective of the interest group a better policy can be obtained when authority is delegated, the additional costs to obtain such a policy may exceed the benefits. Interest groups may thus prefer that policy is set once and for all, discharging them of the need to lobby the bureaucracy. This observation is in line with findings reported in other theoretical studies that interest groups may prefer the absence of possibilities for costly informational lobbying (cf. Chapter 3).

When bg becomes large the interest group certainly prefers delegation over no-delegation. Given the observations made regarding the expected payoffs of the politician (cf. Propositions 6.4 and 6.5), one might conjecture that the interest group does not always benefit from the preferences of the bureaucrat becoming more aligned with its own preferences. This appears not to be the case, though, at least when the delegation decision of the politician is not affected. Given that the politician delegates her policy authority to the bureaucrat, the interest group prefers the bias of the bureaucrat b to be as high as possible. This result follows from the following proposition:

Proposition 6.7
The interest group's net gain of delegation $E_D(U_G;g)-E_N(U_G;g)=X+Y+Z$ is strictly increasing in bias parameter b.   □

Rather intuitive, the interest group prefers a delegate whose interests are aligned with its own interests. Because changes in bias parameter b may affect the delegation decision of the politician, however, an increase in parameter b does not always benefit the interest group. Similar remarks apply to the costs of lobbying the bureaucrat. Given that the politician delegates policy authority to the bureaucrat, the interest group prefers these costs to be that high such that regime U1 applies and information transfer with respect to t cannot occur. As was observed in the previous subsection, though, it is exactly the potential for information transfer about the actual state of the world that induces the politician to delegate. An increase in the costs of lobbying the bureaucrat may therefore induce the politician to delegate less often, and hence does not necessarily benefit the interest group.

6.2.5   The complete game: when is the politician lobbied in equilibrium, and what is the purpose of it?

In the last two subsections the induced preferences of the politician and the interest group were considered in isolation. The results obtained there can be used for the analysis of the complete game by considering these induced preferences simultaneously. We observed that the politician may prefer delegation because the informational gains outweigh the distributional losses. For the interest group the situation appeared to be completely opposite. Delegation is preferred only when the distributional gains outweigh the informational losses. In this section we analyze how the players' induced preferences over delegation and no-delegation affect the first lobbying stage in which the interest group may lobby the politician for (no-)delegation. We focus on two interrelated questions, namely: When does lobbying the politician occur?, and What is the purpose of it?

In order to answer these two questions we have to consider the stage (ii) lobbying strategy of the interest group in equilibrium. In this respect the following observations are useful. Having solved for the equilibria of the two continuation games after stage (iii) (cf. Subsections 6.2.1 and 6.2.2), we are in a position to reduce the complete game to a *reduced game*. This reduced game is obtained by substituting for each of the two possible decisions of P — no-delegation or delegation — the conditional equilibrium payoffs the players obtain in an equilibrium of the continuation game that follows after this decision. These conditional equilibrium payoffs are denoted by $E_N(\cdot)$ and $E_D(\cdot\,|\,g)$, respectively. This reduced game has exactly the setup of the basic signaling game of Chapter 3. To illustrate this, consider Table 6.4 below:

Table 6.4 Gross payoffs over action-state pairs in the reduced game

	$g=g_1$		$g=g_2$	
P chooses No-delegation	$E_N(U_P)$	$E_N(U_G;g_1)$	$E_N(U_P)$	$E_N(U_G;g_2)$
P chooses Delegation	$E_D(U_P\|g_1)$	$E_D(U_G;g_1)$	$E_D(U_P\|g_2)$	$E_D(U_G;g_2)$

The exact expressions for $E_N(U_P)$ and $E_N(U_G;g)$ are given by equations (6.4) and (6.5), and for $E_D(U_P|g)$ and $E_D(U_G;g)$ by Proposition 6.2. The analysis of Chapter 3 is now directly applicable by using the following definitions:

$$x_1 \equiv \text{No-delegation}, \quad x_2 \equiv \text{Delegation}, \quad t_1 \equiv g_1, \quad t_2 \equiv g_2, \quad d_1 \equiv E_N(U_P) - E_D(U_P | g_1)$$

(6.8)

$$d_2 \equiv E_D(U_P | g_2) - E_N(U_P), \quad e_1 = E_D(U_G;g_1) - E_N(U_G;g_1), \quad e_2 = E_D(U_G;g_2) - E_N(U_G;g_2)$$

where the left-hand side of each identity refers to the notation employed in Chapter 3. Using these definitions the payoffs in Table 6.4 can be normalized to obtain payoffs exactly as specified in Table 3.1. Only the additional assumption made in Chapter 3 that $d_1,d_2 > 0$ now has to be dropped.

Using the analysis of Chapter 3, in particular Proposition 3.2, it is fairly straightforward to specify necessary and sufficient conditions on the normalized preference parameters d_i and e_i (i=1,2) defined above for lobbying of P to be an equilibrium phenomenon.[27] Unfortunately, however, these conditions do not reduce to nicely interpretable restrictions on the basic parameters c_B, b, g_1, g_2, z, t_m and p. Similar remarks apply with respect to determining the purpose of lobbying the politician. In order to avoid a vast list of equilibria and conditions under which they exist (cf. Propositions 4.5 and 5.4), we confine ourselves to specifying a number of restrictions on the basic parameters which have a straightforward and interesting interpretation.

First the question is addressed under what circumstances the politician may indeed be lobbied by the interest group. Generally speaking, at least three conditions necessarily have to be satisfied simultaneously for this to happen: (i) whether P prefers no-delegation or delegation must depend on the actual value of g (i.e. $d_1d_2 \geq 0$ is required), (ii) the conditional induced preferences of P and G must not be completely opposed (i.e. $d_1e_1 \leq 0$ or $d_2e_2 \geq 0$ is required), and (iii) the potential gain to G of swaying the decision of P must exceed the costs of lobbying (i.e. $c_P \leq \max\{|e_1|,|e_2|\}$). In case one (or more) of these three necessary conditions is not met the politician is not lobbied in equilibrium. For convenience these observations are summarized in the following proposition:

Proposition 6.8

(a)   When for every value of $g \in \{g_1, g_2\}$ it holds that $E_D(U_P|g) - E_N(U_P) > 0$ or that $E_D(U_P|g) - E_N(U_P) < 0$, the politician is never lobbied in equilibrium.

(b)   When for every value of $g \in \{g_1, g_2\}$ it holds that $(E_D(U_P|g) - E_N(U_P))(E_D(U_G;g) - E_N(U_G;g)) < 0$, the politician is never lobbied in equilibrium.

(c)   When for every value of $g \in \{g_1, g_2\}$ it holds that $c_P > |E_D(U_G;g) - E_N(U_G;g)|$, the politician is never lobbied in equilibrium.   □

The intuition behind Proposition 6.8 is as follows. In the first lobbying stage information about the general characteristics g of the interest group may be transmitted from G to P. In case (a) of Proposition 6.8 information about g has no relevance at all from the politician's point of view, because it will never alter her delegation decision (cf. note 1 to Chapter 3). In that case lobbying P serves no purpose at all and, due to its costs, will not occur. Case (b) in Proposition 6.8 refers to the situation in which there is full conflict of interests between the interest group and the politician (cf. case II in Chapter 3). In such a situation the interest group always has an incentive to misinform the politician on its general characteristics g. By the rational expectations character of an equilibrium, then, the politician will always interpret a costly lobbying message $s_P = c_P$ in a way unfavorable for the interest group. This induces both types of the interest group not to lobby in equilibrium. The last case (c) in Proposition 6.8 simply describes the situation in which lobbying the politician is too costly for the interest group.

Our next step is to translate the conditions specified in Proposition 6.8 into interpretable restrictions on the basic parameters. This simultaneously leads us to consider the purpose of lobbying P. For this latter topic we need some additional notation and definitions. Let $v(s_P)$ denote the probability that P delegates the policy decision to B in case she has received signal $s_P \in \{0, c_P\}$. When a costly lobbying message $(s_P = c_P)$ leads to a higher probability that authority is delegated to the bureaucracy than abstaining from lobbying does $(s_P = 0)$, it is said that, in case the interest group lobbies the politician in equilibrium, it *lobbies for delegation*. In the opposite case — that is, $s_P = c_P$ occurs in equilibrium and $v(c_P) < v(0)$ — it is said that the interest group *lobbies for no-delegation*. Given that the equilibria in the two continuation games after the decision of the politician are unique, and given that $c_P > 0$, it directly follows that when $v(0) = v(c_P)$ lobbying of P does not occur in equilibrium.

It is convenient to distinguish two separate cases, viz. b=0 and b>0. Firstly, consider the case b=0. When the bureaucrat is completely obeying, the politician weakly prefers delegation because she can never lose by doing so; $E_D(U_P|g) \geq E_N(U_P)$ for $g \in \{g_1, g_2\}$ in Table 6.4, and J=Q=0 in expression (6.6). P may only choose not to delegate if she is completely convinced that delegation has no potential informational gain at all. This can only hold if she is completely sure that g has a value such that either regime U1 $(t'=t_m+z)$ or U2 $(t'=t_m-z)$ applies, and no information about t is revealed in the continuation game between B and G. The interest group, on the other hand, always (weakly) prefers no-delegation in that case; $E_D(U_G;g) \leq E_N(U_G;g)$ in Table 6.4.[28] Delegation never leads to a distributional gain (X=0 in expression (6.7)), but yields an informational loss ($Y+Z=Y\leq0$) whenever either regime I or U2 applies. In these two cases (only) the interest group has an incentive to persuade the politician not to delegate policy authority. Taking the above observations together it follows that the politician may only be lobbied by an interest group of type g for which uninformative regime U2 applies. Such a group will try to convince the politician not to delegate policy authority. In summary, when b=0 and the interest group lobbies the politician, it lobbies for no-delegation.

In fact, for the case b=0 necessary and sufficient restrictions on the basic parameters for lobbying of P to occur (with the purpose to induce no-delegation) are relatively easy to derive. These restrictions are summarized in the following proposition.

Proposition 6.9
Suppose b=0. A necessary and sufficient condition for lobbying of P to be an equilibrium phenomenon is $c_P \leq c_B \leq g_2(t_m-z)z$. In all these equilibria $v(c_P)<v(0)$. $\square$

From $c_B>0$ by assumption and Proposition 6.9 above we get that when $t_m=z$ lobbying of P does not occur in equilibrium (recall that by the setup of the model $t_m \geq z$, and that when $t_m=z$ the state of the world t is uniformly distributed on $[0,2t_m]$). Besides, also when $c_P>c_B$ the politician is never lobbied in equilibrium. In that case the potential gain of loss avoidance (c_B) does not outweigh the costs c_P of lobbying P.

Both of the conditions $c_P>c_B$ and $t_m=z$ seem rather plausible. In a loose way, the first condition requires that the interest group has more easy access to the bureaucracy than to the legislature. Especially for an interest group whose interests are explicitly represented by an executive department in the bureaucracy, for instance a Department of Agriculture in case of farmer organizations, this seems a reasonable assumption.[29] Typically, there are more bureaucrats than politicians that the interest group can approach, which is likely to make the former better accessible. Relatedly, politicians usually have a broader orientation than the policy experts in the bureaucracy have. Therefore they will be the potential lobbying target of interest groups in a wide range of different policy areas, and they necessarily have to be more selective in whom they grant access. Also because they face a reelection constraint, politicians tend to be less accessible than bureaucrats (cf. Crain and McCormick, 1984). Finally, access to bureaucrats may be more easy because they are more likely to go through the revolving door in the future. In case the interest group is represented by a former bureaucrat that once went through the revolving door, this representative will also have more easy access to the bureaucracy.

The second plausible restriction $t_m=z$ de facto requires that the actual state of the world may vary substantially. (Note that the variability in t is given by z, which in turn is restricted by its upperbound t_m.) Since the issue of delegation typically incorporates long run considerations, and long-term predictions tend to be less accurate than short-term predictions are, the assumption of substantial ex ante uncertainty seems appropriate. The restriction in particular requires that this ex ante uncertainty is that large such that there always exists circumstances ex post under which the interest group does not find it profitable to lobby the bureaucrat. That is, in the model the restriction excludes the possibility that the uninformative regime U2 applies in the continuation game between B and G (cf. Proposition 6.1). In practice it seems rather unlikely that it always pays for the interest group to lobby the bureaucrat, irrespective the actual state of the world. The restriction $t_m=z$ therefore seems to have some intuitive appeal.[30]

In summary, when the bureaucracy is unbiased ($b=0$) it can be argued that the politician is not likely to be lobbied. When it occurs, however, the interest group always lobbies for no-delegation.

We next consider the case that $b>0$. Based on the conditions specified in Proposition 6.8(a) and 6.8(c) we get the following circumstances under which lobbying the politician does not occur (the condition specified in Proposition 6.8(b) is not easily rewritten in intuitive restrictions on the basic parameters):

Proposition 6.10
Suppose $b>0$. The politician is never lobbied when:
(a) $c_B \leq \tfrac{1}{2} g_1(2+bg_1)(t_m-z)z$ or $c_B \geq \tfrac{1}{2} g_2(2+bg_2)(t_m+z)z$.
(b) $(bg_1)^2 > 4z^2/[4t_m^2+z^2]$.
(c) $c_P > \max\{\tfrac{1}{2}bg^2t_m^2, c_B\}$ for all $g \in \{g_1, g_2\}$. $\square$

In Proposition 6.3 it was already shown that when either (a) or (b) in Proposition 6.10 holds, the politician strictly prefers not to delegate, irrespective the value of g. The first two parts of Proposition 6.10 then immediately follow from Proposition 6.8(a). When it is not possible to sway the delegation decision of the politician, because she can never be persuaded to delegate, the interest group will not undertake costly effort to pursue this. Note that from both part (a) and (b) in Proposition 6.10 together with Proposition 6.3 it follows that when $z=0$ the politician is not lobbied and does not delegate policy authority. Trivially, in case there is no uncertainty about the actual state of the world no informational gains can be obtained by the politician from delegating authority. In that case delegation will not occur when the bureaucrat is biased.

The last part of Proposition 6.10 follows from considering the potential gain to the interest group of persuading the politician to alter her a priori decision. As shown in Subsection 6.2.4 the maximal gain to the interest group from having the politician delegate rather than not delegate is obtained when regime U1 applies and equals the distributional gain X (with $X=\tfrac{1}{2}bg^2t_m^2$, cf. expression (6.7)). The maximal gain from inducing P not to delegate rather than to delegate is given by the maximal loss avoided through no-delegation. From (6.7) immediately follows that the maximal loss never exceeds c_B. When the costs c_P of lobbying P exceed both the maximal gain possible and the maximal loss avoidable, lobbying the politician certainly does not occur. This yields Proposition 6.10(c).

In short, Proposition 6.10 states that for lobbying of P to occur when the bureaucrat is biased it is necessary that G has some, but not too easy lobbying access to B (i.e. c_B is fairly moderate), that the bureaucracy is not too much "captured" by G, and that G has some lobbying access to the politician (c_P is not prohibitively high). Sufficient conditions for lobbying of P to occur, although not difficult to derive given Table 6.4 and the analysis presented in Chapter 3, do not reduce to easily interpretable restrictions on the basic parameters. Therefore, we confine ourselves to showing that lobbying the politician may indeed occur in equilibrium when $b>0$, both in order to induce delegation in some instances, and to pursue no-delegation in other instances. Using numerical examples, the following proposition can be obtained:

Proposition 6.11
Suppose $b>0$. There exist equilibria in which the interest group lobbies the politician for delegation, as well as equilibria in which the interest group lobbies the politician for no-delegation.   □

From Proposition 6.11 it follows that, generally speaking, the politician may be lobbied for delegation as well as no-delegation of policy authority. In order to get an idea about which motive is more likely to occur according to our model, it seems useful to examine the consequences of imposing some plausible restrictions on the basic parameters. As argued previously, the conditions $c_P>c_B$ and $t_m=z$ seem to satisfy this plausibility criterion. The last proposition of this subsection concerns the implications of imposing these restrictions, and, in addition, reports another intuitive observation.

Proposition 6.12
Suppose $b>0$.
(a)   When $c_P \geq c_B$ and the interest group lobbies the politician in equilibrium, it lobbies for delegation.
(b)   When $t_m=z$ and the interest group lobbies the politician in equilibrium, it lobbies for delegation.
(c)   When $bg_1 > \frac{2}{3}$ and the interest group lobbies the politician in equilibrium, it lobbies for delegation.   □

Part (a) and part (c) of Proposition 6.11 follow from considering the induced preferences of the interest group. The only purpose for the interest group to lobby for no-delegation is to avoid the informational loss Y+Z. This potential loss certainly does not exceed c_B (cf. expression (6.7)). Together with $X>0$ when $b>0$ it immediately follows that when $c_P \geq c_B$ the interest group never lobbies for no-delegation. Taking also into account the requirement that for lobbying P to occur the costs of lobbying B necessarily have to be moderate (cf. Proposition 6.10(a)), Proposition 6.12(a) suggests that interest groups typically lobby the legislature for delegation of policy authority. It was already derived in Proposition 6.6 that when $bg_1 > \frac{2}{3}$ the interest group always prefers delegation. Trivially, then, the interest group only lobbies to induce delegation in that case, yielding part (c). Lastly, part (b) of Proposition 6.12 follows from analyzing the congruence of interests between the politician and the interest group. When $t_m = z$ the induced conditional (on g) preferences of P and G never coincide on a preference for no-delegation. As a result, the interest group does not undertake costly effort to pursue this (Proposition 6.8(b)). All in all it is concluded that, under arguably reasonable restrictions on the basic parameters, the interest group typically lobbies the politician to delegate policy authority to a biased bureaucrat.

6.2.6 Summary

When the politician decides to delegate policy authority, a signaling game between the interest group (sender) and the bureaucrat (receiver) is played. The analysis of this signaling game yields that for every specific parameterization a unique universally divine equilibrium exists. For sufficiently low costs of lobbying the bureaucrat the interest group always lobbies him, irrespective of the actual state of the world. When these costs are sufficiently high, the interest group does not lobby the bureaucrat under any circumstances. In both cases, no information about the state of the world is transmitted and the politician suffers from a distributional loss when delegating authority, assuming that her interests and those of the bureaucrat are not fully aligned. The interest group, on the other hand, enjoys a distributional gain. Only when the costs of lobbying the bureaucrat are in between, information transmission between the interest group and the bureaucracy occurs in equilibrium. An informative equilibrium is necessarily a partition equilibrium of size two, in which types knowing that the state of the world is high lobby the bureaucrat, and types knowing that it is low abstain. The politician enjoys an informational gain, but also suffers from a distributional loss. For the interest group the situation is completely opposite, because it receives a distributional gain but has to bear an informational loss.

When the induced preferences of the bureaucrat differ too much from those of the politician, the distributional loss to the politician always exceeds her informational gain from delegation. Roughly put, when the bureaucrat cares too much about his career opportunities outside government, or the stakes of the interest group are relatively high in general, delegation is not profitable for the politician. However, the informational gain to the politician may increase when the preferences of the bureaucrat and the interest group become somewhat more aligned. As a consequence, it is sometimes profitable for the politician that the bureaucrat is somewhat biased, and an increase in the general stake the interest group has in persuading government may in fact benefit the politician.

For lobbying of the politician to be an equilibrium phenomenon there generically must be some potential for information transmission between the interest group and the bureaucrat. This requires that access to the bureaucrat is neither blocked nor too easy for the interest group (i.e. the costs of lobbying him should be moderate). Likewise, the preferences of the bureaucrat must not be too opposed to those of the politician. In general the interest group may prefer delegation as well as no-delegation. Indeed, for specific parameter values the informational loss following delegation outweighs the distributional gain to the group. In that case the interest group prefers to avoid the additional effort of lobbying the bureaucracy. Therefore, in general the interest group sometimes lobbies the politician for no-delegation, and at other instances lobbies for delegation. When either the assumption is made that the actual state of the world may vary substantially, or that the costs of lobbying politicians exceed the costs of lobbying bureaucrats, only lobbying for delegation can occur in equilibrium, though. These assumptions seem rather plausible, hence our model suggests that interest groups will typically lobby politicians to induce delegation.

6.3   GENERALIZATIONS AND ALTERNATIVE MODELING ASSUMPTIONS

In this section first two generalizations of the model analyzed in the previous section are discussed. Subsection 6.3.1 elaborates on the somewhat restrictive result obtained in Subsection 6.2.4 that the interest group always suffers from an informational loss, while in Subsection 6.3.2 the consequences of endogenizing the costs of lobbying are considered. Subsequently, Subsection 6.3.3 briefly analyzes an alternative setup of the model. In that section a model is considered in which the continuation game between B and G is envisaged as a persuasion game rather than as a signaling game (cf. Section 3.3).

6.3.1   Delegation necessarily leads to an informational loss for the interest group?

A somewhat peculiar feature of the setup of the model analyzed in Section 6.2 is that it always results in an informational loss to the interest group ($Y+Z\leq0$, cf. Proposition 6.6). As a consequence, when the bureaucrat has the same preferences as the politician (b=0), the interest group always (weakly) prefers no-delegation. The relevance of this result may be questioned, because one might reasonably expect that in reality there are circumstances under which the interest group prefers delegation even when b=0, to reap some benefits from the transfer of information. The assumptions driving the "always-an-informational-loss" result relate to the prior distribution of the actual state of the world and the preferences of the players. Alternative assumptions about these aspects of the model may qualify this result. For instance, one might conjecture that a different result is obtained when using a distribution of t-types which is skewed to the right rather than symmetric as the assumed uniform distribution is. Loosely speaking, when the distribution of the actual state of the world t is skewed to the right low values of t are more likely than high values of t. When deciding on which policy to implement the politician estimates the mean value of t which, due to the overrepresentation of low types, will be rather low.[31] Thus, without additional information a rather "low" policy will be chosen. Information transmission then becomes more attractive to the high types, because more can be gained by separating themselves out. This may lead to a net informational gain. The following proposition shows that this may indeed be the case.

Proposition 6.13
When the distribution of t-types is skewed to the right it may occur that the interest group benefits from an informational gain in equilibrium.   □

Proposition 6.13 is proved by means of an example in which $b=0$, $g_1=g_2=1$, $t_m=z=\frac{1}{2}$, and where the distribution function of the actual state of the world is given by $F(t;r)=t^r$ $(r>0)$. For $r=1$ this distribution reduces to the uniform distribution used in the model of Section 6.2, while for $r<1$ $(r>1)$ the distribution is skewed to the right (left). It appears that when $F(t;r)$ is skewed to the left, the interest group necessarily suffers from an informational loss, whereas in case $F(t;r)$ is skewed to the right it may occur that the interest group benefits from information transmission per se in equilibrium.[32]

The general assumptions made in the model analyzed in Section 6.2 concerning the preferences of the players seem rather plausible. These general properties are that (i) G is really a special interest group in that it prefers policy to be as extreme as possible, (ii) G's preference intensity does depend on the actual state of the world, though, and (iii) the politician prefers policy to equal some state-dependent ideal policy. While maintaining these three properties, alterations in the specific functional forms of the players' preferences may also yield the result that the interest group gets an informational gain under certain circumstances. Consider, for instance, the utility function $U_G(x;t,g)$ of G. One somewhat restricting feature of the specific functional form used is its linearity in x (cf. expression (6.2)). This makes the interest group, conditional on g and t, risk neutral in policy.[33] It may be conjectured, though, that when the interest group becomes risk loving in policy, an informational gain may be obtained by G. When the interest group is risk loving in policy, it is more likely that it prefers the bet (lottery) on the uncertain policy outcome after information transmission over the certain policy outcome when no information about t is transmitted. The next proposition concerns this possibility.

Proposition 6.14
When the interest group is risk loving in policy conditional on the values of t and g, it may reap an informational gain in equilibrium.   □

Proposition 6.14 is proved by means of an example in which the interest group's utility over public policy is given by $U_G(x;t,g)=gtx^r$, with $r>0$, rather than by (6.2). The parameter r represents the rate of return on public policy, and thereby the risk attitude of the interest group. When $r>1$ the group is risk loving in policy, while in case $r<1$ the group is risk averse (conditional on t and g). For the case $b=0$, $t_m=z=\frac{1}{2}$, and $g_1=g_2=1$ it can be shown that when $r=2$ the interest group still suffers from an informational loss, but when $r=3$ it may benefit from an informational gain (depending on the value of c_B, see the proof of Proposition 6.14). As expected, when the interest group becomes more risk loving in policy, it may indeed sometimes benefit from information transmission per se.

A companion question to the one posed in this subsection is: Does delegation necessarily lead to a distributional gain to the interest group? By the assumption that $b\geq0$ a distributional loss is not possible in the model of Section 6.2. But, when $b<0$ is allowed for, such a distributional loss may indeed occur in equilibrium. It must be noted, however, that a negative value of b is not in the spirit of the model. The relationship between the bureaucrat and the interest group is modeled partly in reduced form, departing from the presumption that the existence of revolving doors brings the preferences of the bureaucrat more in line with those of the interest group. Although $b<0$ is formally possible in our model, such an assumption is inconsistent with this point of departure. Only when there are multiple groups with competing interests, it seems reasonable to assume that in effect the bureaucrat sometimes cares less about the interests of a specific group than the politician does (see Section 6.4 below).

We conclude that the observation that the interest group always suffers from an informational loss does not hold true for any choice of the prior distribution of t and functional forms for the players' preferences. From that point of view, the specific assumptions made in Section 6.2 are somewhat restrictive. They seem to lead to a conservative estimate of the attractiveness of delegation to the interest group. The observation that in general delegation may also lead to an informational gain to the interest group rather than an informational loss, only strengthens the prediction that interest groups will typically lobby for delegation.

6.3.2   Endogenous lobbying costs

In Section 3.3 it was observed that when the costs of lobbying are endogenous rather than fixed, the scope for information transfer increases. This result also holds true in the context of the model analyzed in this paper. When the costs of lobbying the bureaucrat can be determined by the interest group itself, an equilibrium of the continuation game between B and G takes a different form. Let $c_B(t;g)$ be the equilibrium lobbying strategy of the interest group with general stake g, knowing that the actual state of the world equals t. That is, $c_B(t;g)$ denotes the amount spent by G on lobbying B in equilibrium, given the values of t and g. With this additional notation at hand, the following proposition can be obtained:

Proposition 6.15

In case the costs of lobbying B can be freely determined by G, the unique universally divine equilibrium path of the continuation game between B and G is given by:

$$c_B(t;g) = \tfrac{1}{4} g(2+bg)[t^2 - (t_m - z)^2] \quad \text{and} \quad x_B(c_B;g) = [c_B(2+bg)/g + \tfrac{1}{4}(2+bg)^2(t_m - z)^2]^{\tfrac{1}{2}}$$

After having received equilibrium message $s_B = c$ the posterior beliefs of the bureaucrat are completely concentrated on the type t that solves $c_B(t;g) = c$. ☐

The equilibrium of Proposition 6.15 is completely separating because all types $t \in T$ employ a different level of lobbying effort $c_B(t;g)$. After being lobbied by G, the bureaucrat knows the exact value of t and implements his optimal policy $x_B = \tfrac{1}{2}(2+bg)t$.

     With the use of Proposition 6.15 the expected equilibrium payoffs of the players in the continuation game between B and G can be calculated. Employing the same type of notation as in Subsection 6.2.3 (cf. expression (6.6)) the following expression for the payoffs of the politician is obtained:

$$E_D(U_P | s_P) = -\tfrac{1}{3}z^2 - \tfrac{1}{4}b^2 t_m^2 E(g^2 | s_P) + \tfrac{1}{3}z^2 - (1/12)b^2 z^2 E(g^2 | s_P)$$

$$= E_N(U_P) - \quad J \quad + K - \quad Q \qquad (6.9)$$

Once again, there is a pure distributional loss J of delegation, and an (impure) informational effect K-Q. From comparing term K-Q with term J, and using the assumption $z \le t_m$, it follows that when $bg_1 > 1$ delegation is not an equilibrium phenomenon. In addition, the informational effect is now strictly decreasing in b and g. Thus, when lobbying costs are endogenous the politician prefers a completely obeying (unbiased) agency, and an interest group which interests are aligned with those of the politician.[34] These latter results contrast with the conclusions obtained for the model with exogenously fixed costs of lobbying (cf. Propositions 6.4 and 6.5). They follow from the fact that changes in b and g do not affect the level of information transmission between B and G, because the equilibrium in the continuation game is always completely separating (g>0 by assumption). However, changes in b and g do affect the distributional effect. Hence, when b=0 the politician surely wants to delegate.

     Turning now to the expected equilibrium payoffs for the interest group it is obtained that:

$$E_D(U_G;g) = gt_m^2 + \tfrac{1}{2}b g^2 t_m^2 + [\tfrac{1}{3}gz^2 - \tfrac{1}{2}g(2+bg)(t_m - \tfrac{1}{3}z)z] + (1/6)bg^2 z^2$$

$$= E_N(U_G;g) + X + \quad [Y1 - Y2] \quad + Z \qquad (6.10)$$

Also with endogenous lobbying costs the impure informational effect Y+Z is necessarily negative, since $t_m \ge z$. Consequently, the interest group may only prefer delegation due to the distributional effect. It turns out that in case $bg_1>6/5$ the interest group strictly prefers delegation. When bg is sufficiently low, however, the interest group in fact prefers no-delegation (specifically, when $bg<[2z(t_m-\frac{2}{3}z)]/[t_m^2-z(t_m-\frac{2}{3}z)]$). Taking all terms in (6.10) in which b appears together, it follows that the net gain of delegation is increasing in bias parameter b. As was the case when the costs of lobbying are fixed, the interest group prefers a biased delegate over an unbiased one (cf. Proposition 6.7).

The first lobbying (the politician) stage can be conveniently analyzed using expressions (6.9) and (6.10) above. In case the bureaucrat is unbiased (b=0) the politician always prefers to delegate, and the interest group never spends a positive amount on lobbying P in equilibrium. In case b>0 it is easy to derive necessary and sufficient conditions for lobbying of P to occur in equilibrium. The following proposition provides these conditions. It also indicates that when the costs of lobbying are endogenous, the interest group only spends money on lobbying P to further no-delegation.

Proposition 6.16
Suppose that lobbying costs are endogenous and that b>0. Generic necessary and sufficient conditions for an equilibrium to exist in which the interest group spends a positive amount of money on lobbying the politician, are given by:
(i)   Both types g_1 and g_2 prefer no-delegation: $g_2<[2z(t_m-\frac{2}{3}z)]/b[t_m^2-z(t_m-\frac{2}{3}z)]$.
(ii)   Conditional on the value of g, the politician prefers delegation when $g=g_1$ and no-delegation when $g=g_2$: $g_1<2z/b\sqrt{(3t_m^2+z^2)}<g_2$.
(iii)   Type g_2 gains more from persuading P not to delegate than type g_1 does: $E_D(U_G;g_2)-E_N(U_G;g_2)<E_D(U_G;g_1)-E_N(U_G;g_1)$.
In addition, making expenditures on lobbying P in equilibrium increases the probability that the politician does not delegate. $\square$

When analyzing the model with fixed lobbying costs the consequences of imposing two, in our view, plausible restrictions were identified. The first, $c_P>c_B$, does not seem to have a natural counterpart when lobbying costs are endogenous. The second condition $t_m=z$ is still applicable when considering the case with endogenous costs. From Proposition 6.16 above it follows that imposing this restriction excludes the possibility that the politician is lobbied in equilibrium, because condition (i) and (ii) cannot hold simultaneously when $t_m=z$.

The analysis presented in this subsection qualifies some of our conclusions obtained previously. When lobbying costs are endogenous it still holds true that delegation only occurs if bg is sufficiently low. Contrary to the fixed lobbying costs situation, however, the politician now always prefers an unbiased delegate over a biased one. The interest group typically does not lobby the politician, but when it does, it lobbies for no-delegation. The last result challenges our suggestion in Section 6.2 that interest groups often lobby the legislature for delegation of policy authority. However, as holds for the case of fixed lobbying costs, the assumption that these costs can be picked from a continuum should be considered as a benchmark case. In reality, the situation will be somewhere in between. The results of this subsection suggest that the more precisely the lobbying effort can be determined by the interest group and acknowledged by the bureaucrat, the less likely it becomes that the politician is lobbied for delegation, and the more attractive an unbiased bureaucrat becomes to her.

6.3.3 Modeling the continuation game between B and G as a persuasion game

Up till now we have assumed that the communication structure between the politician and the bureaucrat, on the one hand, and the interest group on the other hand, bears the characteristics of a signaling game. However, the possibilities for information transfer can be modeled in different ways. One alternative way is to envisage the continuation game between the bureaucrat and the interest group as a persuasion game. As discussed in Section 3.3, in a persuasion game (i) senders have to select a costless but truthful message from the set of available messages, and (ii) all messages that are truthful are available to them. In some instances it may be reasonable to assume that due to their expertise and long term relationship with interest groups, bureaucrats cannot be fooled by them, inducing interest groups to report truthfully. For instance, the expertise of bureaucrats may enable them to instantaneously and costlessly verify the reports sent by interest groups. Politicians, on the other hand, do not have the expertise to establish whether the claims interest groups make concerning the actual state of the world are correct or not, so proving them with untruthful technical information is possible.

In this subsection we briefly analyze the consequences of assuming that the continuation game between B and G fits the setup of a persuasion game. A lobbying message now has the following characteristics: (i) it is costless, and (ii) an interest group of type t can send all messages $s_B(t) \subseteq T$ that satisfy $t \in s_B(t)$, but cannot send messages for which $t \notin s_B(t)$. As expected, compared with the signaling setup of Section 6.2 with fixed lobbying costs, also in this case the scope for information transmission increases. Specifically, in equilibrium full disclosure about the actual state of the world necessarily occurs (cf. Proposition 6.17 below).

Proposition 6.17
When the continuation game between B and G is modeled as a persuasion game, the equilibrium of this game is necessarily completely separating and all information about t is revealed (unravelling). □

When the politician does not delegate policy authority, no information about the state of the world is revealed. On the other hand, when authority is delegated, all information about t is revealed. It is immaterial to the politician whether this full disclosure occurs either because lobbying costs are endogenous (cf. Proposition 6.15), or because lying to the bureaucrat is not possible (cf. Proposition 6.17). So, also in the latter case expression (6.9) applies. For the interest group there is a difference, though, because when messages are costless no lobbying expenditures have to be made. It immediately follows that in the case of a persuasion game only term Y2 drops out of expression (6.10). We get $E_D(U_G;g)=E_N(U_G;g)+X+Y1+Z$, with X, Y1 and Z as defined in (6.10). The interest group always prefers delegation, and always benefits (ex ante) from an informational gain of delegation.

In order to specify equilibria for the complete model, assumptions have to be made about how information transmission from G to P with respect to the general characteristics g takes place. We consider two possibilities, viz. a signaling game setup with fixed lobbying costs c_P (cf. the model of Section 6.2) and a persuasion game setup. The results for both cases are summarized in the last proposition of this chapter, which is presented below:

Proposition 6.18
(a)   When the lobbying-the-politician stage takes the form of a signaling game with fixed lobbying costs c_P, the politician is not lobbied in equilibrium and information transfer about g does not occur.
(b)   When the lobbying-the-politician stage takes the form of a persuasion game, full information revelation about g is always an equilibrium, and necessarily occurs when $E(g^2)>4z^2/(3t_m^2+z^2)b>(g_1)^2$.   □

In summary, when the interest group cannot lie to the bureaucracy about the actual state of the world, but can report untruthfully to the politician about the value of t, there is also a rationale for delegation of policy authority based on expertise asymmetries.

6.4   CONCLUDING DISCUSSION

In this chapter a stylized game-theoretic model is analyzed that relates the potential influence of interest groups to the internal organization of government. In the model an interest group may try to lobby the legislature, the bureaucracy, or both. Instead of taking government as a black box, the relationship between politicians and bureaucrats is explicitly modeled. Politicians may prefer to delegate policy authority to the bureaucracy because bureaucrats are better informed and are better able to understand technical information coming from the interest group. However, in case of a conflict of interests with bureaucrats, delegation also leads to bureaucratic drift. With their option to delegate authority or not, politicians affect the way in which interest groups try to influence public policy. Rather than acquiescing in the possibility that the bureaucracy is "captured" by organized interests, politicians take account of the potential influence of interest groups when deciding on the delegation of authority. The fact that in some cases this influence is beneficial to the legislature, due to the increased scope for information revelation, may cause the politician to opt for delegation.

In Section 6.1 the specific modeling assumptions were informally introduced by means of the following application of the model. A firm (the interest group) wants to obtain a subsidy from government. The firm prefers the subsidy to be as high as possible. The legislature, on the other hand, prefers the subsidy to serve its economic goals, like a decrease in unemployment. The appropriate subsidy level depends on the economic "health" of the firm, which is only known to the firm itself. Because the bureaucracy cares about both the welfare of the politicians and the firm (due to career opportunities within and outside government), it prefers to give the latter a somewhat higher subsidy than the politicians would do on the basis of the same information. Only bureaucrats have enough expertise to interpret lobbying information coming from the firm about its economic prospects. For that reason, the legislature may delegate to them the authority to decide on the level of subsidies. In turn, the firm may try to influence the delegation decision of the politicians by lobbying them.

   In Section 6.2 the equilibrium analysis of the model was presented in order to shed light on the four questions posed in the introductory paragraphs to this chapter. Firstly, it was derived that the politician prefers to delegate policy authority when the informational gains outweigh the distributional losses. This can only occur when the bureaucrat is not too much biased in favor of the interests of the interest group, and when the general stakes of the interest group in persuading government are not too high. It was observed, however, that the politician may benefit from the preferences of the interest group and the bureaucrat being aligned. Hence, in relation to the second question posed in the introduction, it is obtained that the politician may prefer a moderately biased bureaucrat over an unbiased one. This result is in line with a number of theoretical studies, which were briefly reviewed in Subsection 6.2.3. Thirdly, in case policy authority is delegated to bureaucrats, the interest group lobbies the bureaucracy only when the state of the world is such that the benefits of public policy are large compared with the costs of lobbying. Politicians are typically only lobbied when information about the general characteristics of the interest group may sway their delegation decision.

   Regarding the fourth and last question posed, in general the interest group may lobby politicians for delegation as well as no-delegation of policy authority. Delegation yields the interest group a distributional gain when the preferences of the bureaucracy are somewhat biased towards the group. In addition, delegation also has an informational effect because the bureaucracy is able to understand messages coming from the interest group containing technical, policy relevant information. In the basic setup of the model (Section 6.2) the interest group always loses from information transmission per se. It is not the case that the interest group prefers (in expected payoff terms) to conceal information about the actual state of the world, but rather the assumption that information transmission is costly that causes this result. Whether the interest group overall prefers delegation or no-delegation depends on whether the distributional gain or the informational loss is larger. Taking account of the incentives of the politician as well, it is derived that when the costs of lobbying the politician exceed the costs of lobbying the bureaucracy, the former is only lobbied to induce delegation. The same conclusion is obtained when it is assumed that there is a priori substantial uncertainty about the actual state of the world. To the extent that these two assumptions appear empirically plausible, our model predicts that interest groups typically lobby politicians for delegation of policy authority.

In Section 6.3 it is shown that different assumptions concerning the distribution of the uncertainty parameter or alternative functional forms for the preferences of the interest group may destroy the result that the informational effect is always harmful to the group. This observation indicates that the basic model of Section 6.2 is somewhat biased against delegation. Therefore, the model tentatively suggests that interest groups typically will lobby the legislature for delegation of policy authority. However, the analysis of Section 6.3 also indicates that when the scope for information transfer between the bureaucrat and the interest group increases — either because lobbying costs are endogenous, or lying is prohibited — lobbying for delegation becomes less likely. Also the result that the politician may prefer a biased bureaucrat has to be qualified in this case.

A strong feature of the model presented in this chapter is that it incorporates two levels of information asymmetry, both having some empirical appeal. Interest groups are better informed than bureaucrats about the exact need for certain policies, and bureaucrats are better informed than politicians about the general characteristics of the interest group. Such a feature is, for instance, lacking in the model of Austen-Smith (1993; see Section 2.2 for an informal description of this model). In the context of this chapter his model can be interpreted in the following way. A bureaucrat (the committee, i.e. a specific group of politicians, in his model) has the sole right to propose alternative policies, but these policies have to be approved by the politician (the House). In case the politician does not approve, the status quo policy is maintained. There is uncertainty on the side of both the bureaucrat and the politician about how policies map into consequences. A priori the bureaucrat and the politician possess the same information. An interest group (lobbyist), on the other hand, has the opportunity to acquire this information, and to use it to lobby the bureaucrat (agenda setting stage) and/or the politician (vote stage). The equilibrium results of Austen-Smith (1993) then indicate that, ceteris paribus, lobbying the bureaucrat is more effective in promoting the interests of the group than lobbying the politician. Of course, these results are dictated by the assumption that the politician can only choose between the proposal of the bureaucrat and the status quo. Under an open rule, whereby the politician can freely amend the proposal of the bureaucrat, "the distinction between the agenda-setting stage and the vote stage is blurred." (Austen-Smith, 1993, p. 813). Although strictly speaking incorrect from a formal juridical viewpoint, the closed rule setup as a description of the relationship between politicians and bureaucrats seems to correspond with existing practice rather well.[35]

Now a relevant question regarding the Austen-Smith (1993) model is: Why should the politician (the House) delegate proposal-making power to a bureaucrat (the committee) when the latter has no a priori informational advantage over the former? This question is of special interest, because the model of Austen-Smith is inspired by Gilligan and Krehbiel (1987, 1989a, 1989b, 1990), whose most pertinent claim is that proposal-making power is delegated by House members to the committee in order to reap the benefits from the committee's informational lead. Incidentally, the analysis presented in this chapter provides an answer. In view of the potential for interest group influence, there may be reasons to delegate (proposal-making) authority to bureaucrats even in the absence of an a priori informational advantage. Our claim that the politician may prefer a biased bureaucrat over an unbiased one is namely based on a numerical example in which there is no uncertainty about the general characteristics of the interest group (i.e. $g_1=g_2=g$; cf. Proposition 6.4). Hence, ex ante there is no information asymmetry between the bureaucrat and the politician. Still, in the specific parameterization of the example, the politician prefers a somewhat biased bureaucrat over an unbiased one. This occurs because in the first case information transfer between the interest group and the bureaucrat about the actual state is possible, whereas in the second case this appears not to be the case. So, even without an a priori informational asymmetry the politician may want to delegate policy authority. This is not immediately clear from the original Austen-Smith model, because there the delegation decision is not explicitly modeled; the relationship between the politician (the House) and the bureaucrat (the committee) is taken as given (viz. closed rule).[36] The explicit incorporation of the delegation decision in our model is considered to be another asset of the model.[37]

The model presented in this chapter can be extended in various ways. A first, obviously very interesting, extension of the present model would be to introduce a competing interest group. It seems reasonable to expect that the presence of an opposing interest group makes delegation more attractive to the politician. Firstly, as already indicated in Section 3.3 the literature on signaling games suggests that the scope for information transmission to the bureaucrat increases. Secondly, the presence of a competing group may affect the induced preferences of the bureaucrat, bringing them more in line with those of the politician.[38] These two effects most likely harm the (initially single existing) interest group, so it seems that delegation is preferred less when there are competing interests. This suggests that under severe interest group competition, interest groups lobby less for delegation.

A second interesting extension of the model would be to take account of the fact that in reality delegation is not an all-or-nothing decision (cf. Bawn, 1995, Aghion and Tirole, 1997). The present setup could be altered such as to allow the politician to restrict the level of discretion the bureaucrat has (cf. Epstein and O'Halloran, 1994). For instance, the politician could put an upperbound on the policy the bureaucrat can implement when authority is delegated. An interesting question then becomes whether in equilibrium the politician will set this upperbound always (weakly) below her most extreme potentially optimal policy (t_m+z in the model), or whether she will allow the bureaucrat to implement policies that are even more extreme in order to benefit from information transmission between the bureaucrat and the interest group. Again, it stands to reason to expect that when the politician decides on an upperbound on policy, delegation becomes less attractive to the interest group. In that case the group will lobby less for delegation.

The model presented here focuses on only one aspect of the relationship between (the influence of) interest groups and the internal organization of government. Other aspects of this relationship are ignored, and provide interesting avenues for future research. For instance, one could try to model the idea that the composition of government may provide an interest group with a coalitional force within government. In such a model interest groups either try to infiltrate government to alter its composition (cf. Vogel, 1978), or try to attract former bureaucrats with inside information and a "network" of informal contacts to obtain "insider access". It would be interesting to verify whether such a model could replicate the "stylized fact" derived from the empirical literature that coalitional forces within government are helpful in promoting a group's interests (cf. Subsection 2.1.3). The model could also provide a micro-foundation for the use of reduced form preferences for bureaucrats, as is done here.

APPENDIX 6.A PROOFS OF THE PROPOSITIONS

In this appendix the proofs of the propositions presented in the main text are given. All propositions relate to (perfect Bayesian) equilibrium results. Therefore, we first give a definition of a perfect Bayesian equilibrium for the game considered in Section 6.2. The definition of an equilibrium for the game with endogenous lobbying costs (Subsection 6.3.2), as well as for the game where the continuation game between B and G is envisaged as a persuasion game (Subsection 6.3.3), is a straightforward adaption of the definition given below, and is therefore omitted. With $\lambda_P(g)$ we denote the probability that an interest group with general stake g sends a costly lobbying message $s_P=c_P$ to P. Similarly, $\lambda_B(t;g)$ denotes the probability that the interest group with general stake g sends $s_B=c_B$ after observing that the actual state of the world equals t. The probability that the politician delegates policy authority after receiving signal s_P is denoted with $v(s_P)$. With x_P we denote the policy chosen by P when she decided not to delegate policy authority, whereas $x_B(s_B;g)$ reflects the policy chosen by the bureaucrat after receiving s_B. Lastly, $q_P(s_P)$ is used to reflect the posterior belief of P that $g=g_2$ after having received lobbying signal s_P, $q_B(t|s_B)$ gives B's posterior belief that the actual state of the world equals the value t after he has received signal s_B.

Now a perfect Bayesian equilibrium is a set of strategies $((\lambda_P(\cdot),\lambda_B(\cdot)),v(\cdot),x_P, x_B(\cdot))$ and a set of beliefs $(q_P(\cdot),q_B(\cdot))$ that satisfy the following conditions:

(e1)   for $g\in\{g_1,g_2\}$, if $\lambda_P(g)>0$ (<1), then $[E_D(U_G|\lambda_B(t;g),x_B(s_B;g))-E_N(U_G;g)][v(c_P)-v(0)] \geq (\leq) c_P$.

(e2)   for $s_P\in\{0,c_P\}$, if $v(s_P)>0$ (<1), then $(1-q_P(s_P))\cdot \int_T [-(x_B(c_B;g_1)-t)^2 \lambda_B(t;g_1)-(x_B(0;g_1)-t)^2 (1-\lambda_B(t;g_1))]p(t)dt + q_P(s_P)\cdot \int_T[-(x_B(c_B;g_2)-t)^2 \lambda_B(t;g_2)-(x_B(0;g_2)-t)^2 (1-\lambda_B(t;g_2))]p(t)dt \geq (\leq) \int_T -(x_P-t)^2 p(t)dt.$

(e3) $x_P=\text{argmax}_x \int_T -(x-t)^2 p(t)dt.$

(e4)   for $g\in\{g_1,g_2\}$, if $\lambda_B(t;g)>0$ (<1), then $gt(x_B(c_B;g)-x_B(0;g)) \geq (\leq) c_B$.

(e5)   for $s_B\in\{0,c_B\}$, $x_B(s_B;g)=\text{argmax}_x \int_T [-(x-t)^2+bgtx]q_B(t|s_B)dt.$

(e6) $q_P(c_P)=p\lambda_P(g_2)/[p\lambda_P(g_2)+(1-p)\lambda_P(g_1)]$, $q_P(0)=p(1-\lambda_P(g_2))/[p(1-\lambda_P(g_2))+(1-p)(1-\lambda_P(g_1))]$, $q_B(t|c_B)=p(t)\lambda_B(t;g)/[\int_T p(t)\lambda_B(t;g)dt]$, and $q_B(t|0)=p(t)(1-\lambda_B(t;g))/[\int_T p(t)(1-\lambda_B(t;g))dt]$ whenever the denominators are positive.

Here $E_D(U_G|\lambda_B(t;g),x_B(s_B;g))$ denotes the expected payoff of the interest group with general characteristics g after delegation, given that the group uses lobbying strategy $\lambda_B(t;g)$ and the bureaucrat uses reaction strategy $x_B(s_B;g)$. $E_N(U_G;g)$ denotes the interest group's expected payoff when the politician chooses not to delegate, and subsequently chooses a policy that maximizes her own expected payoff (i.e., $x_P=t_m$). Lastly, $p=P(g=g_2)$ and p(t), with $p(t)=1/2z$ when $t\in[t_m-z,t_m+z]$ and p(t)=0 otherwise, denote the prior beliefs on g and t, respectively.

The equilibrium refinement used in the analysis of the continuation game between B and G is the universal divinity refinement already extensively discussed in previous chapters. Explicit definitions of this concept are therefore omitted. For the continuation game between B and G the refinement is described together with its use in the proof of Proposition 6.1. Similar remarks apply to the application of universal divinity in the context of the reduced game of Subsection 6.2.5.

Proof of Proposition 6.1.
Given s_B, we can write the expected payoff of B in the continuation game after choosing policy x as $\int_T [-(x-t)^2+bgtx]q_B(t|s_B)dt=-(x-E(t|s_B))^2-var(t|s_B)+bgxE(t|s_B)$. Hence the equilibrium policy of B equals $x_B(s_B;g)=\frac{1}{2}(2+bg)E(t|s_B)$ (cf. (e5) above). Pooling on no-lobbying ($s_B=0$) always constitutes a PBE; $t'=t_m+z$ and $x_B(0;g)=\frac{1}{2}(2+bg)t_m$. Out-of-equilibrium beliefs and strategies that support this equilibrium path are not unique, but a belief that $t=t_m-z$ after receiving $s_B=c_B$ and taking an action $x_B(c_B;g)=\frac{1}{2}(2+bg)(t_m-z)$ accordingly always supports the path.

From $x_B(s_B;g)=\frac{1}{2}(2+bg)E(t|s_B)$ follows that the gain for type t from pooling with all higher types $(t,t_m+z]$ rather than with all lower types $[t_m-z,t)$ is given by $H(t)=\frac{1}{2}gt(2+bg)z$. Note that $H(t)$ is monotonically increasing in t (when $z>0$). Now, if $H(t_m+z)<c_B$ neither type wants to lobby B, and only pooling on no-lobbying can occur (U1). Because there exist no (best) response of B to $s_B=c_B$ for which any type wants to deviate from $s_B=0$, universal divinity does not restrict out-of-equilibrium beliefs after $s_B=c_B$. Next, consider the case $c_B<H(t_m+z)$. Pooling on no-lobbying yields the t-type a utility of $gt\cdot\frac{1}{2}(2+bg)t_m$. Let $R_t(c_B)$ be the set of best responses of B to signal $s_B=c_B$ for which type t has an incentive to deviate. We get $R_t(c_B)=\{x\in[\frac{1}{2}(2+bg)(t_m-z),\frac{1}{2}(2+bg)(t_m+z)]| \quad x>(c_B/gt)+\frac{1}{2}(2+bg)t_m\}$. This yields $R_{t_m+z}(c_B)\neq\varnothing$, and $R_t(c_B)\subset R_{t_m+z}(c_B)$ for $t<t_m+z$. By universal divinity then, the bureaucrat must belief after $s_B=c_B$ that $t=t_m+z$ for sure, and all types t for which $c_B<H(t)$ wants to deviate. In short, when $c_B<H(t_m+z)$ lobbying message $s_B=c_B$ necessarily occurs with positive probability in a universally divine equilibrium. Now when $H(t_m-z)<c_B$ it follows that not all types are prepared to lobby, and pooling on lobbying cannot occur. By $H(t_m-z)<c_B<H(t_m+z)$ and $H(t)$ continuous and monotonically increasing there is a unique $t'\in T$ such that $H(t')=c_B$. Specifically, $t'=2c_B/[g(2+bg)z]$. This yields the two-partition equilibrium specified for regime I. Lastly, when $H(t_m-z)>c_B$ only two PBE exist, viz. pooling on $s_B=0$ and pooling on $s_B=c_B$. From the above we already know that pooling on $s_B=0$ is not universally divine. This yields regime U2. It is easily verified that in the pooling on $s_B=c_B$ equilibrium universal divinity requires B to belief that $t=t_m-z$ after receiving out-of-equilibrium message $s_B=0$.[39]

From the above directly follows that the equilibria on the edges $H(t_m+z)=c_B$ and $H(t_m-z)=c_B$ are all characterized by $t'=t_m+z$ and $t'=t_m-z$, respectively.   *QED.*

Proof of Proposition 6.2.
$E_D(U_P|g)=[-(x_B(0;g)-E(t|0))^2-var(t|0)]P(t<t')+[-(x_B(c_B;g)-E(t|c_B))^2-var(t|c_B)]P(t>t')$. Now from Proposition 6.1 $x_B(s_B;g)=\frac{1}{2}(2+bg)E(t|s_B)$, $E(t|0)=(t_m-z+t')/2$, $var(t|0)=(t'-(t_m-z))^2/12$, $P(t<t')=(t'-(t_m-z))/2z$, $E(t|c_B)=(t_m+z+t')/2$, $var(t|c_B)=((t_m+z)-t')^2/12$, and $P(t>t')=(t_m+z-t')/2z$. This yields $E_D(U_P|g)=[-b^2g^2(t_m-z+t')^2/16-(t'-$

$t_m+z)^2/12]P(t<t')+[-b^2g^2(t'+t_m+z)^2/16-(t_m+z-t')^2/12]P(t>t')$. Rewriting this and collecting terms the expression for $E_D(U_P|g)$ is obtained. Similarly, $E_D(U_G;g)=[gE(t|0)x_B(0;g)]P(t<t')+[gE(t|c_B)x_B(c_B;g)-c_B]P(t>t')$. With the equilibrium strategy $x_B(s_B;g)$ we get $E_D(U_G;g)=[\frac{1}{2}g(2+bg)(E(t|0))^2]P(t<t')+[\frac{1}{2}g(2+bg)(E(t|c_B))^2-c_B]P(t>t')$. With this expression the result follows. $\qquad$ QED.

Proof of Proposition 6.3.
The first part trivially follows from equation (6.6).
(a) When $b>0$ it holds that $J>0$. Now, when $c_B\geq\frac{1}{2}g_2(2+bg_2)(t_m+z)z$ regime U1 applies for $g\in\{g_1,g_2\}$, $t'=t_m+z$ and thus $K=Q=0$. Similarly, when $c_B\leq\frac{1}{2}g_1(2+bg_1)(t_m-z)z$ regime U2 applies for $g\in\{g_1,g_2\}$, $t'=t_m-z$ and $K=Q=0$. In both cases $K-Q=0<J$, and the result follows from equation (6.6).
(b) Rewrite (6.6) to obtain $E_D(U_P|s_P)=E_N(U_P)+\frac{1}{4}E(\{(1-\frac{1}{4}b^2g^2)[z^2-(t_m-t')^2]-b^2t_m^2g^2\}|s_P)$. When $bg_1\geq2$ the term within brackets $\{\cdot\}$ becomes negative, so consider the case $bg_1<2$. Subterm $[z^2-(t_m-t')^2]$ is maximized for $t'=t_m$. Hence for the term within brackets $\{\cdot\}$ holds that $\{\cdot\}\leq(1-\frac{1}{4}b^2g^2)z^2-b^2t_m^2g^2$. Now $(1-\frac{1}{4}b^2g^2)z^2<b^2t_m^2g^2$ when $(bg)^2>4z^2/[4t_m^2+z^2]$. So, when $(bg)^2>4z^2/[4t_m^2+z^2]$ for $g\in\{g_1,g_2\}$ it holds that $E(\{\cdot\}|s_P)<0$ for $s_P\in\{0,c_P\}$, and P always prefers not to delegate. $\qquad$ QED.

Proof of Proposition 6.4 by example.
Let $t_m=z=\frac{1}{2}$ such that t is uniformly distributed on [0,1]. Moreover, let $c_B=\frac{1}{2}$ and $g_1=g_2=1$; there is no uncertainty about the general stake the interest group has. Hence we can write $E_D(U_P)$ instead of $E_D(U_P|s_P)$. In the continuation game between B and G we get $t'=2/(2+b)$. The net gain of delegation then equals $E_D(U_P)-E_N(U_P)=\frac{1}{4}b[(1-b-\frac{1}{4}b^2)/(2+b)]$. From $\partial(E_D(U_P)-E_N(U_P))/\partial b=\frac{1}{4}\{[2-2b-\frac{1}{2}b^2]/(2+b)^2-\frac{1}{2}b\}$ follows that the net gain is increasing in b for b close to 0.
To support the second claim of the proposition, when $b=0$ we get $t'=t_m+z=1$. So, no information is revealed and $E_D(U_P)=E_N(U_P)=-1/12$. In case $b=\frac{1}{2}$ we have $t'=4/5$ and $E_D(U_P)=E_N(U_P)+7/320$ ($J=1/64$, $K=1/25$ and $Q=1/400$). Consequently, the politician prefers a biased bureaucrat ($b=\frac{1}{2}$) over an unbiased one ($b=0$). Besides, when $b=\frac{1}{2}$ the politician strictly prefers to delegate, whereas when $b=0$ she is indifferent between delegation and no-delegation. $\qquad$ QED.

Proof of Proposition 6.5 by example.
Let $t_m=z=\frac{1}{2}$, $c_B=7/20$, and $b=4/5$. Then from Proposition 6.1 follows that $t'=\min\{1,7/[g(10+4g)]\}$. When $t'=1$ we have $K=Q=0$. From J strictly increasing in g follows that when $g\leq\frac{1}{4}[(\sqrt{53})-5]$ (i.e. when $t'=1$) the net gain of delegation is negative and decreasing in g. When $g=1$ informative regime I applies in the continuation game between B and G, with $t'=\frac{1}{2}$. In that case the pure distributional loss of delegation J equals 1/25, the informational effect K-Q equals 21/400. Hence $E_D(U_P)-E_N(U_P)=1/80$, the net gain from delegation is positive, and the politician prefers delegation. From $E_D(U_P)-E_N(U_P)$ continuous in g then follows that the net gain of delegation must be increasing in g on some interval within $[\frac{1}{4}\{(\sqrt{53})-5\},1]\approx[0,57,1]$. P prefers $g=1$ over all values $g<0,57$. $\qquad$ QED.

Proof of Proposition 6.6.
Using expression (6.7) and Proposition 6.1 the total impure informational effect
$Y+Z$ equals $-c_B$ when $t'=t_m-z$, 0 when $t'=t_m+z$, and $Y+Z=-\frac{1}{8}g(2+bg)[t_m^2-(t'-z)^2]$
when $t' \in (t_m-z,t_m+z)$. (Substituting $t'=t_m-z$ ($t'=t_m+z$) in the latter expression for $Y+Z$
yields $Y+Z=-c_B$ ($Y+Z=0$), indicating that there are no jumps in the interest group's
expected utility at the boundaries of the regimes U1, I and U2.) Hence $Y+Z\leq0$,
and the interest group always suffers from an (impure) informational loss. The
second part then trivially follows from equation (6.7). With the above expression
for $Y+Z$ at hand, it follows that $-\frac{1}{8}g(2-3bg)t_m^2\leq X+Y+Z$. The last part of the
proposition follows immediately.   *QED.*

Proof of Proposition 6.7.
From equation (6.7) and the expression for $Y+Z$ obtained in the proof of
Proposition 6.6 it follows that in case regime U1 or U2 applies
$\partial(X+Y+Z)/\partial b=\partial X/\partial b=\frac{1}{2}g^2t_m^2>0$. In case regime I applies $\partial(X+Y+Z)/\partial b=\frac{1}{8}g^2[3t_m^2-
(t'-z)(t'+z)]$. By $t_m-z\leq t'\leq t_m+z$ $\partial(X+Y+Z)/\partial b$ is at a minimum for $t'=t_m+z$, with
$\partial(X+Y+Z)/\partial b=\frac{1}{8}[2t_m(t_m-z)]\geq0$ (due to $t_m\geq z$). For $t'<t_m+z$ the derivative is strictly
positive. Only on the boundaries of the regimes — that is, when the knife-edge
case $b=(2c_B/g^2z(t_m-z))-2/g$ or $b=(2c_B/g^2z(t_m+z))-2/g$ applies — the expression for
$E_D(U_G;g)-E_N(U_G;g)$ is not differentiable in b. Right derivatives and the left
derivative at $b=(2c_B/g^2z(t_m-z))-2/g$ are always positive, though, the left derivative
at $b=(2c_B/g^2z(t_m+z))-2/g$ is non-negative, and equals 0 only when $t_m=z$.   *QED.*

Proof of Proposition 6.8.
(a) When $E_D(U_P|g)>E_N(U_P)$ for all g, $v(0)=v(c_P)=1$ by equilibrium condition (e2).
In case $E_D(U_P|g)<E_N(U_P)$ for all g we have $v(0)=v(c_P)=0$ by (e2). From equilibrium
condition (e1) and $c_P>0$ then follows that in both cases $\lambda_P(g)=0$ for all $g\in\{g_1,g_2\}$.

(b) From (e1), when $\lambda_P(g)>0$ for some g in equilibrium we necessarily have
$v(0)\neq v(c_P)$. First, suppose $v(0)>v(c_P)$. Only types g for which $E_D(U_G;g)\leq E_N(U_G;g)-
c_P/(v(0)-v(c_P))$ will choose $s_P=c_P$ in equilibrium. For all these types
$E_D(U_P|g)>E_N(U_P)$ by the assumption in the proposition. By (e2) then $v(c_P)=1$,
contradicting $v(0)>v(c_P)$. Next, suppose $v(0)<v(c_P)$. Then only types for which
$E_D(U_G;g)\geq E_N(U_G;g)+c_P/(v(c_P)-v(0))$ choose $s_P=c_P$. By the assumption of the
proposition and equilibrium condition (e2) $v(c_P)=0$, contradicting $v(0)<v(c_P)$.

(c) The statement follows from (e1) and the fact that $[v(c_P)-v(0)]\in[-1,1]$ by
definition.   *QED.*

Proof of Proposition 6.9.
When $b=0$ we have $E_D(U_P|g)\geq E_N(U_P)$ and $E_D(U_G;g)\leq E_N(U_G;g)$ for all $g\in\{g_1,g_2\}$.
Under regime U1 $E_D(U_G;g)=E_N(U_G;g)$, hence for type g to lobby P regime I or U2
has to apply. Only when $E_D(U_P|g)=E_N(U_P)$ for all types g that lobby P with
positive probability, $v(c_P)<1$ is possible (cf. equilibrium condition (e2)). Hence, for
types g lobbying P regime U2 has to apply. By $g_1\leq g_2$ and Proposition 6.1 it

follows that regime U2 necessarily has to apply for g_2. Hence $c_B \leq g_2(t_m-z)z$ is necessarily required (cf. Proposition 6.1). We get $|E_D(U_G;g_1)-E_N(U_G;g_1)| \leq c_B$, and $|E_D(U_G;g_2)-E_N(U_G;g_2)|=c_B$ from expression (6.7). By Proposition 6.8(c) it follows that $c_P \leq c_B$ is necessarily required as well. To establish sufficiency, consider the following separating equilibrium of the reduced game (i.e. taking the equilibrium path after delegation as in Proposition 6.1, and $x_P=t_m$): $\lambda_P(g_1)=0$, $\lambda_P(g_2)=1$, $v(0)=1$, and $v(c_P)=1-c_P/c_B$; $q_P(0)=0$, $q_P(c_P)=1$.

Lastly, $v(c_P)<v(0)$ when $s_P=c_P$ occurs with positive probability follows from equilibrium condition (e1) and $E_D(U_G;g) \leq E_N(U_G;g)$ for all g when $b=0$. QED.

Proof of Proposition 6.10.
(a) and (b) From Proposition 6.3 it directly follows that $E_D(U_P|g)<E_N(U_P)$ for $g \in \{g_1,g_2\}$. Using Proposition 6.8(a) the result follows.

(c) From expression (6.7) and $Y+Z \leq 0$ (cf. Proposition 6.3) follows $E_D(U_G;g)-E_N(U_G;g) \leq \frac{1}{2}bg^2t_m^2$. Likewise, $E_D(U_G;g)-E_N(U_G;g) \geq -c_B$ from (6.7), and $X,Y1,Z \geq 0$ and $Y2 \leq c_B$. With Proposition 6.8(c) the result follows. QED.

Proof of Proposition 6.11 by example.
We first provide an example of lobbying the politician for delegation. Let $t_m=z=\frac{1}{2}$, $b=4/5$, $c_B=7/20$, $g_1=\frac{1}{4}$, $g_2=1$, and $c_P=1/100$. Then from Proposition 6.1 regime I applies for type g_2 of G, with $t'=\frac{1}{2}$. With respect to type g_1 regime U1 applies ($t'=1$). It is easily calculated that the normalized preference parameters defined in expression (6.8) take the following values in this case: $d_1=1/400$, $d_2=1/80$, $e_1=1/160$, $e_2=1/80$. Given $c_P=1/100$ and Proposition 3.2(ii), a separating equilibrium exists in the reduced game: $\lambda_P(g_1)=0$, $\lambda_P(g_2)=1$, $v(0)=0<1=v(c_P)$; $q_P(0)=0$, $q_P(c_P)=1$.

Next, consider the following example of lobbying the politician for no-delegation. Let $t_m=1$, $z=3/5$, $b=\frac{1}{2}$, $c_B=9/20$, $g_1=\frac{1}{2}$, $g_2=1$, and $c_P=1/20$. From Proposition 6.1 regime I applies for both type g_1 ($t'=4/3$) and g_2 ($t'=3/5$). With (6.8) we get: $d_1=-73/1600$, $d_2=-1/64$, $e_1=-1/400$, $e_2=-1/16$. With $c_P=1/20$ a separating equilibrium exists in the reduced game: $\lambda_P(g_1)=0$, $\lambda_P(g_2)=1$, $v(0)=1$ and $v(c_P)=0$; $q_P(0)=0$, $q_P(c_P)=1$. QED.

Proof of Proposition 6.12.
(a) When $b>0$ we get $X+Y+Z>-c_B$ from expression (6.7). Proposition 6.8(c) now yields the result.

(b) We first show that for all $g \in \{g_1,g_2\}$ the weak inequalities $E_D(U_P|g) \leq E_N(U_P)$ and $E_D(U_G;g) \leq E_N(U_G;g)$ cannot hold simultaneously. When $t_m=z$ regime U2 is not possible for any value of g (cf. Proposition 6.1). When regime U1 applies, $E_D(U_P|g)<E_N(U_P)$ and $E_D(U_G;g)>E_N(U_G;g)$. So, only regime I remains to be considered. When $bg \geq 2$ $E_D(U_G;g)>E_N(U_G;g)$ by the proof of part (c) below. So, let $bg<2$. From Proposition 6.2 it follows that $E_D(U_P|g) \leq E_N(U_P) \Leftrightarrow t'(2t_m-t') \leq 4b^2g^2t_m^2/(4-b^2g^2)$. Similarly, from (6.7) and the expression for $Y+Z$ given in the proof of Proposition 6.6 it follows that $E_D(U_G;g) \leq E_N(U_G;g) \Leftrightarrow t'(2t_m-$

t')$\geq$4bgt$_m^2$/(2+bg). For both P and G to (weakly) prefer no-delegation (conditional on the value of g) it thus certainly must hold that 4bgt$_m^2$/(2+bg)$\leq$4b^{2}g^{2}t$_m^2$/(4-b^{2}g^2). Rewriting this it follows that bg(1+bg)$\geq$2 is required. Using b>0 and g>0 this implies that bg$\geq$1 must hold. Now when bg$\geq$1 we have that 4bgt$_m^2$/(2+bg)$\geq$4t$_m^2$/3. By t'(2t$_m$-t')$\leq$t$_m^2$ it then cannot be the case that t'(2t$_m$-t')$\geq$4bgt$_m^2$/(2+bg). In short, the situation that both P and G (weakly) prefer no-delegation cannot occur.

Next it is shown that when s$_P$=c$_P$ is an equilibrium signal, necessarily v(c$_P$)>v(0). Suppose to the contrary that v(c$_P$)$\leq$v(0). When v(c$_P$)=v(0) it follows from equilibrium condition (e1) that λ_P(g)=0 for all g, contradicting the presumption that s$_P$=c$_P$ is an equilibrium signal. In case v(c$_P$)<v(0) all types g$_i$ for which e$_i$>0 (i=1,2) will use strategy λ_P(g$_i$)=0. Hence only (a subset of) those types of G who prefer no-delegation will choose s$_P$=c$_P$. Using the result obtained in the first part of the proof it follows that for those values of g the politician prefers delegation, implying v(c$_P$)=1, a contradiction.

(c) From Proposition 6.6 it follows that when bg$_i$>⅔ the interest group always prefers delegation (e$_i$>0 for i=1,2, with e$_i$ defined in (6.8)). Hence, v(c$_P$)>v(0) necessarily when s$_P$=c$_P$ is sent by the interest group (cf. equilibrium condition (e1)).   *QED.*

Proof of Proposition 6.13 by example.
Let b=0, t$_m$=z=½, and g$_1$=g$_2$=1. The distribution of t is now given by F(t;r)=t^r, with r>0.[40] For r=1 F(t;r) reduces to the uniform distribution used in the analysis of Section 6.2. The median of F(t;r) equals (½)$^{1/r}$ and E(t)=r/(r+1) under F(t;r). We first show that mean<median when r>1, and that mean>median when r<1. Let h(r)=ln(r)-ln(r+1)+(1/r)ln(2), such that sign(mean-median)=sign h(r). From h(1)=0 and (∂h(r)/∂r)<0 when r<k$\equiv$ln(2)/(1-ln(2)) directly follows that mean>median when r<1, and mean<median when 1<r<k. From h(r) monotonically increasing when r>k and lim$_{r\to\infty}$h(r)=0 it is obtained that mean<median also when r>k. In short, when r>1 the distribution of t is skewed to the left, in case r<1 the distribution of t is skewed to the right.

Cut-off type t' is now defined by t'[E(t|t>t')-E(t|t<t')]=c$_B$. (In general, this equation may have several solutions and hence uniqueness of the two-partition equilibrium is not guaranteed (cf. Potters, 1992).) By b=0 no distributional effects effect occur. G thus obtains an informational gain whenever E$_D$(U$_G$;1)>E$_N$(U$_G$;1). Using x$_P$=E(t), x$_B$(0;1)=E(t|t<t') and x$_B$(c$_B$;1)=E(t|t>t') this inequality reduces to: (E(t|t<t'))^{2}P(t<t')+[(E(t|t>t'))^2-c$_B$]P(t>t') > [E(t)]^2. When F(t;r)=t^r we have E(t)=r/(r+1), E(t|t<t')=t'E(t), and E(t|t>t')=[(1-(t')$^{r+1}$)/(1-(t')^r)]E(t). Using the defining equation of t' and the equation E(t)=E(t|t<t')P(t<t')+E(t|t>t')P(t>t'), the inequality can be rewritten as [r/(r+1)](t')$^{r-1}$(1-t')>(1-(t')) for t' $\in$(0,1). From 0$\leq$t'$\leq$1 this condition certainly cannot hold when r>1. On the other hand, for any value of r<1 there exists values of t', and thus values of c$_B$, such that the condition holds (this observation follows from lim$_{r\downarrow0}$ (t')$^{r-1}$=∞ and lim$_{r\downarrow0}$ (1-(t'))=1). Hence when r$\geq$1 the interest group necessarily suffers from an informational loss, when r<1 not necessarily.   *QED.*

Proof of Proposition 6.14 by example.
Let the interest group's utility over policy be given by $U_G(x;t,g)=gtx^r$, with $r>0$. In addition, let $b=0$, $t_m=z=\frac{1}{2}$, and $g_1=g_2=1$. When $r>1$ ($r<1$) $U_G(x;t,g)$ is convex (concave) in x. Cut-off type t' is now given by the solution of $t'(t'+1)^r-(t')^{r+1}=c_B 2^r$. So, when $r\geq 1$ and a two-partition equilibrium exists, it is unique. With respect to the expected payoffs of the interest group it holds that $E_N(U_G;g)=(\frac{1}{2})^{r+1}$ and $E_D(U_G;g)=[E(t|t<t')]^{r+1}\cdot P(t<t')+[(E(t|t>t'))^{r+1}-c_B]P(t>t')=(\frac{1}{2})^r[\frac{1}{2}(t')^{r+2}+(1-t')(\frac{1}{2}(1-t')(1+t')^r+(t')^{r+1})]$. For $r=2$ the latter expression reduces to $E_D(U_G;g)=E_N(U_G;g)-\frac{1}{4}(t')^2(1-t')$. Thus, when $r=2$, there is still always an informational loss for G. However, for $r=3$ it holds that $E_D(U_G;g)=E_N(U_G;g)+(1/16)t'(1-t')[1-t'-3(t')^2]$. In that case an informational gain is obtained whenever $t'<(\sqrt{13}-1)/6\approx 0,43$, and c_B can be chosen as to obtain such a value. *QED.*

Proof of Proposition 6.15.
The proof follows along exactly the same lines as for the continuous-type Spence job market signaling model, see Banks (1991, pp. 18-20). *QED.*

Proof of Proposition 6.16.
Using expression (6.9) we get $E_D(U_P|g)-E_N(U_P)=\frac{1}{3}z^2-(1/12)b^2g^2(3t_m^2+z^2)$. So, $E_D(U_P|g)-E_N(U_P)>0 \Leftrightarrow g<2z/b\sqrt{(3t_m^2+z^2)}\equiv P^*$. Similarly, we get from expression (6.10) that $E_D(U_G;g)-E_N(U_G;g)>0 \Leftrightarrow g>2z(t_m-\frac{2}{3}z)/b[t_m^2-z(t_m-\frac{2}{3}z)]\equiv G^*$. When $g>P^*$ ($g<P^*$), the politician prefers not to delegate (prefers to delegate). In case $g>G^*$ ($g<G^*$) the interest group prefers P to delegate (not to delegate). It can easily be derived that $P^*=G^*=1/b$ when $t_m=z$, and that $P^*<G^*$ in case $t_m>z$. So, P's conditional preferences can only coincide with those of G on a preference for no-delegation.

 We do not consider non-generic knife-edge cases, so assume $g_i\neq P^*$ and $g_i\neq G^*$ for $i=1,2$. Using Proposition 6.8(a) and (b) (the proposition still applies because the reasoning when c_P can be chosen by the interest group is exactly the same as when c_P is fixed), it follows that for costly lobbying of P to occur necessarily $g_1<P^*<g_2<G^*$ must hold. This immediately yields conditions (i) and (ii). Now suppose (i) and (ii) are satisfied, but that $e_1<e_2<0$ (with e_i as defined in (6.8)). Suppose type g_1 mixes between $s_P=c^l$ and $s_P=c^h$, with $c^l<c^h$. Then necessarily $v(c^l)=v(c^h)+(c^l-c^h)/e_1$, and $v(c^l)>v(c^h)$. But then g_2 prefers sending $s_P=c^l$ over $s_P=c^h$, and will not send $s_P=c^h$. Hence $s_P=c^h$ is a clear signal of being of type g_1, and $v(c^h)=1$ by $g_1<P^*$. This contradicts $v(c^l)>v(c^h)$. Therefore, g_1 plays a pure strategy in equilibrium. A similar reasoning yields that type g_2 plays a pure strategy as well. Now suppose g_1 and g_2 pool on $s_P=c$. Then necessarily $v(c)=0$ (the knife-edge case $p=d\equiv d_1/(d_1+d_2)$ is excluded, with the d_i's defined in expression (6.8)). Type g_i wants to deviate to $s_P=0$ when $v(0)<(-c)/e_i$. Hence type g_2 has a stronger incentive to deviate to $s_P=0$, by universal divinity $v(0)=0$, and both types want to deviate. Pooling on lobbying is not universally divine. Lastly, it is easily seen that separation, type g_1 chooses $s_P=c_1$ and type g_2 chooses $s_P=c_2$ with $c_1\neq c_2$, cannot occur in equilibrium.

In order to show that the conditions are sufficient assume these conditions to hold and consider the following universally divine equilibrium of the reduced game: $c_P(g_1)=0$, $c_P(g_2)=-e_1$, $v(-e_1)=0$, $v(c)=1$ for $c \neq -e_1$; $q_P(-e_1)=1$ and $q_P(c)=0$ for any $c \neq -e_1$. Here $c_P(g)$ denotes the amount spent by type g of G on lobbying P, and $q_P(c)$ denotes the posterior belief of P that G is of type g_2 after observing that the group spent c on lobbying her.   *QED.*

Proof of Proposition 6.17.
Let $s_B(t) \subseteq T$ denote the signal type t sends to B in equilibrium. Because lying is excluded in a persuasion game, necessarily $t \in s_B(t)$. The equilibrium choice of B after receiving signal s_B equals $x_B(s_B;g)=\frac{1}{2}(2+bg)E(t|s_B)$ (cf. Proposition 6.1). Now suppose there is a type t_1 that sends signal $s_B(t_1)$ for which $t_1 > E(t|s_B(t_1))$. Given $x_B(s_B;g)$, this type could gain by sending $s_B=\{t_1\}$ instead because $E(t|\{t_1\})=t_1$, thus it wants to deviate. Hence for all $t_1 \in T$ necessarily $t_1 \leq E(t|s_B(t_1))$. Suppose $t_1 < E(t|s_B(t_1))$ for some t_1, then there must be a $t_2 > t_1$ sending $s_B(t_1)$ as well such that $t_2 > E(t|s_B(t_2))$, a contradiction. Consequently, $t_1 = E(t|s_B(t_1))$ for all $t_1 \in T$, and thus only completely separating equilibria exist.   *QED.*

Proof of Proposition 6.18.
(a) Information about g is only relevant to P when condition (ii) in Proposition 6.16 holds; when $g=g_1$ P prefers delegation, when $g=g_2$ P prefers no delegation ($d_1<0$ and $d_2<0$ in the reduced game of Subsection 6.2.5). From X+Y1+Z positive and increasing in g it follows that the interest group always prefers delegation, and that type g_2 always gains more than type g_1 ($0<e_1<e_2$). Consequently, the sorting condition ($e_1>e_2$) is not satisfied and information transfer about g cannot occur. Using the universal divinity refinement in the same way as in the proof of Proposition 6.16 it follows that P is not lobbied in equilibrium.

(b) The message space is now given by $S=\{\{g_1\},\{g_2\},\{g_1 \text{ or } g_2\}\}$ (cf. Section 3.3). The following separating path always constitutes an equilibrium: let type g_i send message $s_P=\{g_i\}$, after $s_P=\{g_1 \text{ or } g_2\}$ the politician beliefs that G is of type g_2 for sure and acts accordingly (the interest group always prefers P to delegate, and delegation is less attractive to P when she beliefs that $g=g_2$). When $E(g^2)>4z^2/(3t_m^2+z^2)b>(g_1)^2$, the politician prefers to delegate when $g=g_1$, and prefers not to delegate when $g=g_2$. Based on her prior belief about g she chooses not to delegate. Type g_1 then has a strong incentive to separate itself out, and is able to do this by sending message $s_P=\{g_1\}$. Suppose $s_P=\{g_1 \text{ or } g_2\}$ induces P to delegate (with positive probability). Then, given G's incentives and that message $s_P=\{g_2\}$ induces P not to delegate, type g_2 will send $s_P=\{g_1 \text{ or } g_2\}$ for sure. But, then $q_P(\{g_1 \text{ or } g_2\}) \geq p$ by Bayes' rule, and by the starting assumption that P prefers not to delegate based on her prior beliefs, she will choose not to delegate after $s_P=\{g_1 \text{ or } g_2\}$. A contradiction. So, after $s_P \neq \{g_1\}$ P chooses not to delegate, after $s_P=\{g_1\}$ she chooses to delegate. Then type g_1 chooses $s_P=\{g_1\}$ for sure, and only full separation can occur in equilibrium.   *QED.*

NOTES

1. The model presented in this chapter is based on Sloof (1998).

2. A similar remark is made by Riker (1985, p. 198) in his comment on Fiorina (1985): "The problem with an economic description of regulation is that it focuses on the wrong people. To adapt a methaphore I have used elsewhere, the whole action of regulation is up on the acropolis, while the figures in economic theory are down in the agora. The one quasi-political economic theory, that is, the capture theory tells a story about people from the agora going up to the acropolis and bribing venal politicians to send some soldiers down to regulate the agora for the advantage of the bribers. The problem with this theory is that it locates all the initiative in the agora, so that the outcome is bound to reflect only the concerns of the people below."

3. Calvert et al. (1989) forcefully argue that the appointment stage serves as the primary source of the politician's influence over ultimate policy outcomes, compared with opportunities for oversight and control. Using a simple formal model they show that in a complete information environment the final policy choice of the delegate is completely traceable to the preferences of the appointee (here the politician), although the delegate solely decides on policy. In other words, incomplete information on the side of the politician concerning some relevant aspect is necessarily needed for some delegate discretion to persist. As noted, in Legros (1993) the politician has incomplete information about the preferences of the delegate. In the models of Epstein and O'Halloran (1994) and Bawn (1995) to be discussed in the next paragraphs, the politician is uncertain about how policy maps into consequences.

4. The theoretical result that pooling sets can overlap — i.e. types in $[0,a) \cup (b,1]$ choose the same signaling strategy, and types in $[a,b]$ choose a different common strategy — is also reported by Austen-Smith (1995) for a model of campaign contributions and lobbying. In his model the result can be interpreted as saying that observed contributions are not a good indicator of the "average" preferences of the interest group seeking access for lobbying activities.

5. A considerable weakness of the model of Epstein and O'Halloran (1994) is that the veto decision of the politician is based on complete information. Before her choice between the status quo and the policy alternative proposed by the bureaucrat, the politician becomes fully informed on the consequences of every possible policy. In such a setting there seems to be no rationale for delegation in the first place. Why should the politician restrict her possible choices through delegation if it does not lead to additional information at the actual vote (decision) stage?

6. In another paper Epstein and O'Halloran (1996) consider the same simultaneous choice, albeit in a somewhat different setting. In their model Congress decides on the level of discretion an agency gets, whereas the President decides on the preferences of the agency.

7. Aghion and Tirole (1997) discuss the same trade-off in a somewhat different setting. In their model delegation of authority from a principal to an agent increases the agent's incentives to acquire information. When authority is delegated the principal cannot overrule the agent, and the latter takes more initiative to become informed (informational gain). The cost of delegating formal authority is the principal's loss of control over the choice of projects (distributional loss). Aghion and Tirole (1997, p. 10) refer to this as "The Basic Trade-off between Loss of Control and Initiative."

8. A nice illustration that this is a reasonable assumption is the following quote from a former U.S. trade negotiator cited in Choate (1990, p. 39): "When people in government get ready to leave, they know where the money is. It's with the Japanese. Nobody who's looking at an opportunity to make

$200,000 or more a year representing a Japanese company is going to go out of their way to hurt them while in office."

Of course, in practice also politicians may sometimes use the revolving door. As the empirical evidence of Salisbury and Johnson (1989) based on a sample of 776 representatives indicate, however, far more people going through the door had executive experience rather than experience in the legislature (i.e. a former member of Congress). Moreover, interest groups indicated to value executive branch experience more highly than legislative experience. Another interesting observation of these authors is that "what you know" is more important than "who you know", so former civil servants are mainly employed because of their knowledge of the policy process and their issue familiarity, rather than for their contacts within government. It appears that Congressional experience is more likely to provide an understanding of the policy process, whereas executive experience confers issue familiarity.

9. In view of the description given in the text, the fact that the utility function of G is increasing in t is somewhat peculiar; the more severe the economic prospects of the firm are, the higher the firm's utility. The expression given in (6.2) can therefore be better seen as part of a separable utility function of the form $U_G^\circ(x;t,g)=U_G(x;t,g)-f(t)$, with $f(t)$ such that U_G° is decreasing in t for $t \in T$ (this requires $(\partial f/\partial t)>gx$ for all g and x, and thus an upperbound on the values g and x can take). Because the second part $f(t)$ is assumed to be independent of g and x, it is irrelevant for the equilibrium behavior of the interest group. Hence, in our analysis we can work with U_G rather than with U_G°.

10. In the example of a subsidy to a firm, only bureaucrats working in the Department of Economic Affairs are able, and have time, to judge reports about the economic prospects of the firm.

11. Some support for these assumptions can be found in the following quote from Chubb (1983, pp. 9-10): "Interest groups and bureaucratic agencies have important sources of natural affinity for each other. Compared to other major actors in the political system — congressmen, the courts, the president and his executive office — agencies and groups are policy specialists. On a day-to-day basis they are the participants most involved with particular policies. Also, by virtue of formal organization they bring to the process greater expertise and permanence than do other actors. Theoretically they constitute the durable core of any policy arena."

12. The choice between delegation and no-delegation in the model can loosely be interpreted as a choice between administrative, or legislative forms of government intervention (cf. Fiorina, 1982, 1985), or alternatively, between broad delegation of policy to an administrative, "soft-wired" agency, and narrow delegation to a housekeeping, "hard-wired" agency (cf. Epstein and O'Halloran, 1994). Empirically, among distinct policy areas different distributions of policy authority between legislators and bureaucrats can be observed, ranging from areas in which bureaucrats hardly have any policy discretion to policy areas where bureaucrats make the final decision concerning which policy to implement (cf. Fiorina, 1982, 1985).

13. Rather than by (6.3), the preferences of the bureaucrat could also be specified as $U_B^\circ=U_P+b(U_G-c_P I_P-c_B I_B)$, where $I_i=1$ if G lobbies i and 0 otherwise, for $i=P,B$. That is, the bureaucrat could also care about the amount spent by G on lobbying. Due to the separability of U_B° in x and (c_P,c_B) such an alteration does not affect the analysis at all. Thus, contrary to for instance Lagerlöf (1997), here the same results are obtained when the bureaucrat also takes the costs of lobbying for the interest group into account.

14. The assumption that $g \in \{g_1,g_2\}$ is made for expositional reasons. All main results obtained in this chapter remain qualitatively valid for any assumption made about the distribution of g (as long as g>0).

15. This assumption closely resembles one made in Austen-Smith (1995). He assumes that an interest group becomes informed on the consequences of certain policies only *after* campaign contributions are made (meant to buy access). Austen-Smith observes that when policy consequences are known *before* contributions are made, the amount of campaign contributions alone would already be capable of revealing all information, and subsequent lobbying (modeled as cheap talk messages) would serve no purpose. In the model of Austen-Smith the assumption can be motivated by arguing that buying access is typically meant to secure an option to talk to the legislator should the need arise (and by the observation that most details of legislative agendas only emerge over the course of the legislative session). This motivation resembles the first reason mentioned in the main text. See also Section 2.2 for an informal description of the Austen-Smith (1995) study.

16. As in Chapter 5 we focus on universal divinity. The consequences of imposing the other refinement concept frequently mentioned in Chapters 3 and 4, viz. CFIEP, are briefly discussed in Appendix 6.A. As already mentioned in Appendix 5.B, our choice for universal divinity is guided by the observations that contrary to CFIEP: (i) the existence of a universally divine equilibrium is guaranteed, (ii) the universal divinity concept is theoretically related to strategic stability, and (iii) it is widely used in applications.

17. In this way the refined equilibrium concept employed in this chapter exactly fits in the STABAC refinement procedure as discussed and applied in Chapter 5.

18. Strictly speaking, the equilibrium is not unique under regime I because cut-off type t' may take any randomizing behavior between $s_B=0$ and $s_B=c_B$. The actual equilibrium behavior of type t', however, is inessential for the analysis of the complete game. In that sense the equilibrium is *essentially unique*.

19. In the first case the focus is on changes in c_B, b, g, z, and t_m that affect equilibrium policy $x_B(s_B;g)=\frac{1}{2}(2+bg)E(t|s_B)$ via their impact on the term $\frac{1}{2}(2+bg)$. The indirect impact is given by their influence on $E(t|s_B)$.

20. When b=0 an increase in g always lead to a decrease in $x_B(c_B;g)$. So, for b=0 this comparative statics result corresponds with the one obtained for the basic signaling game of Chapter 3 (cf. Result 3.1(d)); the impact of lobbying is decreasing in the stake the interest group has in persuading the bureaucrat.

21. The expression for $U_B(x;t,g)$ given in (6.3) can be easily rewritten as $U_B(x;t,g)=-[x-\frac{1}{2}(2+bg)t]^2+bg(1+\frac{1}{4}bg)t^2$. Because the second term is independent of x, we can conveniently work with $U_B(x;t,g)=-[x-\frac{1}{2}(2+bg)t]^2$ to derive equilibrium results (save, of course, when we have to compute the expected payoffs of B). From the latter expression it is obvious that B's optimal policy equals $\frac{1}{2}(2+bg)t$. In case bg>0 P and B only agree on optimal policy when t=0, and the level of disagreement about the best policy available is increasing in the state of the world t; $\frac{1}{2}(2+bg)t-t=\frac{1}{2}bgt$. This feature, inter alia, distinguishes our setup of the much copied setup of Gilligan and Krehbiel (1987), where the level of disagreement between the House (politician) and the committee (bureaucrat) is independent of the actual state of the world.

22. Instead of using Table 6.2 an alternative way to disentangle distributional and informational effects is to look for the *direct* presence of b and t', respectively, in the various terms (cf. expression (6.6)). Term J indicates a pure distributional effect because only b is present, term K a pure informational effect because only t' is present, and term Q represents a combined effect because both b and t' directly appear in the expression. Note that in term K bias parameter b only appears indirectly through its appearance in t'.

23. The middle case of K-Q=0 generically only occurs when for all $g \in \{g_1, g_2\}$ an uninformative regime (either U1 or U2) applies in the continuation game between B and G. Then no information about the state of the world t is transmitted from G to B, and we get K=Q=0 (recall that when U1 applies $t'=t_m+z$, when U2 applies $t'=t_m-z$).

24. As already touched upon in Section 3.3 of Chapter 3, Austen-Smith and Wright (1992, p. 229) conclude in their study of competitive informational lobbying that "..the more important is an issue to a special interest group, the more likely is the legislator to make the correct full information decision." Although this comparative statics result indeed holds for some equilibria, it is shown in Sloof (1997) that the result is not general because it does not hold true for *all* plausible equilibria.

25. The conclusion that reduction of the stakes of the interest group is always profitable for the politician is unavoidable in the principal-agent setting of Laffont and Tirole (1991), because in their model the potential influence of the interest group (regulated firm) on the bureaucracy (agency) is always harmful to the politician (Congress); the regulated firm may bribe the agency to retain information, and the firm has a larger incentive to do so when its stake in regulation is higher.

26. Again, instead of using Table 6.3 the effects can be identified from noting the direct presence of b in X and Z, and the direct presence of t' in Y and Z.

27. Specifically, necessary and sufficient conditions are given by each of the four following combinations of restrictions (neglecting ties): (i) $d_1, d_2 > 0$, $e_1 < e_2$, $e_2 > 0$, and $c_P < e_2$; (ii) $d_1, d_2 > 0$, $e_1 < e_2$, $e_1 < 0$, and $e_1 < -c_P$; (iii) $d_1, d_2 < 0$, $e_1 > e_2$, $e_2 < 0$, and $e_2 < -c_P$; and (iv) $d_1, d_2 < 0$, $e_1 > e_2$, $e_1 > 0$, and $c_P < e_1$. The "sorting condition" restrictions, i.e. the inequalities comparing e_1 with e_2, are a consequence of imposing the universal divinity equilibrium refinement in the STABAC procedure. They are not needed when this refinement concept is not used (cf. Propositions 3.1 and 3.2).

28. When b=0 we thus have $(E_D(U_P|g)-E_N(U_P))(E_D(U_G;g)-E_N(U_G;g)) \le 0$ for all g. From Proposition 6.8(b) it follows that only when the weak inequality becomes an equality for some $g \in \{g_1, g_2\}$, the politician may be lobbied in equilibrium.

29. There may be important cross country differences, though, as Wilson (1981, p. 126) indicates: "It certainly seems that the close links between interest groups [..] and Executive Departments, are weaker, rather than stronger, in the United States than in many European countries such as Britain." Indeed, from an American viewpoint the assumption may seem less plausible: "In this case [of access to the executive branch], however, access — at least to the president and sometimes to his aides— may be considerably more limited than access to congressional policymakers" (Schlozman and Tierney, 1986, p. 323), and "These results [..] indicate [..] that contacts in Congress are easier to obtain than contacts in the executive branch." (Salisbury and Johnson, 1989, p. 190).

30. When $t_m=z$ the actual state of the world may potentially be such that government policy is of no importance to the group (viz. t=0). However, because t is a continuous random variable this event does not occur with positive probability.

31. If one has the subsidy example of Section 6.1 in mind, this situation seems to have some practical appeal; in general, for instance because they run counter to the market mechanism, firm specific subsidies are not considered to be needed (the state of the world t is rather low). But, there may be exceptional circumstances — demand shocks, for instance — under which a (high) subsidy is needed.

32. The results of this example can be loosely interpreted as follows. When the distribution is skewed to the left, the uninformed policymaker is more or less biased in favor of the interest group and the group prefers a "formal" law, completely fixing policy on the basis of coarse information. In the

opposite case, the uninformed policymaker is biased against the group, and the latter prefers a more flexible "material" law (delegation) which allows for adaption to specific circumstances (cf. Van Schendelen, 1988).

33. Related to the topic of the risk attitude of the players, in the model delegation of policy authority only occurs by virtue of the assumption that reduction of uncertainty about the actual state of the world t benefits the politician. That is, given a specific policy x, the politician prefers a certain outcome of the state of the world t° to a lottery whose expected outcome is t°.

34. Note the assumption that $g>0$. The second statement more formally reads: given that $g>0$, the politician prefers g as low as possible. Note that the case $g=0$ is of no interest because then we have $U_G \equiv 0$.

35. For instance, according to Chubb: "In most industrial nations the bureaucracy's impact on policy formulation is as unmistakable as its effect on implementation. Even the United States Congress, with its comparative political independence, has come to rely on the executive branch for policy proposals." (Chubb, 1983, p. 20), and "In practice this [policy-making] process often entails rather sharp shifts in institutional participation. The executive branch frequently dominates policy formulation and implementation, while the legislative branch controls legitimation." (Chubb, 1983, p. 37).

36. Models that explicitly incorporate procedural endogeneity with respect to the rules governing the relationship between the House and the committee are provided by Gilligan and Krehbiel (1987, 1989b). Contrary to the first model, in the latter the House cannot commit to a specific procedure before the committee makes a proposal. Relatedly, Gilligan and Krehbiel (1990) incorporate institutional endogeneity with respect to the preferences of the committee and its resources. In this model the House chooses the composition of the committee and the resources that are allocated to it before the committee decides whether to specialize or not. Subsequently, proposals are made under open rule. The costs of specialization are assumed to depend on the preferences of the committee and the resources allocated to it. It appears that a necessary condition for assigning *preference outliers* to the committee is that the ability to acquire expertise at low cost is associated with extremity of preferences. The idea that biased committee members have lower costs in acquiring policy relevant information nicely links up with the result obtained in this chapter that biased bureaucrats may appear to be better suited to obtain policy relevant information from interest groups.

37. Similar remarks apply to the signaling model with multiple receivers of Farrell and Gibbons (1989). In this model there is only one level of information asymmetry (between a sender and two receivers). Although one may be inclined to interpret the model of Farrell and Gibbons as a model of lobbying in which the interest group chooses between lobbying the bureaucrat and/or the politician, the relationship between the two receivers is not made explicit. One of the main points of the analysis presented in this chapter is that it is exactly this relationship that governs an interest group's decision which level(s) of government to lobby.

38. For instance, suppose there is an opposing interest group H with preferences $U_H(x;t,g)=-htx$. Now, when the outside career opportunities of the bureaucrat are spread over the two groups (several revolving doors), the induced preferences of B may take the form $U_B(x;t,g)=-(x-t)^2+b_1 gtx-b_2 htx$, with $b_1,b_2>0$. In such a specification the preferences of B are more aligned with those of P compared with at least one of the cases where only group G or group H exists.

39. Unlike universal divinity the CFIEP refinement concept (cf. Chapter 3) never deletes the pooling on no-lobbying ($s_B=0$) equilibrium. To the contrary, it is easily seen that the pooling on $s_B=c_B$ equilibrium is deleted by the pooling on $s_B=0$ PBE (cf. Appendix 4.B of Chapter 4, proof of Proposition 4.6). In order to proof that when $H(t_m-z)<c_B<H(t_m+z)$ the pooling on $s_B=0$ equilibrium is not deleted by the two-partition equilibrium, we show that types $t>t'$ (with t' from the two-partition equilibrium)

always weakly prefer the pooling on $s_B=0$ equilibrium. Suppose to the contrary that for some type $t>t'$ holds that $gt[\frac{1}{2}(2+bg)t_m]<gt[\frac{1}{4}(2+bg)(t'+t_m+z)]-c_B$. If this inequality holds for some type t, then it necessarily holds for all types in $[t,t_m+z]$ because larger types gain more by deviating to $s_B=c_B$. The inequality can be rewritten to $c_B(t-2z)>H(t)(t_m-z)$. Taking $t=t_m+z$ we must have $c_B(t_m-z)>H(t_m+z)(t_m-z)$, contradicting $c_B<H(t_m+z)$. Hence the inequality does not hold for the highest type $t=t_m+z$, and thus for no type t in T. In summary, when CFIEP is used as refinement concept $t'=t_m+z$ is always an equilibrium, $t'=2c_B/[g(2+bg)z]$ only when $t' \in (t_m-z,t_m+z)$. The equilibrium of the continuation game is thus not always unique, and pooling on lobbying $(s_B=c_B)$ cannot occur.

40. For the numerical example considered here it can be shown that when the alternative skewed distribution $F(t;r)=1-(1-t)^r$ (with $r>0$) is used, no informational gain can be obtained by the interest group.

REFERENCES

Aghion, P. and J. Tirole, 1997, Formal and real authority in organizations, *Journal of Political Economy* 105, 1-29.

Austen-Smith, D., 1993, Information and influence: Lobbying for agendas and votes, *American Journal of Political Science* 37, 799-833.

Austen-Smith, D., 1995, Campaign contributions and access, *American Political Science Review* 89, 566-581.

Austen-Smith, D. and J.R. Wright, 1992, Competitive lobbying for a legislator's vote, *Social Choice and Welfare* 9, 229-57.

Banks, J.S., 1991, Signaling games in political science (Harwood, Chur).

Bawn, K., 1995, Political control versus expertise: Congressional choices about administrative procedures, *American Political Science Review* 89, 62-73.

Breton, A., 1995, Organizational hierarchies and bureaucracies: An integrative essay, *European Journal of Political Economy* 11, 411-440.

Calvert, R.L., 1985, The value of biased information: A rational choice model of political advice, *Journal of Politics* 47, 530-555.

Calvert, R.L., McCubbins, M.D. and B.R. Weingast, 1989, A theory of political control and agency discretion, *American Journal of Political Science* 33, 588-611.

Che, Y.K., 1995, Revolving doors and the optimal tolerance for agency collusion, *Rand Journal of Economics* 26, 378-397.

Choate, P., 1990, Agents of influence (Simon and Schuster, New York).

Chubb, J.E., 1983, Interest groups and the bureaucracy (Stanford University Press, Stanford).

Crain, W.M., and R.E. McCormick, 1984, Regulators as an interest group, in: J.M. Buchanan and R.D. Tollison, eds., The theory of public choice II (The University of Michigan Press, Ann Arbor) 287-304.

Epstein, D. and S. O'Halloran, 1994, Administrative procedures, information and agency discretion, *American Journal of Political Science* 38, 697-722.

Epstein, D. and S. O'Halloran, 1996, Divided government and the design of administrative procedures: A formal model and empirical test, *Journal of Politics* 58, 373-397.

Farrell, J. and R. Gibbons, 1989, Cheap talk with two audiences, *American Economic Review* 79, 1214-1223.

Fiorina, M.P., 1982, Legislative choice of regulatory forms: Legal process or administrative process?, *Public Choice* 39, 33-66.

Fiorina, M.P., 1985, Group concentration and the delegation of legislative authority, in: R.G. Noll, ed., Regulatory policy and the social sciences (University of California Press, Berkeley) 175-197.

Gilligan, T.W. and K. Krehbiel, 1987, Collective decisionmaking and standing committees: An informational rationale for restrictive amendment procedures, *Journal of Law, Economics and Organization* 3, 145-193.

Gilligan, T.W. and K. Krehbiel, 1989a, Asymmetric information and legislative rules with a heterogeneous committee, *American Journal of Political Science* 33, 459-490.

Gilligan, T.W. and K. Krehbiel, 1989b, Collective choice without procedural commitment, in: P.C. Ordeshook, ed., Models of strategic choice in politics (The University of Michigan Press, Ann Arbor) 295-314.

Gilligan, T.W. and K. Krehbiel, 1990, Organization of informative committees by a rational legislature, *American Journal of Political Science* 34, 531-564.

Laffont, J.-J. and J. Tirole, 1991, The politics of government decisionmaking: A theory of regulatory capture, *Quarterly Journal of Economics* 106, 1089-1127.

Lagerlöf, J., 1997, Lobbying, information, and private and social welfare, *European Journal of Political Economy* 13, 615-637.

Legros, P., 1993, Information revelation in repeated delegation, *Games and Economic Behavior* 5, 98-117.

Letterie, W. and O.H. Swank, 1997, Learning and signalling by advisor selection, *Public Choice* 92, 353-367.

Mazza, I, and F. van Winden, 1998, An endogenous policy model of hierarchical government, Working Paper TI 98-007/1 (Tinbergen Institute, Amsterdam).

Milner, H.V. and B.P. Rosendorff, 1996, Trade negotiations, information and domestic politics: The role of domestic groups, *Economics and Politics* 8, 145-189.

Mo, J., 1995, Domestic institutions and international bargaining: The role of agent veto in two-level games, *American Political Science Review* 89, 914-924.

Niskanen, W.A., 1971, Bureaucracy and representative government (Aldine Atherton, Chicago).

Peacock, A., 1994, The utility maximizing government economic advisor: A comment", *Public Choice* 80, 191-197.

Potters, J., 1992, Fixed cost messages, *Economics Letters* 38, 43-47.

Riker, W.H., 1985, Comment (on: Group concentration and the delegation of legislative authority, by M.P. Fiorina), in: R.G. Noll, ed., Regulatory policy and the social sciences (University of California Press, Berkeley) 197-199.

Rogoff, K., 1985, The optimal degree of commitment to an intermediate monetary target, *Quarterly Journal of Economics* 100, 1169-1189.

Salant, D.J., 1995, Behind the revolving door: A new view of public utility regulation, *Rand Journal of Economics* 26, 362-377.

Salisbury, R.H. and P. Johnson, 1989, Who you know versus what you know: the uses of government experience for Washington lobbyists, *American Journal of Political Science* 33, 175-195.

Schlozman, K. and J. Tierney, 1986, Organized interests and American democracy (Harper and Row, New York).

Sloof, R., 1997, Competitive lobbying for a legislator's vote: A comment, *Social Choice and Welfare* 14, 449-464.

Sloof, R., 1998, Interest group influence and the delegation of policy authority, submitted to: *Economics and Politics*.

Spiller, P.T., 1990, Politicians, interest groups, and regulators: A multiple-principals agency theory of regulation, or "let them be bribed", *Journal of Law and Economics* 33, 65-101.

Tirole, J., 1994, The internal organization of government, *Oxford Economic Papers* 46, 1-29.

Van Schendelen, M.P.C.M., 1988, De markt van politiek en bedrijfsleven (Kluwer, Deventer).

Vogel, D., 1978, Lobbying the corporation. Citizen challenges to business authority (Basic Books, New York).

Wilson, G.K., 1981, Interest groups in the United States (Clarendon Press, Oxford).

7 SUMMARY AND EVALUATION

In this final chapter a summary and evaluation of the monograph are given. Section 7.1 provides a brief, yet comprehensive survey of the main results obtained. Subsequently, Section 7.2 evaluates these results and the method(s) employed to derive them, and advances some ideas for future research.

7.1 SUMMARY

The monograph starts in Chapter 1 with the observation that taking a well-informed position on a number of relevant policy issues requires a basic knowledge and understanding of the "how and when" of interest group influence. Already a substantial literature, both empirical and theoretical in nature, concerned with the positive analysis of the behavior of interest groups and their influence exists. Especially a number of recent game-theoretic models have considerably increased our insight in this respect. These game models highlight the strategic issues involved in the interaction between interest groups and governmental decisionmakers. The theoretical models presented in this monograph extends these earlier models in a number of directions, to address theoretical issues that remained up till now largely unresolved. The directions in which the theoretical models are extended are guided by results obtained in empirical studies.

By means of an overview of the empirical findings concerning the influence of interest groups, the first part of Chapter 2 provides some basic knowledge of how interest groups try to influence governmental decision-making, and when these activities are likely to be effective. The general observations that can be made on the basis of this survey are summarized in seven "stylized facts" (cf. Subsection 2.1.3), which will not be repeated here. Moreover, given that the empirical literature is rather well developed, most of the themes that have been brought to the fore in relation to the positive analysis of interest group influence pass in review. In that way the discussion of existing empirical results also provides a general background for the theoretical work presented in later chapters. An empirical observation of particular interest is that interest groups may, and indeed do, use several means of influence, and that these means are likely to interact. As the second part of Chapter 2 indicates, however, theoretical models that analyze how interest groups will choose between the distinct options they have are relatively scarce. Moreover, some of the studies that do focus on this issue are of limited use in trying to uncover the determinants of this choice because they employ a black box influence process. Together, these observations motivated the theoretical research presented in this monograph.

The driving force behind interest group influence is taken to be the strategic transmission of information. Chapter 3 first discusses the basic signaling game that underlies all our subsequent models. In this basic signaling game a privately informed sender may, by means of sending a fixed cost message, try to affect the decision of a less well-informed decisionmaker. It is shown that information transfer — and, thus, actual influence of the sender on final decision-making — only occurs when there is sufficient congruence of interests between the sender and the decisionmaker (sorting condition). In that case the informational value of a costly message lies merely in its costs, and not in its content. Subsequently, these results are interpreted in light of the specific applications of the basic signaling game that are used later on in the monograph. For instance, it is argued that political campaigns can be informative to voters regarding the policy stance of the candidate just by means of the amounts of money spent on the campaign, rather than through its content. Similar remarks apply to direct endorse-ments in support of a particular candidate. Not the content of an interest group's endorsement necessarily matters, but the amount of money involved. Also the application to informational lobbying, for which the basic signaling game was originally introduced, is given some attention. Finally, a number of extensions of the basic signaling game that are of particular interest in view of our three applications are discussed. It is argued that factors like the presence of multiple senders, the possibility of the verification of messages, some discretion in determining the amount of money that is spent on sending messages, and the exclusion of the possibility of explicit lying, all increase the scope for information revelation.

Chapter 4 addresses two main questions. The first question arises from combining the theoretical result that campaign outlays in itself may be informative to voters about the policy position of the candidate, with the empirical observation that campaign chests are largely financed by interest groups. It is analyzed whether the funds provided by interest groups can serve a useful purpose as "indirect" endorsements, and whether the presence of campaign contributions may be crucial for campaign expenditures to transfer some information in equilibrium. In our game-theoretic model where contributions are not directly observable to the voter, but where their existence can be concluded upon in case a costly campaign is observed, this indeed appears to be the case. When the preferences of the candidate are such that (s)he cannot be trusted by the (decisive) voter, it is crucial for campaign costs to carry some information that the campaign is financed by an interest group whose preferences are more aligned from the voter's perspective. In case the electorate has more faith in the donating interest group than in the candidate, campaign contributions indeed serve a useful purpose as indirect endorsements.

The second question addressed in Chapter 4 relates to the main topic of this monograph, the endogenous choice between various means of influence. Specifically, it concerns the choice of an interest group between campaign contributions and direct endorsements to support a particular political candidate. Quite intuitively, our results predict that groups with preferences opposed to those of the electorate typically prefer to use campaign contributions in order to filter their opposing interests. Encompassing (public) interest groups, on the other hand, are more likely to reach out to the electorate directly by means of direct endorsements.

In order to investigate the robustness of our answers to the two questions posed in Chapter 4, also the more realistic situation is analyzed in which contributions are directly observable to the voter. Rather reassuringly, it appears that our answers carry over to this case. As a nice additional result this exercise also yields a theoretical justification for the popular call for full disclosure laws. Even in a modeling environment where it might be conjectured that unobservability of campaign donations in some instances can benefit the electorate by increasing the scope for information transfer (cf. receiver uncertainty discussed in Chapter 3), this result cannot be derived in our model. Although the model seems rather biased towards the potential benefits of no-disclosure, not in the least because it excludes explicit "quid pro quos", it unambiguously points at the desirability of full disclosure.

Chapter 5 analyzes the choice an interest group may have between lobbying and pressure. Lobbying is interpreted as persuasion through the use of "words", pressure as persuasion through the use of "actions". In fact, three different models are analyzed which differ in the opportunities an interest group has for lobbying. The conclusions drawn in this chapter with respect to getting and maintaining an established position are based on an overall evaluation of all three models. Thus, they are robust to changes in the specific assumptions made in regard to the exact opportunities for lobbying. This robustness strengthens our faith in the results obtained.

The main result that follows from the analysis of all three models together is that pressure only occurs in case the "reputation for being strong" of the interest group (the interest group's actual credibility for carrying out a sanction) is sufficiently low. In these cases pressure is typically indispensable for building up a reputation. Roughly put, the models predict that, in building up a reputation, "actions speak louder than words", and that it is necessary to "show them your teeth first!". Once the interest group has arrived at an established position, the group may be forced to maintain this position through lobbying. In case an established interest group fails to lobby, it is sometimes able to regain its reputation by exerting pressure; contrary to other game-theoretic models of reputation building, the models presented in Chapter 5 allow an interest group to make up for a previous loss of reputation. At the same time, these models predict that established interest groups still, once in a while, have to exert pressure to keep their reputation. The analysis of repeated lobbying shows that, only under rather specific circumstances, repeated lobbying may yield the group a high reputation without having to exert pressure. Finally, a case could be made for policymakers to design their institutions in such a manner as to force newly organized interest groups to show their teeth first, before giving these groups lobbying access.

Chapter 6 combines the observation that interest groups typically have an informational advantage over government with the observation that delegation of powers within government are (partly) dictated by information and expertise asymmetries between bureaucrats and politicians. A formal model is presented in which bureaucrats are able to understand and interpret technical, policy relevant information coming from interest groups, whereas politicians are not. Because potential gains can be obtained from information transmission about these technical policy details, politicians may prefer to delegate policy authority to bureaucrats. The incentives to delegate authority are tempered, however, when the preferences of the bureaucrat are not completely aligned with those of the politician. In the latter case the bureaucracy will namely implement policies that would not be chosen by politicians were they to have the same information. In view of this last observation, perhaps a somewhat surprising result derived from the model is that politicians sometimes prefer a (more) biased delegate in the bureaucracy over a less biased, or even unbiased one. The intuition behind this result is that the scope for information transfer between the interest group and the bureaucracy may increase when their preferences become more aligned. So, when the preferences of the bureaucracy move away from the interests of politicians towards the interests of special interest groups, politicians may still gain because the informational effect outweighs the "shift-in-policy-to-be-implemented" effect. Beyond a certain point, of course, the preferences of the bureaucrats and the politicians are too opposed to make delegation of policy authority profitable for the latter.

With their option to delegate policy authority or not, politicians affect the way in which interest groups might try to influence public policy. Instead of standing by helplessly how the bureaucracy is "captured" by organized interests, politicians may actively control the potentially negative impact of interest groups on bureaucrats by giving the latter no discretion at all. The model indicates that — due to differences in their knowledge and decision power — interest groups may advance politicians for different reasons than they approach bureaucrats for. In the model setup of Chapter 6 the interest group may lobby the legislature for delegation or no-delegation of policy authority. Lobbying politicians takes the form of transmitting information about the potential benefits of (no-)delegation.

The equilibrium analysis of the model yields that interest groups try to affect the politician's delegation decision only when interest group access to the bureaucracy is neither blocked nor too easy. This last result is a consequence of the strategic behavior of politicians; when interest groups have no or too easy access to the bureaucracy, no technical policy information is revealed through their interaction, and delegation has no potential benefits at all to the politician. Given that the decision of the politician will not be affected by the lobbying efforts of the interest group, it will not waste resources on these activities. When lobbying costs are assumed to be fixed the theoretical model predicts that interest groups are more likely to lobby for delegation than no-delegation of policy authority. It is shown that this prediction may become stronger when more general preferences and prior beliefs are allowed for, and also when a communication structure is analyzed in which technical policy information coming from an interest group can be verified by bureaucrats at zero costs. However, the conclusion that interest groups mostly pursue delegation of policy authority is not robust to changes in the model that allow the costs of lobbying the bureaucracy to be endogenously determined.

7.2   EVALUATION

In the introductory chapter it was observed that a better understanding of the nature of interest group influence can be of great help when tackling certain real world policy problems (examples of such policy issues that are currently of interest are given in Chapter 1). Therefore, the general purpose of this research project was to shed light on how interest groups try to affect government decision-making, and when these influence activities are likely to be effective. Specifically, the monograph aimed to contribute to the existing game-theoretic research in this area by considering issues largely ignored so far in these models. The monograph has attained this end if its results, summarized in the previous section, are of significance to a better understanding of the phenomena at issue. Although we ourselves are not dissatisfied with them in this respect, the final judgment concerning their significance is up to the reader.

An assessment of the in our opinion strong and weak points of the analysis presented may be helpful to the reader to pass this judgment. We first mention a number of assets of the general method employed here to derive our results. Then we direct attention to the more important shortcomings of our analysis, which also leads to some suggestions for further research.

Given our objective to contribute to the theoretical literature which already has a firm footing, the general theme of this research project as well as the tools of analysis are not original. The theoretical models presented all depart from models already discussed and analyzed at length by other authors. However, since these models have proven to be very helpful in understanding intricate real world phenomena, it is considered to be an advantage that all game-theoretic models presented in this monograph elaborate on the same underlying basic signaling game (cf. Chapter 3). This common structure provides a base for coherence and embeddedness.

The direction in which existing models are extended is supported by both casual and more rigorous empirical observations (cf. Chapter 2). Rather than focusing on topics that are primarily of theoretical interest, extensions are proposed and analyzed that allow the inspection of issues that are of practical relevance. The main characteristic of the game-theoretic models that are analyzed is the endogeneity of the choice an interest group has between several means of influence. Given the focus on strategic information transmission, at first glance a setup starting from an existing multi-dimensional signaling game model would have seemed to be appropriate.[1] This approach is not taken, however, because closer examination of the issues under consideration shows that the interactions between interest groups, politicians, bureaucrats and voters do not directly fit into the format of one of these games. The models presented in this book display an institutional embedding by taking account of some of the peculiarities of these interactions. For instance, in the model of Chapter 5 pressure per se is not directly traded-off against lobbying, because the model does not allow the interest group to choose to exert pressure when choosing whether or not to lobby. In our view such a direct trade-off is not characteristic of the actual choice between lobbying and pressure. It seems rather unrealistic that the politician cannot avoid the costs of being pressured by giving in immediately, or that the interest group can commit itself to exert pressure beforehand (which gives the interest group a direct trade-off between lobbying and pressure before the politician takes her decision). Although our models are directed towards particular relationships in the political arena, they seem to have wider applications, and therefore seem to be of some general interest as well (the model of Chapter 5 could for instance be applied to fraternity raids).

Predictions obtained from a particular equilibrium of a stylized game-theoretic model are typically highly sensitive to changes in specific assumptions made about the order of play, the prior beliefs of the agents, precisely what information a specific agent has, and the structure of communication. In order to arrive at theoretical predictions that do have some general theoretical validity, making them worthwhile to explore by means of empirical tests, it is important to investigate the robustness of these predictions to alterations in specific modeling assumptions. Each of the three models discussed in Chapters 4, 5 and 6, are examined in this respect for at least some of their modeling aspects. The conclusions drawn from these models take these examinations into account. When conclusions are valid for a more general set of assumptions we can have more faith in them. Ultimately, of course, real confidence in conclusions obtained from theoretical models only follows from their apparent empirical validity. In this regard a vast number of empirical illustrations are given throughout the book that suggest the practical relevance of our models. However, possibilities for more rigorous empirical testing also exist (see below).

Besides having a number of attractive properties, the analysis presented in this monograph also has some important shortcomings. Although our models elaborate on existing models and make these more realistic, they are still too simplistic in a number of respects. For instance, the games analyzed in Chapter 4 model the electorate as a single unitary actor (the voter). This assumption is tenable by referring to a reduced form approach to majority vote elections, making the median voter the single decisive voter. But, as the discussion presented in Section 4.5 indicates, considerable gains could be obtained from introducing a spectrum of voters. Various other suggestions for further extensions of our models were given in the concluding discussions of the individual chapters. Except for one recurrent suggestion, they will not be repeated here.

In the final chapter of his book Potters (1992) advances two suggestions for future theoretical research. One of them, the endogenization of the type of influence attempt, is followed here. His second suggestion is to account for the presence of competing interest groups. Indeed, probably one of the weakest points of the game-theoretic models presented here is that they incorporate only one interest group. Although we repeatedly indicated what kind of results are to be expected from such an extension, these were not derived from a formal model incorporating multiple groups. It stands to reason, though, that the choice of a specific means of influence is significantly affected by the comparative advantages a group may have over other groups in this respect, or determined by the competition between interest groups. In this way competition between interest groups may affect their choice of instruments in important ways, and hence its omission as a determinant of this choice may be rather serious.

The reason why multiple interest groups are not formally analyzed here is that allowing for multiple senders in a signaling game context typically introduces severe complexities in the equilibrium analysis of the model. Even a rather straightforward extension of the basic signaling game of Chapter 3 to the case with two competing senders leads to an equilibrium analysis which is "tedious" (Potters, 1992, p. 60). In combination with the technical difficulties the endogenous choice between several means of influence introduces, allowing for multiple groups as well would prima facie lead to extremely complex models.[2]

Given these difficulties, the prospects for a comprehensive model of the political influence of competing interest groups on government policy look rather dim. Potters (1992) cherishes the hope that further work will establish a theoretical micro-foundation for the use of a "composite utility function". The idea is that in equilibrium the behavior of a policymaker would be in line with maximizing a function which has the weighted utilities of the respective interest groups as arguments. Such micro-foundations already exist for influence based on campaign contributions or votes in a probabilistic voting context.[3] A model focusing on influence through information transfer would indeed provide an important complement to the other models that support a "composite utility function". However, a comprehensive micro-foundation, allowing competing interest groups to employ a variety of means of influence, will probably appear to be too ambitious.

Similar remarks as for the single interest group involved in our models apply to the observation that in each chapter only the choice between (at most) two means of influence is studied. In reality interest groups in fact choose from a larger set of possible political activities. For instance, rather than choosing either between campaign contributions and direct endorsements, or between lobbying and pressure, or between lobbying politicians and lobbying bureaucrats, an interest group may choose its means from all these activities together. The increased complexity observed in Chapters 4, 5 and 6, however, may inescapably require a partial analysis when studying the choice of an interest group among a range of possible influence methods. At the expense of partly giving up a base for embeddedness and comparison, partial (equilibrium) models are more likely to yield insightful and interpretable results and testable hypotheses. Apart from the choices considered in this monograph, a number of other choices remain worthwhile to investigate by means of such a partial analysis.

For instance, the game models considered here all focus on interest group influence through the use of different means of transferring information. Although the (strategic) transmission of information is among the most important methods of influence, it is by no means the only one. As already observed in Chapter 2 the supply of votes and the use of campaign contributions as explicit "quid pro quos" are important examples. Models investigating the choice between means which have a different base for exerting influence — such as information transfer versus the supply of votes — may therefore be particularly worthwhile to analyze. Also, it would be interesting to contrast means of influence not directly connected to elections (like lobbying) with means that are more election related (like campaign contributions, mobilizing members, and direct endorsements). The latter methods are, either directly or indirectly, based on the control of voters over politicians.[4] Contrarily, the ways of transferring information studied in Chapters 5 and 6 are more or less independent of voters and elections. Lastly, with the model of Chapter 6 in mind it might be interesting to study the choice between trying to influence different types of agents, rather than focusing on different ways to influence the same agent. For instance, a relevant issue to investigate is an interest group's choice between influencing politicians and/or bureaucrats, on the one hand, and going to court, on the other hand.

A question closely related to how to shape one's strategy in trying to exert political influence is whom to employ for such activities. A relevant research question in this respect is the following: Under what circumstances does (or should) a corporation create its own public affairs department instead of hiring external lobbyists? The study of the political activities of individual corporations raises other interesting issues as well. For instance, how does (should) an individual corporation allocate its resources over, on the one hand, regular economic activities, and, on the other hand, political influence seeking activities? Recent papers start to address these type of questions (see, e.g., Johnson, 1996, and Moloney and Jordan, 1996), but clearly much work remains to be done in this area.

Another aspect missing in the present monograph is a rigorous empirical test of the models that are presented. Although on many occasions empirical findings are mentioned that are highly suggestive with respect to the empirical relevance of our models, "the proof of the pudding is the eating". The conclusions obtained from our models would strongly gain in cogency when they could be sustained by rigorous empirical tests of their validity. The highly stylized nature of these models, relying crucially on notions that are hard to measure in the field (such as prior beliefs, information levels, order of moves, structure of communication, reputation etc.), make them probably more easily testable using laboratory experiments. Indeed, games of asymmetric information of the type considered here are increasingly put to the test in the lab. Surveys of the results obtained so far are provided by Crawford (1995) and Van Winden (1997a). In particular, the following conclusion by Crawford (1995, p. 41), based on a broad review of experimental evidence, is of relevance here: "Most subjects [in the experiments] seem to have some strategic sophistication, but seldom enough to justify an analysis based exclusively on equilibrium, however refined."

The conclusion reached by Crawford calls into question our modeling assumptions about the level of strategic sophistication of the economic agents involved in the interaction. It seems that considerable gains in terms of descriptive accuracy can be obtained by relaxing the assumption of full rationality in our theoretical models. That is, a turn to bounded rationality or adaptive (learning) behavior models may prove to be very useful in understanding actual behavior (cf. Van Winden, 1997b). Exactly how the notion of bounded rationality should be modeled is far from clear, though.[5] In this regard Conlisk (1996, p. 692) argues that: "..a sensible rationality assumption will vary by context, depending on such conditions as deliberation cost, complexity, incentives, experience, and market discipline", and, "..the conditions of a particular context may favor either bounded or unbounded rationality." From this perspective models in which agents are fully rational continue to serve as a very useful benchmark when tackling positive research questions. Of course, unbounded rationality models are of interest from a normative viewpoint as well, because they indicate how fully rational agents ought to behave.

Although the basic ingredients of stylized game-theoretic models are typically difficult to measure in the field, problems are not always unsurmountable. Sometimes creativity and the use of supplementary information from various sources can be very helpful in finding satisfactory proxies for the unobserved quantities. This is, for instance, nicely illustrated in the empirical analysis presented by Austen-Smith and Wright (1994).[6] They test their theory of competitive informational lobbying using the 1987 confirmation vote on the nomination of judge Bork for the U.S. Supreme Court. They make use of a headcount by the American Conservative Union to assess a senator's stand (i.e. his prior beliefs) *before* the onset of any lobbying. Substantial empirical support for the main predictions of the model is reported (cf. Chapter 2). In short, though probably rather difficult, it seems possible to collect field data to test the game-theoretical models at issue here in a satisfactory way.

In fact, the game models presented in this monograph may guide the search for new empirical material, and indicate the (competing) hypotheses that have to be tested. In that way empirically estimated relationships can be given a clear cut interpretation. As observed in Chapter 2, one of the two important shortcomings of the current empirical literature on the influence of interest groups concerns the lack of theory. As a result, the empirical models become rather ad hoc and the estimated relationships difficult to interpret. Theoretical models in the spirit of the models presented in this monograph may prove to be useful here.

Although we mainly focused on game-theoretic models, a substantial part of this monograph reviewed existing empirical results (cf. Chapter 2). Therefore, we end this section with some suggestions for future empirical work. These suggestions relate to the other major shortcoming of the existing empirical research observed in Chapter 2: the lack of data. This shortage constrains the assessment of the influence of interest groups on government policies in important ways. Most empirical studies are directed towards one country (the U.S.), one means of influence (campaign contributions), and one or a small set of narrow vote issues. Due to the abundance of U.S. campaign contribution data, possibilities for empirical inquiry are almost unlimited here. However, the marginal returns to future work in this area are probably very low. It seems unlikely that yet another U.S. oriented study regarding the influence of campaign contributions coming from a particular interest group on the outcome of a certain bill will yield new insights that are of general interest. New empirical research in other areas, on the other hand, appears to be more promising in this respect. Particularly welcome are empirical studies that focus on other countries than the U.S. or means of influence other than campaign contributions, as well as studies that incorporate the dynamic features that are typical of the relationship between interest groups and government.

NOTES

1. See, for instance, Cho and Sobel (1990), Engers (1987), Milgrom and Roberts (1986), Quinzii and Rochet (1985), Ramey, (1996), and Wilson (1985).

2. These models would not only be demanding in regard to the modeler's own analytical capabilities, they would also be very demanding regarding the capabilities of the agents whose behavior they try to describe and predict. Recent results obtained from laboratory experiments put serious doubts on the accuracy of game-theoretic predictions of actual behavior, even for the simple (basic signaling) game presented in Chapter 3. Potters and Van Winden (1996) find that their experimental data for this game are more in line with a theory in which subjects are assumed to behave as non-strategic decisionmakers. Although a follow-up study indicated that professional lobbyists have a somewhat keener eye for the strategic issues involved, they also do not closely follow the predictions of game theory (cf. Potters and Van Winden, 1998).

3. For influence based on the potential supply of votes see Coughlin et al. (1990a, 1990b) and Lindbeck and Weibull (1987), and for influence based on campaign contributions see e.g. Baldwin (1987) and Grossman and Helpman (1996). A severe drawback of the first type of models is that they do not allow interest groups to compete actively for favors (cf. Grossman and Helpman, 1996). A point of critique applying to the second type of models is that the uninformed voters in the model are irrational and suffer from "voter illusion"; they are positively affected by campaigns that are financed at their own expense (Lohmann, 1994, p. 7).

4. Likewise, it would also be worthwhile to merge partial models of interest group influence, elections and policy making "into more complete equilibrium models that allow for interactions between electoral, influence, and legislative behaviour." (Persson, 1998, p. 311)

5. In the words of Milgrom and Roberts (1987, p. 189): "The problem, of course, is that we as yet have little agreement on how to model more descriptively accurate forms of rational behavior, little faith that we can find hypotheses on behavior that will be as tractable and powerful as maximization and equilibrium, and a general fear that by renouncing our standard methods we will forfeit elegance and, in return, get only *ad hockery*." See Conlisk (1996) for a sampler of economic models that incorporate bounded rational agents, and for a more general discussion of the topic of bounded rationality.

6. By taking this study as an example of outstanding scientific (empirical) inquiry we take a position diametrically opposed to Baumgartner and Leech (1996a, 1996b), who take this very same article as a prominent recent example of faulty empirical research: "They [Austen-Smith and Wright] study an unusual case, test an incomplete theory with imperfect measures, and commit a variety of other errors.." (1996a, p. 522). See Austen-Smith and Wright (1996) for an, in our view, convincing rebuttal of this critique.

REFERENCES

Austen-Smith, D. and J.R. Wright, 1994, Counteractive lobbying, *American Journal of Political Science* 38, 25-44.

Austen-Smith, D. and J.R. Wright, 1996, Theory and evidence for counteractive lobbying, *American Journal of Political Science* 40, 543-564.

Baldwin, R., 1987, Politically realistic objective functions and trade policy PROFs and tariffs, *Economics Letters* 24, 287-290.

Baumgartner, F.R. and B.L. Leech, 1996a, The multiple ambiguities of "counteractive lobbying", *American Journal of Political Science* 40, 521-542.

Baumgartner, F.R. and B.L. Leech, 1996b, Good theories deserve good data, *American Journal of Political Science* 40, 565-569.

Cho, I.K. and J. Sobel, 1990, Strategic stability and uniqueness in signaling games, *Journal of Economic Theory* 50, 381-413.

Conlisk, J., 1996, Why bounded rationality?, *Journal of Economic Literature* 34, 669-700.

Coughlin, P.J., Mueller, D.C. and P. Murrell, 1990a, A model of electoral competition with interest groups, *Economics Letters* 32, 307-311.

Coughlin, P.J., Mueller, D.C. and P. Murrell, 1990b, Electoral politics, interest groups, and the size of government, *Economic Inquiry* 29, 682-705.

Crawford, V.P., 1995, Theory and experiment in the analysis of strategic interaction, Discussion Paper 95-37 (University of California, San Diego).

Engers, M., 1987, Signalling with many signals, *Econometrica* 55, 663-674.

Grossman, G.M. and E. Helpman, 1996, Electoral competition and special interest politics, Review of Economic Studies 63, 265-286.

Johnson, P.E., 1996, Corporate political offices in a rent-seeking society, *Public Choice* 88, 309-331.

Lindbeck, A. and J. Weibull, 1987, Balanced budget redistribution as the outcome of political competition, *Public Choice* 52, 273-297.

Lohmann, S., 1994, Electoral incentives, political intransparency and the policy bias toward special interests, Mimeo (University of California, Los Angeles).

Milgrom, P. and J. Roberts, 1986, Price and advertising signals of product quality, *Journal of Political Economy* 94, 796-821.

Milgrom, P. and J. Roberts, 1987, Informational asymmetries, strategic behavior, and industrial organization, *American Economic Review* 77, 184-193.

Moloney, K. and G. Jordan, 1996, Why companies hire lobbyists, *The Service Industries Journal* 16, 242-258.

Persson, T., 1998, Economic policy and special interest politics, *Economic Journal* 108, 310-327.

Potters, J., 1992, Lobbying and pressure. Theory and experiments (Thesis Publishers, Amsterdam).

Potters, J. and F. van Winden, 1996, Comparative statics of a signaling game, *International Journal of Game Theory* 25, 329-353.

Potters, J. and F. van Winden, 1998, The performance of professionals and students in an experimental studying of lobbying, Working paper TI 98-008/1 (Tinbergen Institute, Amsterdam).

Quinzii, M. and J.-C. Rochet, 1985, Multidimensional signalling, *Journal of mathematical economics* 14, 261-288.

Ramey, G., 1996, D1 signaling equilibria with multiple signals and a continuum of types, *Journal of Economic Theory* 69, 508-531.

Van Winden, F., 1997a, Experimental studies of signaling games, Working Paper TI 97-033/1 (Tinbergen Institute, Amsterdam).

Van Winden, F., 1997b, On the economic theory of interest groups: Towards a group frame of reference in political economics, forthcoming in: *Public Choice*.

Wilson, R., 1985, Multi-dimensional signalling, *Economics Letters* 19, 17-21.

AUTHOR INDEX

SUBJECT INDEX